MECHANOBIOLOGY: CARTILAGE AND CHONDROCYTE

Biomedical and Health Research

Volume 52

Earlier published in this series

ISSN: 0929-6743

Mechanobiology: Cartilage and Chondrocyte

Vol. 2

Edited by

J.-F. Stoltz

Angiohématologie–Hémorhéologie, UMR CNRS 7563, Faculté de Médecine, Université Henri Poincaré, Vandœuvre-lès-Nancy, France

IOS Press

Ohmsha

Amsterdam • Berlin • Oxford • Tokyo • Washington, DC

ISBN 1 58603 236 4 (IOS Press)
ISBN 4 274 90524 1 C3047 (Ohmsha)
Library of Congress Control Number: 2002104876

This is the book edition of the journal *Biorheology,* Volume 39, Nos 1,2 (2002), ISSN 0006-355X.

Publisher
IOS Press
Nieuwe Hemweg 6B
1013 BG Amsterdam
The Netherlands
fax: +31 20 620 3419
e-mail: order@iospress.nl

Distributor in the UK and Ireland
IOS Press/Lavis Marketing
73 Lime Walk
Headington
Oxford OX3 7AD
England
fax: +44 1865 75 0079

Distributor in the USA and Canada
IOS Press, Inc.
5795-G Burke Centre Parkway
Burke, VA 22015
USA
fax: +1 703 323 3668
e-mail: iosbooks@iospress.com

Distributor in Germany, Austria and Switzerland
IOS Press/LSL.de
Gerichtsweg 28
D-04103 Leipzig
Germany
fax: +49 341 995 4255

Distributor in Japan
Ohmsha, Ltd.
3-1 Kanda Nishiki-cho
Chiyoda-ku, Tokyo 101-8460
Japan
fax: +81 3 3233 2426

PRINTED IN THE NETHERLANDS

Contents

Author Index

Biorheology 39 (2002) 1
IOS Press

Preface

Research concerned with the influence of mechanical stresses on cells and tissues has advanced considerably during the last two decades, and we can readily trace the way this occurred, beginning with the first congresses of Biorheology and Biomechanics in the years between 1950 and 1960. That was how the concepts of mechanobiology and of cell and tissue engineering were developed. These new approaches, at the crossroads between physical mechanics, chemistry, and clinical and therapeutic science, are recognized today as entire disciplines.

In that respect, if mechanobiology can shed new light on the understanding of pathological phenomena, it is also, through the originality of the problems being faced, a stimulant for work in fundamental cell biology. The main aim of this symposium was to focus on research in the mechanobiology of cartilage and chondrocyte, and to promote the creation of new studies and collaborations in the osteo-articular field.

The proceedings of the first international symposium on mechanobiology, held in Nice in August 1999, were assembled and published in *Biorheology* (**37** (2000), 1–190). This issue of the journal has assembled the papers presented at the second international symposium, held in Paris in April 2001 under the auspices of the French Society of Biorheology, and with the support of the Negma-Lerards Laboratories. This meeting occurred between the first international symposium on mechanobiology and the upcoming 4th World Congress of Biomechanics in Calgary, August 2002. It permitted us to focus on the different mechanisms which arise during cartilage remodeling and the possibilities of clinical applications. Three main themes were developed during the two days work in the Paris symposium: (i) the mechanical properties of cartilage and their modeling; (ii) the importance of the effects of mechanical stress on the physiology of the chondrocyte; and (iii) clinical applications and cartilage engineering. For future work, there is a need to reinforce clinical and pharmacological studies on the importance of mechanical stresses on the properties of cartilage.

In order to ensure the continuity of the two first symposia, a third one will be organized in May 2003. It is proposed to award a prize to reward the research one or more young scientists.

J.-F. Stoltz
Nancy, December 24, 2001

PART I

Osteoarticular Cartilage Mechanics

Biorheology 39 (2002) 5–10
IOS Press

From biomechanics to mechanobiology

J.-F. Stoltz * and X. Wang

Cell and Tissue Engineering and Mechanics, LEMTA - UMR 7563 CNRS-INPL-UHP and IFR 111 Bioengineering CNRS-UHP-INPL-CHU, Faculty of Medicine, Brabois, 54500 Vandoeuvre-lès-Nancy, France

Abstract. Biomechanics can be defined as the application of mechanical concepts to the living world, and various fields of research have been developed such as the mechanics of movement, ergonomics, the mechanical properties of cells and tissues, and the relationship between physiology and applied forces. In this paper, the authors give, through several examples, an outline of these approaches and their potential biomedical applications, as in tissue remodelling, cell and tissue engineering and the development of biotissues.

Keywords: Mechanobiology, biomechanics, stress, cell, tissue, remodelling

"Bold ideas, unjustified expectations and speculations constitute our only means for comprehending nature" K. Kopper

1. Biomechanics: What definitions?

"Everything should be made as simple as possible but not simpler" Albert Einstein

Etymologically, biomechanics may be defined as mechanics applied to the living world. However, this definition is too vast and imprecise. It is also limiting, because it does not include physiological or pathological effects induced by the application of mechanical forces. Three approaches (more or less complete) allow us to define the fields of biomechanics.

The first one proposes to find, by application of mechanical laws, solutions to problems in medicine, biology, ergonomics and athletics.

The second one envisages studies of the mechanical properties of cells and tissues in considering the complexity of the structures being studied, for example, the properties of cardiac muscle, blood vessels and blood in the microcirculation.

The third one, more recent and no doubt more integrative, is interested not only in the mechanical properties of objects being studied through their structures, but also in their biological functions and in physiopathological consequences. This approach demands not only the resolution of fundamental problems and the development of models, but also the most recent knowledge in molecular biology, genomics and cell biology. This new approach which seems to be quite promising for its potential applications in cell and tissue engineering, can be called "mechanobiology".

*Address for correspondence: Prof. J.-F. Stoltz, Cell and Tissue Engineering and Mechanics, LEMTA - UMR 7563 CNRS-INPL-UHP and IFR 111 Bioengineering CNRS-UHP-INPL-CHU, Faculty of Medicine, Brabois, 54500 Vandoeuvre-lès-Nancy, France. Tel.: +33 3 83 59 26 41 / +33 3 83 15 37 79; Fax: +33 3 83 59 26 43 / +33 3 83 15 37 56; E-mail: stoltz@hemato.u-nancy.fr, jf.stoltz@chu-nancy.fr.

2. Biomechanics: An old history but a modern science

"...mechanical science is of all the noblest and most useful, seeing that by means of this all animate bodies which have movement perform all their action." Leonardo da Vinci

Throughout the ages, mankind has exhibited a strong interest in explaining the behaviour of living bodies, no doubt because of the importance of the challenge. Today the development of biomechanics, evoked by Aristotle (384–322 BC) in his work "*De motus animalium*", requires complementary research in physics, chemistry, mechanics, biology, medicine and surgery. In fact, it is very difficult to imagine a theoretical approach without any reference to experiments.

Historically we can distinguish different phases in the development of biomechanics, just as in any other scientific domain. Over more than 20 centuries the studies in biomechanics remained descriptive. Only with the evolution of sciences in 17th and 18th centuries did quantitative approaches and the first applications touching on instrumentation, and the concepts of implantable biomaterials and ergonomics come into being. The revolution of knowledge in biology brought about by genetics and molecular biology at the end of the 20th century, and recent progress in physical instrumentation such as atomic force microscopy, confocal fluorescence microscopy and laser tweezers to name a few, has again raised interest for a new biomechanics and its applications as in biotissue engineering and cell and tissue therapy.

Great names illuminate the history of biomechanics. Without being exhaustive, one can cite Aristotle's Treatise on the movement of animals, Archimedes (287–212 BC), and Galien on anatomy. These people were without any doubt the first biomechanicians.

The Renaissance was one of the most prosperous periods for the sciences. There was Leonardo da Vinci (1452–1519) and his work on human body movement, Galileo (1564–1642), a physician before being a physicist, William Harvey (1578–1658) and his fundamental work on blood circulation, who was also the first to pose the problem of microcirculation when he wrote "*How does the blood manage to traverse the porosities of the flesh on its way from arteries to veins?*"

The 17th and 18th centuries were particularly rich in works on physiomechanics. Malpighi (1628–1694) described the capillaries and red blood cells (at the beginning confused with fat particles), van Leeuwenhoek who completed the work of Malpighi, and Holes (1677–1761) who measured blood pressure.

It was in 19th century that the first physio-mechanical and rheological approaches were described. Thus, Jean Marie Léonard Poiseuille (1799–1869) can be considered as one of the pioneers of the modern physiomechanics. It was on the basis of Poiseuille's experiments that Hagen proposed the classical law on flow in a small circular tube which has generally been used in textbooks of physiology to describe the microcirculation. These studies constitute the beginning of almost one century of research in biorheology and biomechanics, and show the complexity of biological systems, in particular that of blood.

The term biomechanics appears to have been used for the first time in 1887 by M. Benedikt (Über Mathematische Morphologie und Biomechanik). Among the outstanding achievements of this period, we can cite the work of Wolff on the adaptability of a bone tissue, Fåhraeus on microcirculation, Bernstein on movement, and more recently, the work of Alan Burton, Al Copley, George Scott-Blair, Syoten Oka, Alex Silberberg, Gustav Born, Yuan-cheng Fung, Richard Skalak, Shu Chien, Van Mow, Savio Woo, etc.

Finally, the discovery of some important physiological mechanisms (Vane and Moncada for the prostaglandins, and Furchgott for nitric oxide, both works crowned by Nobel prizes) allows us to better understand the role of local mechanical stresses on cell physiology and tissue remodelling.

This brief historical review covering the last 25 centuries shows the evolution of Aristotle's initial preoccupation, which was reconsidered during the centuries leading to the modern concept of bioengineering.

3. Classical biomechanics: Movement and ergonomics

The analysis of bodily motion and its modelling helps us not only to understand the movement of different structures and to design machines to test implantable materials, but also to improve the physical performance of organs through a better understanding of their limitations. Such a biomechanical approach plays an important role, for example, in the re-establishment of motor functions through understanding musculoskeletal problems, and in the redistribution of stresses by using a plantary orthesis to re-establish posture equilibrium. In the workplace, biomechanics touches especially on ergonomics as in the optimisation of work stations. This widely employed biomechanical approach brings together an important scientific community around sports medicine and the technology of healthcare (readaptation, implantable prosthesis, surgery).

4. Tissue and cell mechanics: Physiological models

In the second half of the 20th century, biomechanics was applied to explain and to model the behaviour of organs, tissues, and more recently, cells. It is not surprising that these developments have been slow, because a fluid or a Hookean solid is generally an abstraction when we consider the behaviour of a tissue or a cell. Thus we have long known that blood is a non-Newtonian fluid, but research of physiologically acceptable models has gone beyond the classical knowledge on the behaviour of colloidal suspensions.

4.1. Necessity of models

"Good mathematical models don't start with the mathematics, but with a deep study of certain natural phenomena." Stephen Smale (1930)

The extremely complex biological systems present different types of heterogeneity. On the one hand, there are different shapes, dimensions, and constitutive elements; on the other hand, there is variability in time and in space. But through an appropriate selection of biological parameters of a system to be studied, the contribution of mechanics to the understanding of the physiology of organs, tissues, and cells has been demonstrated. For example in hemomechanics, the particulate nature of blood, a complex suspension of deformable and non-spherical particles, and of its vascular interface have been considered in order to understand complex blood flow in the microcirculation, and mass transfer in the capillary circulation. In large vessels, the role of local singular flows at bifurcations, stenoses and aneurysms have been considered in order to understand the occurrence of the pathological processes of thrombosis and atherosclerosis. Certainly the heterogeneity of scale must also be considered in hemomechanics or in studies of other fluids and tissues. An organ is macroscopic compared to that of its constitutive cells which in their turn are dimensionally different form their constitutive macromolecules or their environmental matrix components.

These problems of scale and structure, known to be fundamental since the 1960s, are inseparable from reliable biomechanical approaches which would permit us to link molecules to microscale structures and to entire tissues. For example, the work of Alan Burton (1959) on enzymatic digestion of arterial tissue helped to elucidate the role of vascular matrix components (elastin and collagen) on its mechanical properties.

According to Fung, research in biomechanics needs the integration of different data:

- Detailed structure of tissue and its matrix;
- Morphometric data on constitutive elements;
- Knowledge of mechanical properties of the constitutive elements and of their structure with and without loading.

It is in this spirit that studies have been developed on pulmonary parenchyma, on heart and blood vessels, and on cartilage and bone.

4.2. Cell mechanics: A developing field

Basically, all tissues and organs are composed of assemblages of cells. Experimental studies of their mechanical properties remain a delicate problem. Recent developments of non or little invasive physical techniques such as the cell scanner, biphotonic confocal microscopy, fluorescence spectroscopy, laser tweezers, and magnetocytometry, permits reliable measurements in cell mechanics to be made. We refer to measurements of cell aggregation and adhesion, local micromechanical behaviour, the effects of the cell cycle and division, and the phenomenon of polarisation.

5. Mechanobiology, mechanotransduction and tissue remodelling

Cells of the living body are always exposed to mechanical stresses which can vary from several Pa (shear stress at the vessel wall) to millions of Pa (stress on hip cartilage). Only very recently has it been admitted that these stresses can influence all organ and cell functions (physiology, synthesis, gene expression), just as biochemical factors do.

One of the early observations which remained forgotten for a long time is that of Wolff, a German surgeon, in 1892 on bone adaptability. He wrote "*every change in the form and function of bone or of their functions alone is followed by certain definitive changes in their internal architecture and equally definitive secondary alterations in their external conformation in accordance with mathematical laws.*" Thus, he defined the phenomenon of today's well known tissue remodelling.

In fact, cell mechanics, being at the interface of physics and biology, has experienced a conceptual revolution in the last 20 years accompanied by the development of molecular biology, genomics, and also bioengineering, with the possibility of measuring forces of the order of pico-Newtons and deformations of the order of nanometers. We are now capable of investigating the relationships between local mechanical parameters and cell functions (the concept of mechanobiology). It has been shown that most cells are not only sensitive to mechanical forces in their environment, but also to the origin and history of the mechanical loading. Although the biological effects of mechanical forces on certain cells are relatively well described, the mechanisms involved in these phenomena still remain unclear. In fact, how does one explain the passage from a mechanical stimulus to a physiological process as in the secretion and expression of a receptor, or activation of a gene, knowing that the effects are not always the same on different cell types, even for the same function? Conceptually, it is accepted today that these phenomena take place in 4 steps:

1. Mechanical coupling, which generally implies the transformation of applied forces into detectable stimuli by cells or the induction of a physical phenomenon. For example, pressure on a bone can induce a fluid flow in the canicular system and an electrokinetic potential.
2. Mechanotransduction corresponding to the action of induced stimuli on specific structures. Today, different hypotheses have been proposed and they constitute the subject of a large number of studies either in mechanics or in biology, such as cytoskeleton restructuring, specific receptor localization and receptors related to functional proteins (G protein, ionic channels, existence of mechanosensitive elements).
3. Signal transduction, which means the passage of intracellular physiological signals.
4. Cell response including gene regulation, release of autocrine or paracrine factors and specific receptor expression.

It should be noted that although steps 3 and 4 have been well elucidated for certain cell types and functions, the understanding of steps 1 and 2 still needs the development of models and specific experimental approaches for every cell type studied.

6. Cell and tissue engineering: Emergent technology and industry

At the present time, research in mechanobiology finds applications mainly in vascular, cardiac and osteoarticular fields. Recent reports show that mechanical consequence is specific for the system being considered. Furthermore, mechanical stresses are involved in tissue physiology, for example, the production of extracellular matrix (cartilage), and some specific secretions such as NO and prostaglandins in endothelial cells under flow.

These new findings can lead to the development of the concept of biotissue which means substituting tissues made *in vitro*, based on bioreabsorbable (or no) scaffolds and on cells cultured in a mechanical environment similar to the *in vivo* physiological conditions (vessel, cartilage, bone). It is estimated that this new biomedical industry will reach a market turnover of more than 50 billion US dollars in the year 2020.

6.1. Cardiovascular tissue engineering

Presently, research in this field is aimed at various targets such as myocardiac grafts and biovessels. In 2000, the autologous graft of leg skeletal myoblast cultures in a patient suffering from a major cardiac insufficiency, has opened the door for new applications of cell therapy. In fact, this first therapy attempted to demonstrate that grafted muscle cells could become functional muscle fibres under the appropriate biochemical and mechanical environment.

In the vascular field, the group of Robert Nerem in Atlanta (Georgia Tech/Emory University) is developing vascular substitutes in vitro using cultures of smooth muscle cells and vascular endothelial cells in a collagen gel under physiological flow conditions.

6.2. Cartilage engineering

Cartilage is an interfacial tissue composed of only one cell type, the chondrocyte, and an autosynthesized matrix. Its properties are a function of its mechanical environment. A large number of studies aim to make cartilage *in vitro* with chondrocytes cultured in an initial matrix of polymers (hyaluronic acid) and under a physiologically compatible pressure. Clinical applications of such cultured tissue could see the light of day in the not too distant future.

There are still many cell and tissue therapies targeted by researchers, which need further investigations on the effect of the mechanical environment and on the quality of cell and tissue grafts (for example: mesenchymal cells, hematopoietic stem cells, bone tissues).

7. Conclusion

"An unique point of view is always wrong." Paul Valéry

Like any other biomedical discipline which wants to be credible, biomechanics must come to a better understanding of pathological situations (e.g., thrombosis, atherosclerosis, arthrosis). That is why future evolution of the science necessitates interdisciplinary links between mechanics, physics, chemistry, molecular biology and genomics. Research in these fields will lead to implantable biotissues and to new cell therapies in the near future.

However, a large number of unknowns still exist, giving full meaning to the thought of the philosopher H. Michaux in the last century:

"Any science creates new ignorance
Any conscious creates a new unconscious
Any new acquisition creates a new void"

General bibliography on biomechanics and mechanobiology

N. Akkas, Biomechanics of active movement and deformation of cells, *NATO ASI series – serie H: Cell Biology*, Springer-Verlag, New York, Berlin **42** (1990), 524 pp.

A.A. Biewener, *Biomechanics – Structures and Systems*, IRL Press at Oxford University Press, New York, 1992, 290 pp.

H.J. Bereiter, O.R. Anderson and W.E. Reif, *Cytomechanics. The Mechanical Basis of Cell Form and Structure*, Springer-Verlag, New York, Berlin, 1987, 294 pp.

C.G. Caro, T.J. Pedley, R.C. Schroter and W.A. Seed, *The Mechanics of the Circulation*, Oxford University Press, New York, Toronto, 1978, 527 pp.

J. Enderle, S. Blanchard and J. Bronzino, *Introduction to Biomedical Engineering*, Academic Press, London, 2000, 1062 pp.

H.M. Frost, *An Introduction to Biomechanics*, Charles C. Thomas, Springfield, 1971, 151 pp.

Y.C. Fung, *Biomechanics: Mechanical Properties of Living Tissues*, Springer-Verlag, New York, Berlin, 1981, 433 pp.

Y.C. Fung, K. Hayushi and Y. Seguchi, *Progress and New Directions of Biomechanics*, Meta Press, Tokyo, 1989, 444 pp.

T. Hianik and V.I. Passechnik, *Biolayer Lipid Membranes: Structure and Mechanical Properties*, Kluwer Academic Publishers, Dordrecht, London, 1995, 436 pp.

M.Y. Jaffrin and F. Goubel, *Biomécanique des Fluides et des Tissus*, Masson, Paris, 1997, 454 pp.

R.P. Lanza, R. Langer and W.L. Chick, *Principles of Tissue Engineering*, Academic Press, London, 1997, 808 pp.

A. Larcan and J.F. Stoltz, *Micocirculation et Hémorheologie*, Masson, Paris, 1970, 273 pp.

R.B. Martin, D.B. Bun and N.A. Sharley, *Skeletal Tissue Mechanics*, Springer-Verlag, New York, Berlin, 1998, 392 pp.

A. Silberberg, ed., *Perspectives in Biorheology (Festschrift for A.L. Copley)*, Pergamon Press, New York, 1981, 422 pp.

J.F. Stoltz and P. Drouin, *Hemorheology and Diseases*, Doin, Paris, 1980, 709 pp.

J.F. Stoltz, New trends in Biorheology, *Biorheology* **30** (1993), 305–322.

J.F. Stoltz, M. Singh and P. Riha, *Hemorheology in Practice*, IOS Press, Amsterdam, 1999, 128 pp.

J.F. Stoltz, ed., Mechanobiology: Cartilage and Chondrocyte, *Biorheology* **37** (2000), 1–190.

J.R. Vane, G.V.R. Born and D. Welzel, *The Endothelial Cell in Health and Disease*, Schattauer, Stuttgart, 1995, 203 pp.

S.L.Y. Woo and Y. Seguchi, *Tissue Engineering*, The American Society of Mechanical Engineers, BED, 1989, 146 pp.

Biorheology 39 (2002) 11–25
IOS Press

The functional environment of chondrocytes within cartilage subjected to compressive loading: A theoretical and experimental approach

Christopher C.-B. Wang, X. Edward Guo, Dongning Sun, Van C. Mow, Gerard A. Ateshian and Clark T. Hung *

Department of Biomedical Engineering, Columbia University, New York, NY 10027, USA

Abstract. A non-invasive methodology (based on video microscopy, optimized digital image correlation and thin plate spline smoothing technique) has been developed to determine the intrinsic tissue stiffness (H_a) and the intrinsic fixed charge density (c_0^F) distribution for hydrated soft tissues such as articular cartilage. Using this technique, the depth-dependent inhomogeneous parameters $H_a(z)$ and $c_0^F(z)$ were determined for young bovine cartilage and incorporated into a triphasic mixture model. This model was then used to predict the mechanical and electrochemical events (stress, strain, fluid/osmotic pressure, and electrical potentials) inside the tissue specimen under a confined compression stress relaxation test. The integration of experimental measurements with theoretical analyses can help to understand the unique material behaviors of articular cartilage. Coupled with biological assays of cell-scale biosynthesis, there is also a great potential in the future to study chondrocyte mechanotransduction *in situ* with a new level of specificity.

1. Introduction

A major motivation for cartilage biomechanics research has been to provide a greater insight into the interaction among the type II collagen meshwork, the fixed negative charges of the entrapped proteoglycan, and the ion-rich interstitial fluid that gives rise to the load-bearing capacity of this specialized connective tissue [24,29,32]. Consisting of 3–10% proteoglycans, 10–30% collagen, 60–80% water, the distribution of each component varies with the depth of the tissue. Although well recognized, the physiologic implications of this inhomogeneity on tissue function and chondrocyte activities remain to be elucidated. The study of cartilage inhomogeneity has been traditionally addressed by slicing the tissue into thin "homogenized" strips to be tested in uniaxial tension or compression [1,19,25,41,62].

There have been significant advances in cartilage mechanical testing by the introduction of various optical-based techniques [18,27,35,36,42,43,58,59]. Similarly, advances in constitutive modeling of hydrated soft tissues (e.g., [12,13,21,31,52]) have contributed to the understanding of cartilage structure-function relationship (Fig. 1). Motivated by these advances, we have developed an automated approach for determination of the intrinsic tissue properties (e.g., intrinsic solid matrix stiffness and intrinsic fixed

*Address for correspondence: Dr. Clark T. Hung, Cellular Engineering Laboratory, Department of Biomedical Engineering, Columbia University, 3513 Engineering Terrace, M.C. 8904, 1210 Amsterdam Avenue, New York, NY 10027, USA. Tel.: +1 212 854 6542; Fax: +1 212 854 8725; E-mail: cth6@columbia.edu.

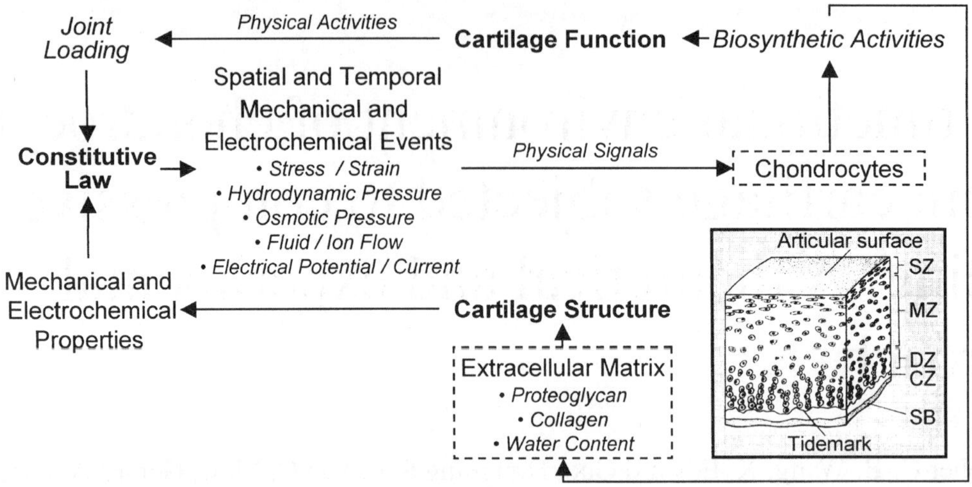

Fig. 1. A schematic of the structure–function relationship of articular cartilage. The extracellular matrix, with associated interstitial fluid, solutes and ions, can collectively be thought of as a mechanical signal transducer that receives mechanical input in the form of joint or explant loading and yields an output of various extracellular signals.

charge density (FCD)) of full-thickness cartilage explants using a combination of applied osmotic chemical loading and mechanical loading. With the knowledge gained from this technique, we are in a position to provide a more refined description of the chondrocyte environment (e.g., hydrostatic pressure, deformation, fluid, electrokinetic, and osmotic pressure fields) *in situ*.

Through these efforts, we can better identify the stimuli and their appropriate levels that the cells may see *in situ*. New information regarding the *in situ* environment of the cell will provide motivation for *in vitro* cell and explant studies to study physical effects on chondrocytes, whose metabolic and catabolic activities regulate cartilage maintenance. The role of a particular stimulus can then be identified using *in vitro* models aimed at simplifying the cell environment or delineating the role of concomitant stimuli [5,7,15,17,22,23,28,37,39,47,48]. Thus, the application of constitutive modeling permits an understanding of the potential contributions of specific stimuli that cannot be directly inferred from experiments alone.

We have previously [56,57] described the effect of an inhomogeneous aggregate modulus on the stress relaxation response of a cartilage explant using a biphasic analysis [31]. In this earlier study, the depth-dependent distribution of tissue stiffness was obtained from the work of Schinagl and co-workers [42,43] using digital video microscopy and fluorescently labeled cell nuclei as fiducial markers. Additionally, we have also considered the effect of an inhomogeneous FCD on the mechanoelectrochemical behavior of cartilage [53] using the triphasic mechano-electrochemical theory which considers the tissue as consisting of solid, fluid and ion phases [21]. The depth-dependent FCD distribution for this analysis was based on the depth-dependent variation of FCD reported in the literature [25]. In the current study, we describe our approach to integrating novel experimental measurements of depth-varying intrinsic tissue stiffness and intrinsic FCD [55] with the triphasic theory to describe the chondrocyte environment within articular cartilage during the stress relaxation response of a tissue explant subjected to loading in confined compression.

2. Methods

2.1. Triphasic mixture model for articular cartilage

In the triphasic mixture theory, the articular cartilage is modeled as a composite material with three interacting incompressible phases: (1) a charged solid phase, which is composed predominately of a densely-woven, strong, collagen fibrillar network enmeshed with a high concentration of charged proteoglycan aggregates; (2) an interstitial fluid phase which is water; and (3) an ion phase of both cation and anion species, which is required to neutralize the charges fixed to the collagen–proteoglycan solid matrix. According to this theory,

1. The net charge at every point within the tissue is zero. This assumption, i.e., the electroneutrality condition, can be expressed in terms of ion concentrations and valance by:

$$\sum_{\alpha} z_\alpha c_\alpha - c^F = 0,$$

where z_α is the valance of the αth ionic species including sign $(+/-)$, c_α (mol/m^3) is the molar concentration of the αth ionic species (per unit of solvent volume) and c^F (mEq/ml) is the effective FCD of the tissue. The effective FCD is related to the intrinsic FCD (c_0^F) by the conservation equation of fixed charges:

$$c^F = \frac{c_0^F}{1 + \varepsilon/\phi_0^w}, \tag{1}$$

where ϕ_0^w is the intrinsic porosity of the tissue[1].

2. The only motive forces for interstitial fluid flow and ion transport are the gradients of their electrochemical potentials; and
3. Frictional forces exist when there is relative motion between these three phases of the tissue; however, the frictional force between the solid matrix and the ions are negligible.

In this triphasic theory, the total stress $\boldsymbol{\sigma}^T$ may be decomposed into partial stresses, an elastic stress $\boldsymbol{\sigma}_E^s$ and the fluid pressure p (which includes the osmotic pressure):

$$\boldsymbol{\sigma}^T = \boldsymbol{\sigma}_E^s - p\mathbf{I}. \tag{2}$$

At equilibrium, the fluid pressure $p = \pi$ is the Donnan osmotic pressure given by van't Hoff pressure law:

$$\pi = RT\left(\sum_{\alpha}(\phi_\alpha c_\alpha) - \sum_{\alpha}(\phi_\alpha^* c_\alpha^*)\right), \tag{3}$$

where R is the universal gas constant, T is the absolute temperature, and ϕ_α and ϕ_α^* are osmotic coefficients[2]; the superscript (*) denotes the quantities associated with the external solutions.

[1] ϕ_0^w is assumed constant through the depth of cartilage in this study.
[2] In the current study we assume $\phi_\alpha \approx \phi_\alpha^* \approx 1$.

The electrochemical potentials of the ions within the interstitium of cartilage are given by:

$$\mu^\alpha = \mu_0^\alpha + \frac{RT}{M_\alpha}\ln(\gamma_\alpha c_\alpha) + z_\alpha \frac{F_c\psi}{M_\alpha}, \tag{4}$$

where μ_0^α are the reference potentials of the αth ionic species, γ_α are the activity coefficients of c_α, M_α are the atomic weight of these ions, F_c is the Faraday constant and ψ is the electric potential.

2.2. One-dimensional (1D) confined compression test

Normal articular cartilage is firmly attached to the impermeable subchondral bone, and articulates against an opposing cartilage surface. Thus one needs to examine how articular cartilage will respond when pressed against a solid material. In a 1D confined compression test, all motions are uniaxial in the direction of the applied load. This test requires a precisely prepared cartilage specimen harvested from an articular layer. The experiment calls for inserting this cartilage specimen into a frictionless, impermeable confining chamber (to prevent lateral expansion and fluid flow), and loading it axially between two rigid solid loading platens. The top platen is porous-permeable to permit the interstitial fluid and ions to flow out of the tissue. The pore pressure within this free-draining loading platen is ambient.

In this study, a stress–relaxation test is analyzed using the triphasic mixture theory to study the mechanical and electrochemical events of a tissue under confined compression. In such a test, the surface-to-surface strain ε_o (say 10%) is imposed in a linear manner over the time duration $0 < t < t_o$ (e.g., 400 seconds). For time $t > t_o$, the strain ε_o is maintained constant for the remainder of the experiment. Under this test configuration, the elastic stress (σ_E^s) is given by

$$\sigma_E^s = H_a\varepsilon, \tag{5}$$

where H_a is the intrinsic stiffness of the solid matrix at equilibrium and ε the strain inside the tissue.

In this 1D configuration, the general equations of the triphasic constitutive law are reduced to a system of partial differential equations with u (matrix displacement) and c^+ (cation concentration) as dependent variables, and z (depth) and t (time) as independent variables. The system of partial differential equations can be solved numerically by finite element or finite difference methods [34,53]. The intrinsic stiffness of the solid matrix (H_a) and the intrinsic FCD (c_0^F) are two of the most important parameters that govern the mechanical and electrochemical behavior of the tissue. Both H_a and c_0^F have been found to vary significantly through the depth (z-direction) of the tissue [25,42]. Consequently, these inhomogeneous material properties will lead to the development of non-uniform mechanical and electrochemical events ($\sigma(z,t)$, $\varepsilon(z,t)$, $\pi(z,t)$, etc.) during the transient phase of loading and at equilibrium. Below, we describe a methodology to determine these important parameters on the same tissue specimen without sectioning the specimen in the axial (or depth) direction.

2.3. Experimental measurement of cartilage properties

Healthy carpometacarpal joints from 1- to 2-month-old calves were obtained from a local slaughterhouse 6–10 hours post-mortem. A steel trephine with a 6mm diameter core was used to harvest cartilage-bone plugs with the core axis perpendicular to the articular surface. Following harvesting, the plugs were rinsed with phosphate buffered saline (PBS) containing protease inhibitors (PI) and frozen at −80°C for storage until the day of use. On the day of testing, plugs were thawed to room temperature by immersion in freshly prepared PBS + PI solution. The plug was microtomed to remove residual subchondral bone and vascularized deep zone only, leaving the articular surface intact. 2 mm-diameter disks were cored out from each plug with a stainless-steel micro-dissecting trephine (Biomedical Research Instruments, Rockville, MA, USA). Before testing, each 2 mm-diameter disk was cut into two semi-cylindrical microcompression specimens with a custom cutter. The remaining tissue annulus for each plug was frozen and saved for biochemical analyses. Each annulus was sectioned axially into three tissue slices of equal thickness and FCD determined using the DMMB dye-binding assay following the method described by Narmoneva et al. [36].

A confined compression microscopy device was designed for this study, which incorporated a semi-cylindrical confining ring with porous platen at bottom and a semi-cylindrical impermeable stainless steel indenter. The indenter was rigidly connected to a precision miniature load cell (Model 31; Sensotec Inc., Columbus, OH, USA; range 0–150 grams) and then to a differential screw translator (M-R-O Industrial Supplies, Manville, NJ, USA; resolution 1 μm). The tissue sample, with the cross section of tissue depth facing down, was confined compressed with the impermeable indenter, while bathed in NaCl solution + PI contained in an immersion chamber with a glass window at bottom. The custom loading device was mounted onto the motorized stage (ProScan H128 Series; Prior Scientific, Rockland, MA, USA) of an inverted microscope (Olympus IX70; Olympus America, Melville, NY, USA) equipped for transmission and epifluorescence microscopy with a 100 W Xenon arc lamp, a DeltaRam high speed monochrometer, a DAPI filter cube, and a Uplan 10× objective (Olympus America, Melville, NY, USA). The optical path of the microscope was adjusted through the immersion chamber and the cross section of tissue depth between the platens was focused and visualized. All images were captured such that the image width contained the central region of the cross section. As shown in Fig. 1, the half-cylindrical specimen was first equilibrated in hypertonic saline (2 M NaCl) for 45 minutes. The initial thickness (h_0) was optically determined by moving the indenter until no space could be observed from either side of the specimen (Fig. 2, I). The specimen was then compressed 10% of h_0 (Fig. 2, II) and permitted to relax

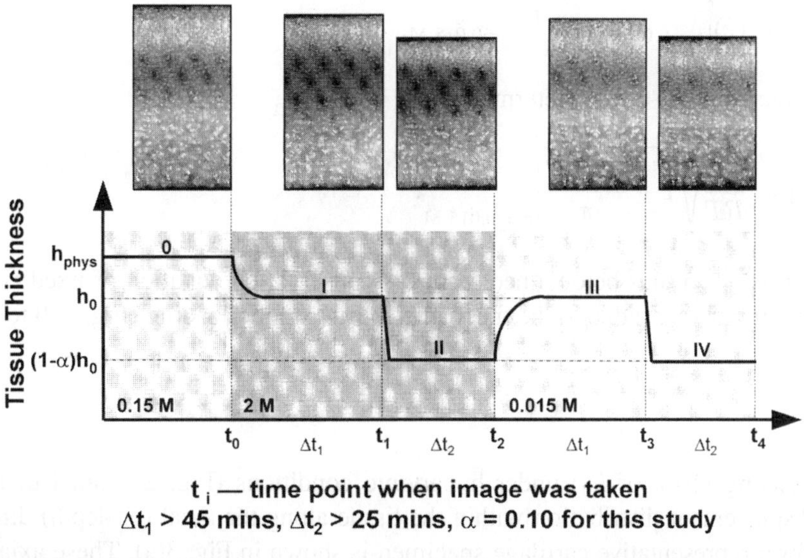

Fig. 2. Experimental protocol for determination of the depth-dependent intrinsic solid matrix stiffness ($H_a(z)$) and intrinsic fixed charge density ($c_0^F(z)$). In the plot of the tissue thickness versus time, h_{phys} is the initial thickness of the tissue specimen in physiological saline (0.15 M NaCl), h_0 is the initial thickness of the tissue specimen in 2M NaCl and $(1-\alpha)h_0$ is the tissue thickness after applied compression.

to equilibrium, and images of the cross section recorded. Next, the distance between the bottom of the confining ring and the indenter was reset to h_0. The bathing solution was then changed to 0.015 M (Fig. 2, III) and the preceding loading protocol repeated (Fig. 2, IV).

The strain distributions resulting from mechanical and/or osmotic loading were determined automatically using an optimized digital image correlation (DIC) technique and a thin-plate spline smoothing algorithm. By simple manipulation of the common Donnan osmotic equation, we get the expression $\pi = RT(\sqrt{c^{F2} + 4c^{*2}} - 2c^*)$ [21]. Thus, under extreme hypertonic conditions when $c^* \gg c^F$ (e.g., $c^* = 2$ M), the osmotic pressure in cartilage is negligible. According to Eqs (2) and (5), at equilibrium,

$$\sigma^T = H_a \varepsilon - \pi. \tag{6}$$

Under extreme hypertonic conditions ($c^* = 2$ M), since $\pi \approx 0$, hence measurements of the depth-dependent strain field ($\varepsilon(z)$) in a specimen subjected to a prescribed surface-to-surface clamping strain (Fig. 2, I to II) provides a measure of the inhomogeneity of the intrinsic solid matrix stiffness ($H_a(z)$), i.e.,

$$H_a(z) \simeq \left.\frac{\sigma^T}{\varepsilon(z)}\right|_{c^*=2\ \mathrm{M}}. \tag{7}$$

If this sample, subjected to the same clamping strain, is then exposed to an isotonic/hypotonic salt environment (Fig. 2, II to IV), its osmotic pressure rises according to the local distribution of effective FCD; if this distribution is inhomogeneous, the depth-dependent strain will change accordingly. Thus, a measure of the change in depth-dependent strain ($\varepsilon(z)$) between hypertonic and hypotonic conditions (e.g., 0.015 M), coupled with knowledge of the clamping stress, can produce a measure of the osmotic pressure distribution ($\pi(z)$), i.e.,

$$\pi(z) = H_a(z) \cdot \varepsilon(z)|_{c^*=0.015\ \mathrm{M}} - \sigma^T|_{c^*=0.015\ \mathrm{M}}, \tag{8}$$

from which the effective FCD can be determined as

$$c^F|_{c^*=0.015\ \mathrm{M}} = \left.\frac{\pi}{RT}\sqrt{1 + \frac{4c^* RT}{\pi}}\right|_{c^*=0.015\ \mathrm{M}}. \tag{9}$$

The intrinsic FCD (c_0^F) can thus be obtained from the conservation equation of fixed charges (Eq. (1)), given the resultant strains ($\varepsilon(z)$) due to the mechanical and osmotic loading (Fig. 2, 0 to IV).

3. Results

Compressive loading (10% of h_0) under hypertonic conditions (Fig. 2, from I to II) resulted in a nonlinear axial displacement distribution within the tissue along the axial (or depth) direction. The displacement field for a representative cartilage specimen is shown in Fig. 3(a). These axial displacements exhibit no significant variation along the radial direction ($p > 0.9$); the optimized DIC also measured an average lateral displacement of 0.14 ± 0.16 μm, which is on the order of the system errors introduced by the image acquisition. Together, these data demonstrate a perfect confinement of the specimen within the

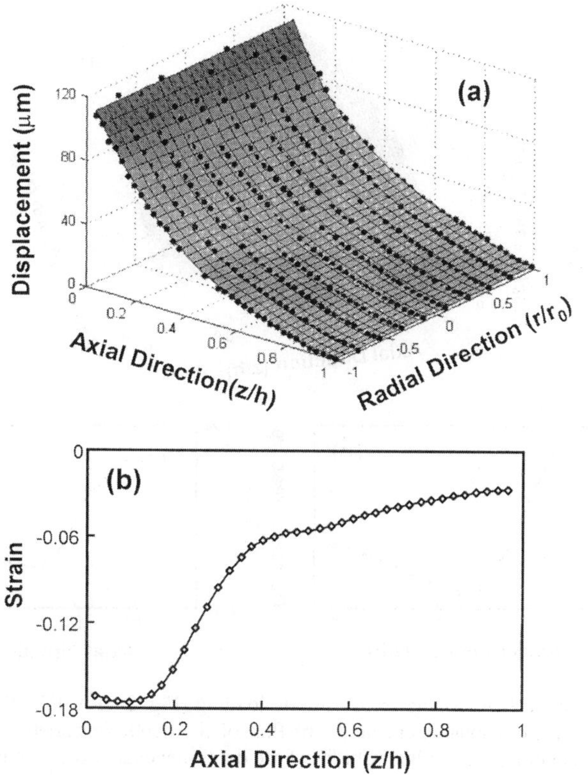

Fig. 3. (a) The resultant displacement field arising from compressive loading in hypertonic (2M NaCl) solution (Fig. 2, I to II). The shaded surface represents the thin-plate smooth fitting spline of the displacement field. (b) Plot of the depth-dependent strain distribution obtained from differentiation of the thin-plate smooth fitting spline.

confining ring. The shaded surface in Fig. 3(a) represents the thin-plate smooth fitting spline of the measured axial displacement fields ($R^2 = 0.99$). Strain analysis performed by differentiating the smoothed surface revealed an inhomogeneous strain distribution along the axial direction, with the magnitude decreasing from the surface layer to the deep layer (Fig. 3(b)). Since the osmotic effects are negligible under the extreme hypertonic condition, the intrinsic matrix stiffness ($H_a(z)$) for the specimen in Fig. 3(a) can be determined from Eq. (7) (Fig. 5(a)).

Under the same clamping strain (10% of h_0), the introduction of osmotic loading via change of bathing solution concentration from 2 M to 0.015 M (Fig. 2, II to IV) yielded a non-uniform displacement field within the same specimen (Fig. 4(a)). Again, no significant variation of the axial displacement was observed along the radial direction ($p > 0.85$). The resultant strain distribution (Fig. 4(b)) was determined from the thin-plate smooth fitting spline of the displacement data ($R^2 = 0.92$). The osmotic effects are no longer negligible under this hypotonic condition; from Eq. (6), the surface traction on the specimen surface arises from both the osmotic pressure and the elastic stress of the solid matrix. With knowledge of the intrinsic matrix stiffness $H_a(z)$ (Fig. 5(a)) and the measured osmotic strain distribution (Fig. 4(b)), the depth-dependent osmotic pressure π was determined (Fig. 4(c)). By further noticing that the osmotic pressure can be expressed as a function of the effective FCD of the tissue as well as the bathing concentration, a depth-dependent effective FCD ($c^F(z)$) can be determined from the osmotic pressure distribution. Furthermore, the intrinsic FCD ($c_0^F(z)$) for the same specimen was determined using Eq. (1) and is plotted in Fig. 5(b).

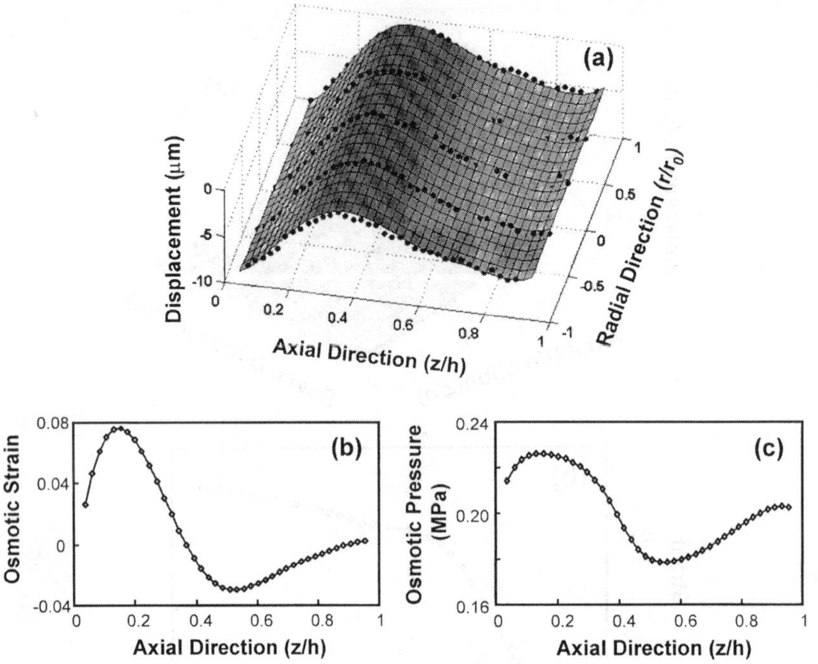

Fig. 4. (a) The resultant displacement field arising from osmotic loading (Fig. 2, II to IV). The shaded surface represents the thin-plate smooth fitting spline of the displacement field. (b) Plot of the depth-dependent strain distribution obtained from differentiation of the thin-plate smooth fitting spline. (c) Plot of the depth-dependent osmotic pressure π.

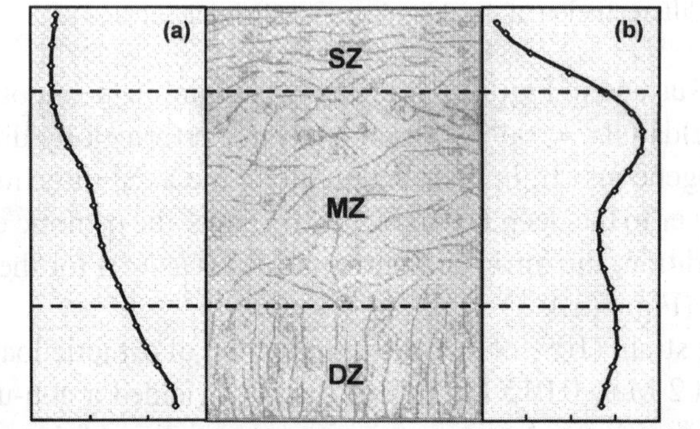

Fig. 5. The depth-dependent intrinsic solid matrix stiffness ($H_a(z)$) and intrinsic fixed charge density ($c_0^F(z)$) determined for a typical young bovine articular cartilage specimen. These properties are incorporated in our theoretical analysis of the tissue behavior under confined compression.

The distributions of intrinsic matrix stiffness ($H_a(z)$) and FCD ($c_0^F(z)$) (Fig. 5) were then incorporated into the triphasic model to analyze the transient environment of chondrocytes within a cartilage explant under confined compression with the articular surface against the porous loading platen. It is noted that for the purpose of comparison, an average intrinsic FCD and an "equivalent" constant intrin-

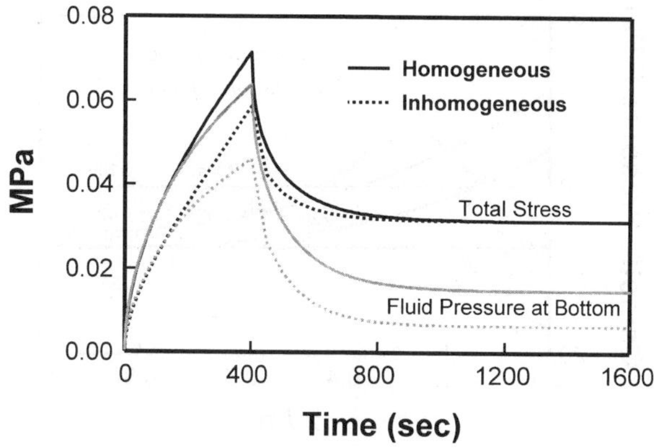

Fig. 6. The stress relaxation response of cartilage specimens with homogeneous and inhomogeneous intrinsic solid matrix stiffness and intrinsic fixed charge density. The fluid pressure determined at the bottom of the specimen ($z = h$) is also depicted to show the degree of load support by the fluid phase.

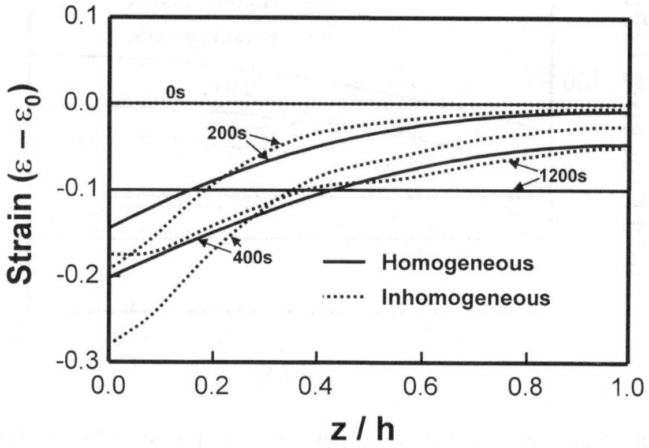

Fig. 7. Plot of the spatial distribution and temporal development of strain for the homogeneous and inhomogeneous tissue subjected to stress relaxation in confined compression. $z = 0$ corresponds to the articular surface.

sic matrix stiffness[3] were used for a homogeneous model in the following results. Comparison of the stress–relaxation curves for the homogeneous and inhomogeneous tissues clearly shows distinct differences in the transient development of stress before equilibrium is reached at 1200 seconds (Fig. 6). The homogeneous model predicts a greater peak stress and lengthier relaxation phase. For both models, the loading-induced tissue deformation proceeds from the surface zone to the deep zone, with the strain eventually being distributed equally throughout the tissue depth for the homogeneous case (Fig. 7). It should be emphasized that during the transient loading phase (i.e., before equilibrium is reached) the peak strains within the tissue for each case can exceed the nominally applied 10% surface-to-surface strain. This result is indicative of the applied loading condition (1D loading with a permeable platen). The inhomogeneity of the tissue amplifies the conditions established by this applied loading condition. Although the stress relaxation test is performed with the zero external applied loading condition as the reference, there exists an initial prestress due to the osmotic pressure within the proteoglycan–collagen

[3]The "equivalence" between the homogeneous tissue and inhomogeneous tissue is defined such that both tissues, which have the same values for all the parameters except $c_0^F(z)$ and $H_a(z)$, should achieve the same equilibrium stress in the stress–relaxation test.

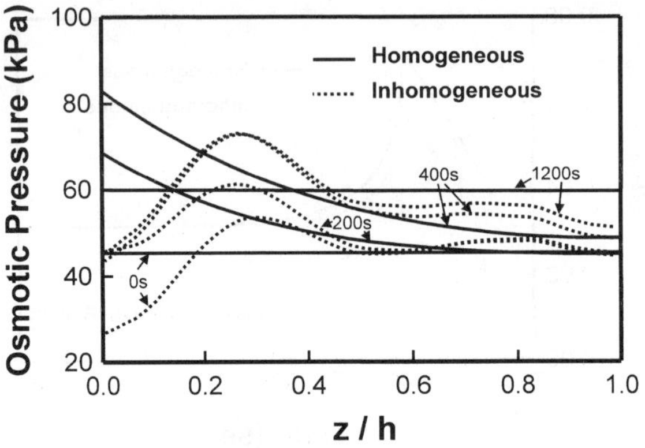

Fig. 8. Plot of the spatial distribution and temporal development of osmotic pressure for the homogeneous and inhomogeneous tissue subjected to stress relaxation in confined compression. $z = 0$ corresponds to the articular surface.

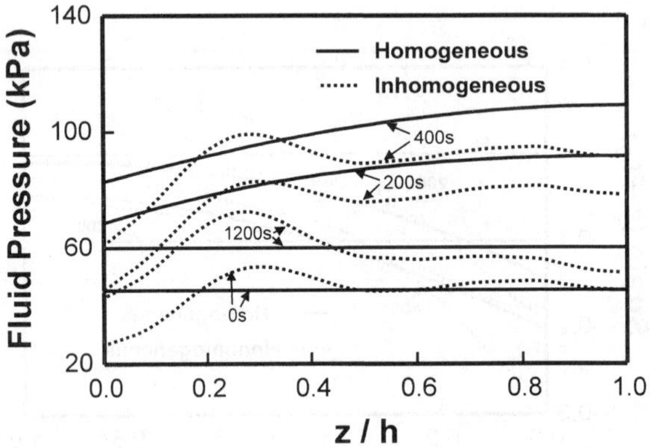

Fig. 9. Plot of the spatial distribution and temporal development of fluid pressure for the homogeneous and inhomogeneous tissue subjected to stress relaxation in confined compression. At $t = 0$ and at equilibrium, the fluid pressure is equal to the osmotic pressure (Fig. 8). $z = 0$ corresponds to the articular surface.

network of the tissue (Fig. 8). This prestress generates an initial free-swelling strain (ε_0) distribution (constant throughout the depth in the homogenous case) within the tissue (Fig. 7).

The FCD and the tissue stiffness govern the osmotic pressure within the tissue. In the case of the homogeneous tissue, the strain distribution alone accounts for the development of non-uniform osmotic pressure in the tissue during the transient phase. Conversely, the existence of an intrinsic inhomogeneous FCD gives rise to an initial non-uniform osmotic pressure (and initial matrix strain) which is then modulated by the subsequent loading-induced matrix strain. From Figs 7 and 8, the spatial and temporal development of solid matrix deformation and osmotic pressure are directly related, with the peak osmotic pressure (and peak compressive matrix deformation) in the tissue corresponding to the surface zone ($z = 0$) at time 400 seconds. Accordingly, the osmotic pressure distribution parallels the uniform strain distribution in the tissue at equilibrium for the homogeneous case.

The load during the stress relaxation test is partitioned between the solid and fluid phases of the tissue, manifested as matrix stress and fluid pressure. In the triphasic model, the fluid pressure derives from both the hydrodynamic fluid pressure and the osmotic pressure. Accordingly, from Fig. 9, the fluid pressure is equal to the osmotic pressure at $t = 0$ and at the equilibrium condition (1200 sec) when there is no longer fluid exudation from the tissue. At equilibrium, the measurable load (e.g., calculated total stress)

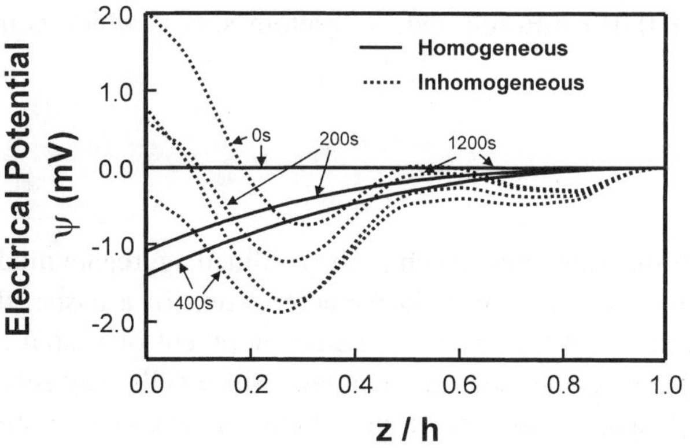

Fig. 10. Plot of the spatial distribution and temporal development of electrical potential for the homogeneous and inhomogeneous tissue subjected to stress relaxation in confined compression. $z = 0$ corresponds to the articular surface.

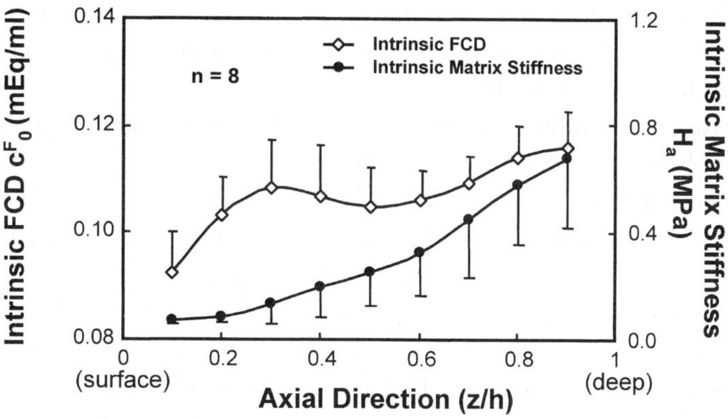

Fig. 11. A summary of the depth-dependent intrinsic solid matrix stiffness ($H_a(z)$) and intrinsic fixed charge density ($c_0^F(z)$) for 8 young bovine articular cartilage specimens determined using the non-invasive methodology described in the text and the experimental protocol depicted in Fig. 2.

reflects the change in osmotic pressure and solid matrix stress. During the loading period, there can be significant load support by the fluid pressure (Fig. 6). This fluid support has been proposed to contribute significantly to cartilage's load-bearing and lubrication capacity [3,29,50].

The electrical potential (Fig. 10) in the homogeneous tissue is zero at $t = 0$ and at $t = 1200$ seconds. This results from the uniform distribution of FCD within the tissue before loading and at equilibrium. In contrast, electrical potential gradients exist in the inhomogeneous case during the entire stress relaxation response due to the intrinsic inhomogeneous FCD distribution. At $t = 0$ and equilibrium, the distribution of the electrical potential resembles that for the FCD distribution within the inhomogeneous tissue. From Fig. 10, there is a sign change in the electrical potential inside the inhomogeneous tissue that is not observed for the homogeneous tissue.

Although the theoretical analysis of the spatial and time-varying mechanical and electrochemical events (Figs 6–10) was performed using parameters derived from a single cartilage specimen, it should be noted that its inhomogeneous properties are representative of seven other samples derived from 4 young bovine carpometacarpal joints. We summarize the intrinsic sold matrix stiffness and intrinsic FCD from all 8 cartilage specimens in Fig. 11. For qualitative comparison, the mean FCD determined from biochemical analyses of the top slices (which contains the surface zone) was 0.098 ± 0.016 mEq/ml, the middle slices was 0.112 ± 0.019 mEq/ml and the bottom slices (which contains the deep zone) was 0.132 ± 0.063 mEq/ml.

4. Discussion

While surface-to-surface measurements (such as the confined aggregate modulus or surface-to-surface applied strain) provide bulk properties or bulk characterization of a tissue, they provide only indirect insights to the internal behavior of the tissue or the development of spatial and temporal gradients of strain, pressure, osmotic pressure and electric potentials. Since cells may sense change as well as stimulus magnitude (e.g., [11]), spatial and time-varying fields or gradients of stimuli may be important in chondrocyte mechanotransduction and related changes to cell biosynthetic and catabolic activities [30]. From our analysis of a homogeneous tissue subjected to the simple loading case of uniaxial confined compression, one can quickly gain an appreciation for the complexity of the environment within the tissue during loading. When a more accurate description of the tissue properties (inhomogeneous intrinsic tissue stiffness and FCD) is incorporated into the analysis, the variation of the cell environment within the tissue becomes even more pronounced.

Using confocal microscopy, Guilak and co-workers [14] demonstrated depth-varying cell deformation in loaded explants, being greatest in the superficial zone. Consistent with this finding is that of a depth-dependent aggregate modulus reported in the literature [42,43,46]. Our measurements also reveal a depth-dependent aggregate modulus, as well as a depth-dependent FCD [55,58] determined through mechanical and osmotic loading. The current investigation represents the first effort to incorporate experimental measurements of inhomogeneous aggregate modulus and FCD into a theoretical analysis of articular cartilage loading behavior. While our experiments can permit determination of the FCD, osmotic pressure, hydrostatic pressure and strain at equilibrium, we are unable to currently apply our technique to the transient phase of the stress–relaxation response. However, with constitutive modeling we can predict the time-varying environment within the tissue.

The novel non-invasive methodology used to determine the intrinsic tissue stiffness and the intrinsic FCD for the same specimen utilized a custom loading device, digital video microscopy, optimized digital image correlation technique and application of the familiar equations for Donnan osmotic pressure [55,58]. The technique reports FCD values that are similar to those reported using conventional techniques for determination of FCD [25,26] and are in qualitative agreement with our own biochemistry results. Thus, our methodology appears to represent an alternative means for determination of FCD in soft hydrated tissues without requiring excessive tissue sectioning. Efforts to further validate this non-invasive methodology are currently underway. Assumptions in the study included the use of a linear stress/strain law to describe the deformation behavior of the solid matrix and ideal Donnan behavior. In the future, non-ideal Donnan behavior may be incorporated, including deviations of the osmotic coefficient from unity at low salt concentrations [6], and configurational entropy effects which produce an osmotic pressure above and beyond Donnan electrostatic interactions [9,20,54]; similarly, other tissue inhomogeneities can be incorporated, theoretically or through measured parameters, into the theoretical framework adopted in this study. Tissue anisotropy [8,9], tension-compression nonlinearity [16,49] and intrinsic viscoelasticity [16,51] are other properties that can be introduced as well.

Coupled with biological assays of cell-scale chondrocyte biosynthesis [38,60,61] and real-time fluorescence probes of cellular activities [40], there is a great potential in the future to study chondrocyte mechanotransduction *in situ* with a new level of specificity. With a better prediction of the chondrocyte environment *in situ*, we may begin to elucidate the mechanisms that underlie normal cartilage maintenance as well as the etiology of osteoarthritis [33]. Similarly, an understanding of the changes to the tissue inhomogeneity that accompany aging and disease may provide clues to the initiation and progression of joint degeneration [2,4,10,41,44,45]. It may also be used to assess the development of inhomogeneity in tissue engineered cartilage constructs, and to explore means of prescribing tissue inhomogeneity in growing tissues via modulation of the physical environment (e.g., loading, chemical mediators etc.) in custom bioreactors.

Acknowledgements

This study was supported by the National Institute of Arthritis and Muscoloskeletal and Skin Diseases (NIAMS) of the National Institutes of Health (AR 46568, AR 41913, AR 46532 and AR 45832). The authors also would like to thank Ms. Nadeen O. Chahine for help in the experiments.

References

[1] S. Akizuki, V.C. Mow, F. Muller, J.C. Pita, D.S. Howell and D.H. Manicourt, Tensile properties of human knee joint cartilage: I. Influence of ionic concentrations, weight bearing and fibrillation on the tensile modulus, *J. Orthop. Res.* **4** (1986), 379–392.

[2] C.G. Armstrong and V.C. Mow, Variations in the intrinsic mechanical properties of human articular cartilage with age degeneration and water content, *J. Bone Jt. Surg.* **64A** (1982), 88–94.

[3] G.A. Ateshian, H. Wang and W.M. Lai, The role of interstitial fluid pressurization and surface porosities on the boundary friction of articular cartilage, *J. Tribology* **120** (1998), 241–248.

[4] J.A. Buckwalter, K.E. Kuettner and E.J.-M. Thonar, Age-related changes in articular proteoglycans: Electron microscopic studies, *J. Orthop. Res.* **3** (1985), 251–257.

[5] M.D. Buschmann, Y.A. Gluzband, A.J. Grodzinsky and E.B. Hunziker, Mechanical compression modulates matrix biosynthesis in chondrocyte/agarose cultures, *J. Cell Sci.* **108** (1995), 1497–1508.

[6] M.D. Buschmann and A.J. Grodzinsky, A molecular model of proteoglycan-associated electrostatic forces in cartilage mechanics, *J. Biomech. Eng.* **117** (1995), 179–192.

[7] P.-H.G. Chao, R. Roy, R.L. Mauck, W. Liu, W.B. Valhmu and C.T. Hung, Chondrocyte translocation response to direct current electric fields, *J. Biomech. Eng.* **122** (2000).

[8] B. Cohen, W.M. Lai, G.S. Chorney, H.M. Dick and V.C. Mow, Unconfined compression of transversely-isotropic biphasic tissue, *Adv. Bioengng. Trans. ASME* **BED-19** (1992), 187–190.

[9] S. Ehrlich, N. Wolff, R. Schneiderman, A. Maroudas, K.H. Parker and C.P. Winlove, The osmotic pressure of chondroitin sulphate solutions: Experimental measurements and theoretical analysis, *Biorheology* **35:6** (1998), 383–397.

[10] D.R. Eyre, I.R. Dickson and K.P. Van Ness, Collagen cross-linking in human bone and articular cartilage. Age related changes in the content of matrix hydroxypyridinium residues, *Biochem. J.* **252** (1988), 495–500.

[11] J. Frangos, T. Huang and C.B. Clark, Steady shear and step changes in shear stimulate endothelium via independent mechanisms – superposition of transient and sustained nitric oxide production, *Biochem. Biophys. Res. Commun.* **224** (1996), 660–665.

[12] E.H. Frank and A.J. Grodzinsky, Cartilage electromechanics-II. A continuum model of cartilage electrokinetics and correlation with experiments, *J. Biomechanics* **20** (1987), 629–639.

[13] W.Y. Gu, W.M. Lai and V.C. Mow, A mixture theory for charged-hydrated soft tissues containing multi-electrolytes: passive transport and swelling behaviors, *J. Biomech. Eng.* **120** (1997), 169–180.

[14] F. Guilak, A. Ratcliffe and V.C. Mow, Chondrocyte deformation and local tissue strain in articular cartilage: a confocal microscopy study, *J. Orthop. Res.* **13** (1995), 410–422.

[15] F. Guilak, H.P. Ting-Beall, W.R. Jones, G.M. Lee and R.M. Hochmuth, Mechanical properties of chondrocytes and chondrons, *Adv. Bioengng. Trans. ASME* **BED-33** (1996), 253–254.

[16] C.-Y. Huang, A. Stankiewiecz, G.A. Ateshian, E.L. Flatow, L.U. Bigliani and V.C. Mow, Anisotropy, inhomogeneity and tension–compression nonlinearity of human glenohumeral cartilage in finite deformation, *Trans. Orthop. Res. Soc.* **24** (1999), 95.

[17] C.T. Hung, D. Henshaw, C.C.-B. Wang, R.L. Mauck, F. Raia, G. Palmer, V.C. Mow, A. Ratcliffe and W.B. Valhmu, Mitogen-activated protein kinase signaling in bovine articular chondrocytes in response to fluid flow does not require calcium mobilization, *J. Biomechanics* **33** (2000), 73–80.
[18] J.S. Jurvelin, M.D. Buschmann and E.B. Hunziker, Optical and mechanical determination of Poisson's ratio of adult bovine humeral articular cartilage, *J. Biomechanics* **30** (1997), 235–241.
[19] G.E. Kempson, Mechanical properties of articular cartilage, in: *Adult Articular Cartilage*, M.A.R. Freeman, ed., Pitman Medical, Kent, England, 1979, pp. 333–414.
[20] I.S. Kovach, The importance of polysaccharide configurational entropy in determining the osmotic swelling pressure of concentrated proteoglycan solution and the bulk compressive modulus of articular cartilage, *Biophys. Chem.* **53** (1995), 181–187.
[21] W.M. Lai, J.S. Hou and V.C. Mow, A triphasic theory for the swelling and deformational behaviors of articular cartilage, *J. Biomech. Eng.* **113** (1991), 245–258.
[22] M.J. Lammi, R. Inkinen, J.J. Parkkinen, T. Jakkinen, M. Jortikka, L.O. Nelimarkka, H.T. Jarvelainen and M.I. Tammi, Expression of reduced amounts of structurally altered aggrecan in articular cartilage chondrocytes exposed to high hydrostatic pressure, *Biochem. J.* **304** (1994), 723–730.
[23] D.A. Lee and D.L. Bader, Compressive strains at physiological frequencies influence the metabolism of chondroctyes seeded in agarose, *J. Orthop. Res.* **15** (1997), 181–188.
[24] A.I. Maroudas, Balance between swelling pressure and collagen tension in normal and degenerate cartilage, *Nature* **260** (1976), 808–809.
[25] A.I. Maroudas, H. Muir and J. Wingham, The correlation of fixed negative charge with glycosaminoglycan content of human articular cartilage, *Biochim. Biophys. Acta* **177** (1969), 492–500.
[26] A.I. Maroudas, Physicochemical properties of articular cartilage, in: *Adult Articular Cartilage*, M.A.R. Freeman, ed., Pitman Medical, Kent, UK, 1979, pp. 215–290.
[27] I. Martin, B. Obradovic, L.E. Freed and G. Vunjak-Novakovic, Method for quantitative analysis of glycosaminoglycan distribution in cultured natural and engineered cartilage, *Ann. Biomed. Eng.* **27** (1999), 656–662.
[28] R.L. Mauck, M.A. Soltz, C.C.-B. Wang, D.D. Wong, P.-H.G. Chao, W.B. Valhmu, C.T. Hung and G.A. Ateshian, Functional tissue engineering of articular cartilage through dynamic loading of chondrocyte-seeded agarose gels, *J. Biomech. Eng.* **122** (2000), 252–260.
[29] V.C. Mow and G.A. Ateshian, Lubrication and wear of diarthrodial joints, in: *Basic Orthopaedic Biomechanics*, V.C. Mow and W.C. Hayes, eds, 1997, Lippincott-Raven, Philadelphia, pp. 273–315.
[30] V.C. Mow, N.M. Bachrach, L.A. Setton and F. Guilak, Stress, strain, pressure and flow fields in articular cartilage and chondrocytes, in: *Cell Mechanics and Cellular Engineering*, V.C. Mow, F. Guilak, R. Tran-Son-Tay and R.M. Hochmuth, eds, 1994, Springer-Verlag, New York, pp. 345–379.
[31] V.C. Mow, S.C. Kuei, W.M. Lai and C.G. Armstrong, Biphasic creep and stress relaxation of articular cartilage in compression: theory and experiments, *J. Biomech. Eng.* **102** (1980), 73–84.
[32] V.C. Mow and A. Ratcliffe, Structure and function of articular cartilage and meniscus, in: *Basic Orthopaedic Biomechanics*, V.C. Mow and W.C. Hayes, eds, 1997, Lippincott-Raven, Philadelphia, pp. 113–177.
[33] V.C. Mow, C.C.-B. Wang and C.T. Hung, The extracellular matrix, interstitial fluid and ions as a mechanical signal transducer in articular cartilage, *Osteoarthritis Cartilage* **7** (1999), 41–59.
[34] V.C. Mow and C.C.-B. Wang, Some bioengineering considerations for tissue engineering of articular cartilage, *Clin. Orthop.* **367** Suppl. (1999), S204–S223.
[35] D.A. Narmoneva and L.A. Setton, Measurement of nonuniform swelling strains in full-thickness articular cartilage, *Trans. Orthop. Res. Soc.* **22** (1997), 81.
[36] D.A. Narmoneva, J.Y. Wang and L.A. Setton, Nonuniform swelling-induced residual strains in articular cartilage, *J. Biomechanics* **32** (1999), 401–408.
[37] J.J. Parkkinen, J. Ikonen, M.J. Lammi, J. Laakkonen, M. Tammi and H.J. Helminen, Effects of cyclic hydrostatic pressure on proteoglycan synthesis in cultured chondrocytes and articular cartilage explants, *Arch. Biochem. Biophys.* **300** (1993), 458–465.
[38] T. Quinn, A. Grodzinsky, M. Buschmann, Y. Kim and E. Hunziker, Mechanical compression alters proteoglycan deposition and matrix deformation around individual cells in cartilage explants, *J. Cell Sci.* **111** (1998), 573–583.
[39] P.M. Ragan, V.I. Chin, H.H. Hung, K. Masuda, E.J. Thonar, E.C. Arner, A.J. Grodzinsky and J.D. Sandy, Chondrocyte extracellular matrix synthesis and turnover are influenced by static compression in a new alginate disk culture system, *Arch. Biochem. Biophys.* **383** (2000), 256–264.
[40] S. Roberts, M. Knight, D. Lee and D. Bader, Mechanical compression influences intracellular Ca^{2+} signaling in chondrocytes seeded in agarose constructs, *J. Appl. Physiol.* **90** (2001), 1385–1391.
[41] V. Roth and V.C. Mow, The intrinsic tensile behavior of the matrix of bovine articular cartilage and its variation with age, *J. Bone Jt. Surg.* **62A** (1980), 1102–1117.

[42] R.M. Schinagl, D. Gurkis, C.C. Chen and R.L.-Y. Sah, Depth-dependent confined compression modulus of full-thickness bovine articular cartilage, *J. Orthop. Res.* **15** (1997), 499–506.
[43] R.M. Schinagl, M.K. Ting, J.H. Price and R.L. Sah, Video microscopy to quantitate the inhomogeneous equilibrium strain within articular cartilage during confined compression, *Ann. Biomed. Eng.* **24** (1996), 500–512.
[44] L.A. Setton, D.M. Elliot and V.C. Mow, Altered mechanics of cartilage with osteoarthritis: human osteoarthritis and an experimental model of joint degeneration, *Osteoarthritis Cartilage* **7** (1999), 2–14.
[45] L.A. Setton, V.C. Mow, F.J. Muller, J.C. Pita and D.S. Howell, Mechanical behavior and biochemical composition of canine knee cartilage following periods of joint disuse and disuse with remobilization, *Osteoarthritis Cartilage* **5** (1997), 1–16.
[46] D. Shin, J.-H. Lin and K. Athanasiou, Microindentation of the individual layers of articular cartilage, *Adv. Bioengng. Trans. ASME* **BED-36** (1997), 155–156.
[47] R.L. Smith, B.S. Donlon, M.K. Gupta, M. Mohtai, P. Das, D.R. Carter, J. Cooke, G. Gibbons, N. Hutchinson and D.J. Schurman, Effects of fluid-induced shear on articular chondrocyte morphology and metabolism *in vitro*, *J. Orthop. Res.* **13** (1996), 824–831.
[48] R.L. Smith, S.F. Rusk, B.E. Ellison, P. Wessells, K. Tsuchiya, D.R. Carter, W.E. Caler, L.J. Sandell and D.J. Schurman, In vitro stimulation of articular chondrocyte mRNA and extracellular matrix synthesis by hydrostatic pressure, *J. Orthop. Res.* **14** (1996), 53–60.
[49] M.A. Soltz and G.A. Ateshian, A conewise linear elasticity mixture model for the analysis of tension–compression nonlinearity in articular cartilage, *J. Biomech. Eng.* **122** (2000), 576–586.
[50] M.A. Soltz and G.A. Ateshian, Experimental verification and theoretical prediction of cartilage interstitial fluid pressurization at an impermeable contact interface in confined compression, *J. Biomechanics* **31** (1998), 927–934.
[51] J.K. Suh and S. Bai, Finite element formulation of biphasic poroviscoelastic model for articular cartilage, *J. Biomech. Eng.* **120** (1998), 195–201.
[52] D.N. Sun, W.Y. Gu, X.E. Guo, W.M. Lai and V.C. Mow, A mixed finite element formulation of triphasic mechano-electrochemical theory for charged, hydrated biological soft tissues, *Int. J. Numer. Meth. Engng.* **45** (1999), 1375–1402.
[53] D.N. Sun, W.Y. Gu, X.E. Guo, W.M. Lai and V.C. Mow, The influence of inhomogeneous fixed charge density of cartilage mechano-electrochemical behaviors, *Trans. Orthop. Res. Soc.* **23** (1998), 484.
[54] J.P.G. Urban, A. Maroudas, M.T. Bayliss and J. Dillon, Swelling pressures of proteoglycans at the concentrations found in cartilaginous tissues, *Biorheology* (1979), 447–464.
[55] C.C.-B. Wang, X.E. Guo, J.J. Deng, V.C. Mow, G.A. Ateshian and C.T. Hung, A novel non-invasive technique for determining distribution of fixed charge density within articular cartilage, *Trans. Orthop. Res. Soc.* **26** (2001), 129.
[56] C.C.-B. Wang, C.T. Hung and V.C. Mow, Analysis of the effects of depth-dependent aggregate modulus on articular cartilage stress–relaxation behavior in compression, *J. Biomechanics* **34** (2000), 75–84.
[57] C.C.-B. Wang and V.C. Mow, Effects of aggregate modulus inhomogeneity on cartilage compressive stress-relaxation behavior, *Adv. Bioengng. Trans. ASME* **BED-39** (1998), 261–262.
[58] C.C.-B. Wang, M.A. Soltz, R.L. Mauck, W.B. Valhmu, G.A. Ateshian and C.T. Hung, Comparison of equilibrium axial strain distributions in aricular cartilage explants and cell-seeded alginate disks under unconfined compression, *Trans. Orthop. Res. Soc.* **25** (2000), 131.
[59] M. Wong, M. Ponticiello, V. Kovanen and J.S. Jurvelin, Volumetric changes of articular cartilage during stress relaxation in unconfined compression, *J. Biomechanics* **33** (2000), 1049–1054.
[60] M. Wong, P. Wuethrich, M. Buschmann, P. Eggli and E. Hunziker, Chondrocyte biosynthesis correlates with local tissue strain in statically compressed adult articular cartilage, *J. Orthop. Res.* **15** (1997), 189–196.
[61] M. Wong, P. Wuethrich, P. Eggli and E. Hunziker, Zone-specific cell biosynthetic activity in mature bovine articular cartilage: a new method using confocal microscopic stereology and quantitative autoradiography, *J. Orthop. Res.* **14** (1996), 424–432.
[62] S.L.-Y. Woo, W.H. Akeson and G.F. Jemmott, Measurements of nonhomogeneous directional mechanical properties of articular cartilage in tension, *J. Biomechanics* **9** (1976), 785–791.

Biorheology 39 (2002) 27–37
IOS Press

Proteoglycan deposition around chondrocytes in agarose culture: Construction of a physical and biological interface for mechanotransduction in cartilage

T.M. Quinn [a,b,c], P. Schmid [c], E.B. Hunziker [c] and A.J. Grodzinsky [b,*]
[a] *Biomedical Engineering Laboratory, Swiss Federal Institute of Technology, Lausanne, Switzerland*
[b] *Center for Biomedical Engineering, Massachusetts Institute of Technology, Cambridge, MA, USA*
[c] *M.E. Mueller Institute for Biomechanics, University of Bern, Switzerland*

Abstract. With a view towards the development of methods for cartilage tissue engineering, matrix deposition around individual chondrocytes was studied during *de novo* matrix synthesis in agarose suspension culture. At a range of times in culture from 2 days to 1 month (long enough for cartilage-like material properties to begin to emerge), pericellular distributions of proteoglycan and matrix protein deposition were measured by quantitative autoradiography, while matrix accumulation and cell volumes were estimated by stereological methods. Consistent with previous work, tissue-average rates of matrix synthesis generally decreased asymptotically with time in culture, as *de novo* matrix accumulated. Cell-scale analysis revealed that this evolution was accompanied by a transition from predominantly pericellular matrix (within a few μm from the cell membrane) deposition early in culture towards proteoglycan and protein deposition patterns more similar to those observed in cartilage explants at later times. This finding may suggest a differential recruitment of different proteoglycan metabolic pools as matrix assembly progresses. Cell volumes increased with time in culture, suggestive of alterations in volume regulatory processes associated with changes in the microphysical environment. Results emphasize a pattern of *de novo* matrix construction which proceeds outward from the pericellular matrix in a progressive fashion. These findings provide cell-scale insight into the mechanisms of assembly of matrix proteins and proteoglycans in *de novo* matrix, and may aid in the development of tissue engineering methods for cartilage repair.

1. Introduction

Articular cartilage serves biomechanical roles in load bearing and joint lubrication in synovial joints. Its specialized extracellular matrix contains primarily cartilage-specific collagens and charged proteoglycans in a water-rich gel. A range of molecular properties which include electrostatic, electrokinetic, and transport phenomena allow proteoglycans to contribute to cartilage function in many different ways [10, 14]. Chondrocytes continually synthesize cartilage matrix, and can alter their rates [35] and spatial patterns [31] of pericellular matrix deposition in response to mechanical loads applied to the tissue. Although collagen deposition and remodelling is relatively slow in adult cartilage (requiring years [26]), proteoglycans are turned over much more rapidly (over the course of weeks [6]) and therefore represent a means of primary importance for the chondrocyte response to mechanical loads and matrix damage.

[*]Address for correspondence: Prof. A.J. Grodzinsky, MIT, Room 38-377, Cambridge, MA 02139, USA. Tel.: +1 617 253 4969; Fax: +1 617 258 5239; E-mail: alg@mit.edu.

Assembly of cell-associated proteoglycan-rich matrix and the chondrocyte metabolic response to mechanical compression are therefore of significant interest in cartilage physiology, biomechanics, and tissue engineering.

Chondrocytes in agarose represent a well-characterized model cell culture environment within which cell-associated matrix assembly can be observed during the progression from essentially no matrix (cells only) to the establishment of a biomechanically functional [3] "artificial tissue". The chondrocyte phenotype is well-preserved in agarose culture [1,17], making it valuable for the study of *de novo* matrix deposition by chondrocytes [21], and the influence of matrix accumulation on synthesis. Such information is of significant interest for the development of tissue engineering methods of cartilage repair. Indeed, agarose has even been used *in vivo* with some success as a "scaffold" for chondrocytes in an experimental cartilage repair application [34]. Additionally, since the cell response to compression has been characterized and shown to evolve as a function of time in agarose culture [2,9,20,22,27], more detailed study of matrix structure in this system could further elucidate relationships between mechanical compression, cell–matrix interactions, and pathways of the biological response.

Matrix proteoglycans (PGs) are abundant macromolecular constituents of the chondrocyte pericellular matrix which may play particularly important roles in mechanotransduction due to the wide range of physical phenomena in which they participate. PGs are the main source of resistance to solute transport and fluid flow in cartilage matrix [24] due to the dense polymeric meshwork they are believed to form *in situ* [28]. They are largely responsible for cartilage electrokinetic phenomena (such as streaming potentials) arising from flow-induced separation of electrical charges fixed to the PG polymeric structure from counter-ions dissolved in the matrix fluid phase [7,10]. They contribute significantly to cartilage mechanical stiffness, in large part due to double layer-mediated electrostatic repulsion between their constituent glycosaminoglycan (GAG) side chains [4]. Proteoglycans strongly influence cartilage mechanical function, and play central roles in the transduction of microphysical "signals" to chondrocytes during tissue compression. The pericellular distribution and evolution of matrix proteoglycan deposition during *de novo* matrix synthesis are therefore of significant interest for understanding the roles of proteoglycans in cartilage repair and the biological response to mechanical compression.

In an established chondrocyte-agarose culture system, we studied *de novo* matrix deposition around primary calf chondrocytes over a culture period long enough for cartilage-like material properties to begin to emerge. At a range of times in culture from 2 days to 1 month, spatial distributions of proteoglycan and matrix protein deposition around individual cells were measured by quantitative autoradiography, while matrix accumulation and cell volumes were estimated by stereological methods. Our objectives were to elucidate the roles of specific matrix components (particularly proteoglycans) in cartilage responses to compression and injury, insofar as *de novo* matrix synthesis in agarose culture represents a model for chondrocyte-mediated cartilage repair.

2. Methods

Under continual irrigation with phosphate buffered saline (PBS; Gibco) supplemented with 100 U/ml penicillin and 100 μg/ml streptomycin, articular cartilage was obtained from the distal femur (patellar groove) of freshly slaughtered 2-week old calves. Cartilage was sliced into small pieces of $\sim$1 mm characteristic size with a scalpel, and left overnight in culture medium (for each gram of tissue, 8 ml of DMEM (Gibco) high glucose, no HEPES, supplemented with 0.1 mM nonessential amino acids, 0.4 mM proline, 2 mM glutamine, 100 U/ml penicillin, 100 μg/ml streptomycin, 10% fetal bovine serum

(Hyclone), and 50 μg/ml ascorbate (Sigma)) in a tissue culture incubator. The next day, the tissue matrix was digested by sequential treatment with 56 U/ml pronase E (Sigma) for 1–1.5 hrs and 752 U/ml collagenase II (Sigma) for 3–4.5 hrs. Liberated chondrocytes were then isolated by passing digestion products through a 100 μm and then a 20 μm pore-size filter (Millipore), with centrifugation ($\sim$10 min at 100 g) and washing (removal of supernatant and resuspension in $\sim$50 ml culture media) after filtrations to remove enzymes. Isolated cells were then left overnight in culture media at 4°C. The next day, chondrocytes were resuspended and an aliquot was counted in a Coulter counter. Cells were then mixed with low melting temperature agarose (SeaPlaque, Sigma) and cast in 1 mm sheets of 3% agarose at a density of 2×10^7 cells/ml. 3 mm diameter disks were punched from these sheets and maintained in culture for up to 32 days with daily changes of culture media (80 μl per cell-gel disk).

On days 2, 6, 11, and 32 after the initiation of cell-gel culture, disks were incubated for 1.5 hrs in medium containing (nominally) 50 μCi/ml of either ^{35}S-sulfate (for labelling of newly-synthesized proteoglycans) or ^{3}H-proline (for labelling of newly-synthesized collagen and matrix proteins). After a 1 hr wash in DMEM (to remove unincorporated radiolabel), disks were chemically fixed in PBS containing 5% glutaraldehyde and 0.05 M Na-cacodylate (Sigma) and equilibrated in graded series of (respectively) water/ethanol, ethanol/propylene oxide, propylene oxide/Epon 812. After embedding in Epon 812 (Fluka), ultrathin (1 μm) sections taken axially through a disk diameter were exposed to an autoradiographic emulsion (Kodak NTB-2) for $\sim$1 week. Following emulsion development, sections were stained with Toluidine Blue O.

Cell volume and spatial patterns of cell-associated matrix deposition (autoradiography grain density as a function of distance from the cell membrane) were measured by image analysis as previously described [31]. Sections were systematically, randomly sampled [11] at $\sim$20 different locations and examined for cells with a well-defined nucleus–cytoplasm interface (indicating a section taken through a cell "diameter" such that distances from the cell membrane as seen on sections were representative of actual distances in three dimensions). Color images were centered on identified cells. Digitized high power light microscope (Olympus Vanox) images, 100 μm $\times$ 75 μm in total area at 6 pixels/μm resolution, were captured using a CCD color video camera (Sony), frame grabber (RasterOps XLTV) and microcomputer (Macintosh). Using an image processing program (IPLab Spectrum, Signal Analytics Corp.), the cell–matrix boundary was traced and autoradiography grains were identified by green intensity thresholding. The physical space in each image was then parameterized in terms of radial position relative to the traced cell–matrix boundary. Previously developed methods of calculating autoradiography grain density [5] were employed within regions of space defined by concentric annuli, 1 μm in breadth and conforming to individual cell shapes with an angular resolution of $\pi/12$ rad, at increasing distances from the cell membrane (cell-length scale grain distributions). Local "tissue-average" grain densities, calculated without regard for location relative to any cell membranes, were obtained using the entire 100 μm $\times$ 75 μm images. Traced cell–matrix interfaces from grain density measurements were used for estimation of cell volume with the nucleator sizing principle, using the average pixel location of the trace as reference point and sine-weighted directional probes for isotropic sampling within vertical sections [12]. Point counting of independently-acquired images was also used for the identification of volume fractions of cells and Toluidine Blue-staining matrix within cell–gel constructs. Although the Toluidine Blue stain appeared to function only above a certain threshold concentration of matrix PGs (data not shown) this measurement provided a useful indication of the accumulation of *de novo* matrix within cell–gel constructs during culture.

For each measured parameter, one-way ANOVA with post hoc Tukey testing was used to identify differences between radiolabelling times [8]. Results were considered to be significant for $p < 0.05$. Data are reported as mean $\pm$ sem.

3. Results

Spatial patterns of deposition of newly synthesized proteoglycans in the cell-associated matrix around chondrocytes were similar to those observed in calf cartilage explants [32], with the most rapid rates of deposition occurring in the immediately pericellular matrix (Fig. 1). In contrast, the visual impression of ^{3}H-proline histological autoradiography suggested that matrix protein deposition was somewhat more uniformly distributed throughout the matrix (Fig. 2), throughout the culture period.

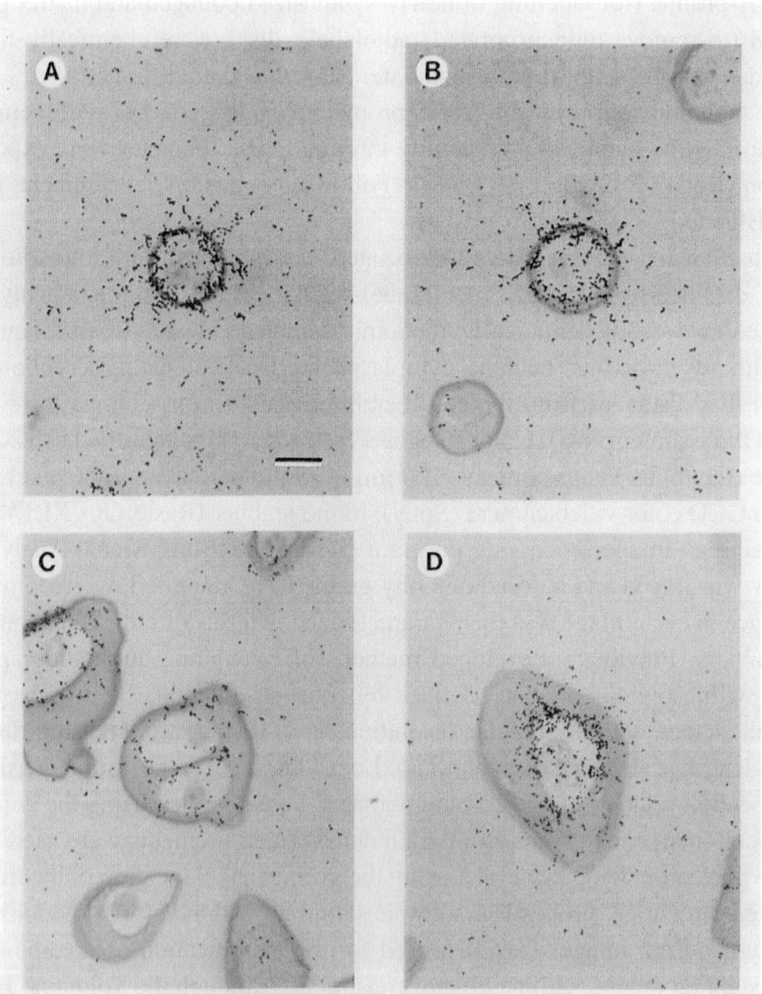

Fig. 1. Histological appearance of chondrocytes in agarose prepared for ^{35}S-sulfate autoradiography after (a) 2, (b) 6, (c) 11, and (d) 32 days of culture. Autoradiography grains represent proteoglycan deposition in the extracellular matrix.

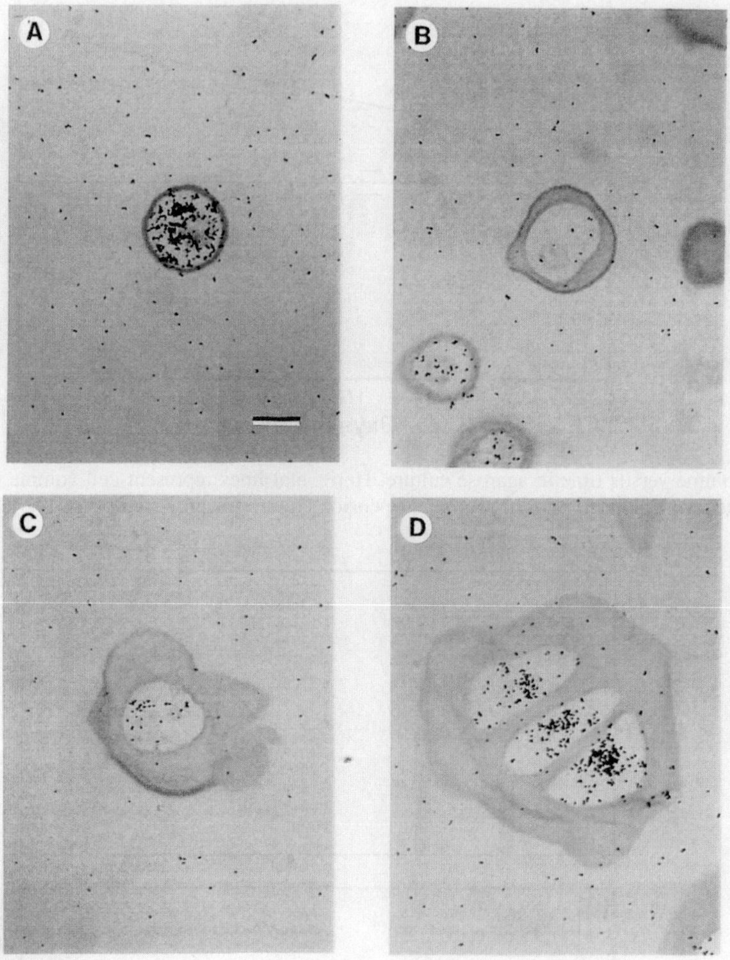

Fig. 2. Histological appearance of chondrocytes in agarose prepared for ^{3}H-proline autoradiography after (a) 2, (b) 6, (c) 11, and (d) 32 days of culture. Autoradiography grains represent deposition of collagen and matrix proteins in the extracellular matrix.

Over the culture period, chondrocyte volumes increased dramatically from values similar to those found in normal calf cartilage [32] on day 2 to almost double this at day 32 (Fig. 3). Therefore, isolation of chondrocytes from calf cartilage and casting in agarose appeared to have negligible effects on cell volumes, but subsequent culture for 32 days (during which the synthesis and deposition of *de novo* cartilaginous matrix proceeded rapidly) was associated with large increases in cell volumes.

The volume fraction of Toluidine Blue-staining matrix increased monotonically with time in culture (Fig. 4a), indicating the steady accumulation of *de novo* matrix. Tissue-average rates of proteoglycan and collagen/matrix protein deposition, as evidenced by ^{35}S-sulfate and ^{3}H-proline autoradiography, were highest during the first week of culture (Fig. 4b), consistent with previous observations [3]. In general, rates of matrix synthesis were highest on day 2 and then decreased monotonically to day 11, after which a steady-state or slightly increasing rate of deposition was evident between days 11 and 32 (Fig. 4b). These

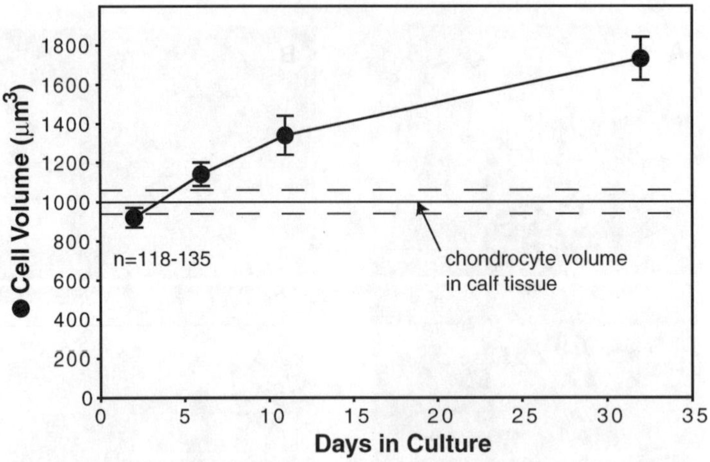

Fig. 3. Chondrocyte volume versus time in agarose culture. Horizontal lines represent cell volume in calf articular cartilage from which chondrocytes were isolated prior to agarose suspension culture (mean ± sem, $n = 142$ [32]).

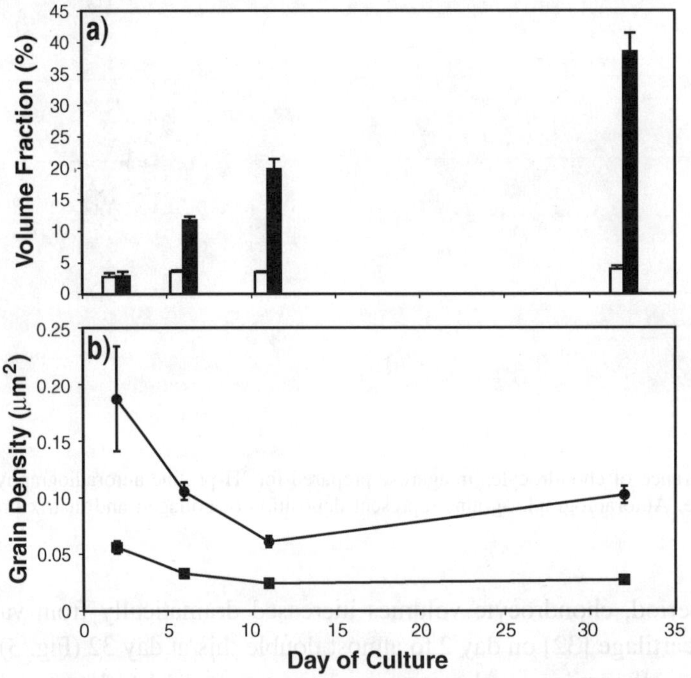

Fig. 4. (a) Volume fractions of chondrocytes (□) and Toluidine Blue-staining matrix (■) versus time in agarose culture. (b) Tissue-average densities of ^{35}S-sulfate (●) and ^{3}H-proline (■) autoradiography grains (representing proteoglycan and matrix protein deposition, respectively) appearing on histological sections.

"steady-state" deposition rates were comparable to those observed in calf articular cartilage explants under similar culture conditions [31,32].

Normalizing the spatial distributions of ^{35}S-autoradiography grains to their values just outside the cell membrane revealed that PG deposition was relatively more restricted to the pericellular matrix on day 2 than on day 32 (Fig. 5a). Similarly, collagen and matrix protein deposition as evidenced by ^{3}H-proline

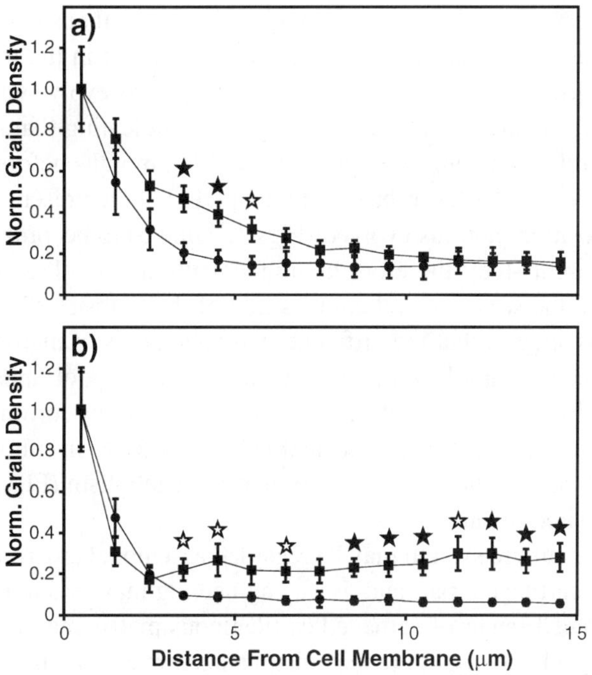

Fig. 5. Patterns of cell-associated matrix deposition as determined by image analysis of histological autoradiography. Extracellular grain densities (measured as a function of "radial" distance from the cell membrane) on days 2 (●) and 32 (■) have been normalized to those just outside the cell membrane. (a) ^{35}S-sulfate autoradiography grains reflect patterns of pericellular PG deposition ($n = 21$). (b) ^{3}H-proline autoradiography grains reflect patterns of deposition of collagen and other matrix proteins ($n = 21$). ☆: $p < 0.05$; ★: $p < 0.01$.

autoradiography was emphasized in the pericellular matrix on day 2 (Fig. 5b). However, with time in culture, the spatial distribution of matrix protein deposition included relatively more of the further-removed matrix by day 32. At this later time in culture, patterns of matrix protein deposition therefore appeared to be similar to cell-associated ^{3}H-proline deposition in calf cartilage, where deposition rates are greatest outside of the pericellular matrix [32].

4. Discussion

Consistent with previous findings, results indicate that assembly of *de novo* cell-associated matrix around chondrocytes in agarose gel culture progresses from the pericellular matrix outward. This appears to be consistent with understanding of cartilage matrix assembly and the role of cell–matrix interactions in "anchoring" both the proteoglycan and collagen network constituents of the extracellular matrix [14, 16]. Combined with previous results indicating rapidly turning over proteoglycan pools in the pericellular matrix [23], these observations suggest that pericellular matrix deposition and remodelling may represent a leading priority activity of chondrocytes, particularly within a damaged extracellular matrix.

Assuming a constant density (2×10^7 ml^{-1}) of spherical cells in a cubic lattice arrangement, the time in culture at which Toluidine Blue-staining cell-associated matrices around adjacent cells "contacted" one another was likely between days 6 and 11 (Figs 3 and 4a). This contiguity of deposited matrix would seem to mark an important point in the transition from a collection of individual cells in suspension culture to the establishment of a functional cartilage-like tissue. Previous studies have shown that the chondrocyte biosynthetic response to mechanical compression strengthens over a similar time period [2], suggesting a mediating role for the cartilaginous extracellular matrix in the transduction of tissue compression to cell activities. However, cartilage-like material properties evolve within chondrocyte–agarose cultures over somewhat longer times (more then 70 days [3]), indicating that *de novo* matrix deposition and remodelling may be a relatively long-term process requiring months before a biomechanically static, cartilage-like tissue is "regenerated". Over the one month period of the present study, cell-associated proteoglycan and protein deposition patterns evolved (Figs 1, 2 and 5) to become similar to those observed in calf tissue [32]. The present study therefore highlights patterns of cell-associated matrix deposition which depend heavily upon the structure and composition of the "substrate" to which *de novo* matrix is added. Interestingly, results suggest that the different metabolic pools of matrix proteoglycans may have been differentially affected: the rapidly turning-over "pericellular" pool appeared to be preferentially deposited at early times within a sparse matrix, while the more slowly turning over "further-removed" pool [15] formed a greater proportion of the deposited PG at later times in culture (Fig. 5a). These associations were consistent with the "tissue-scale" rates of matrix metabolism (Fig. 4b) which were relatively rapid early in culture compared to later times.

The chondrocyte-agarose culture system may be regarded as a model environment within which events important to cell-mediated cartilage repair and tissue engineering may be studied. Present results emphasize that chondrocyte-mediated repair of damaged cartilaginous matrix may follow a "scar-remodelling" temporal sequence similar to bone or skin repair: disorganized matrix is initially deposited rapidly, and then longer-term remodelling may continue at slower metabolic rates. Therefore, the high level of structural organization of adult articular cartilage might best be regarded as a long-term goal of cartilage tissue engineering, with the establishment of a minimally-functional, less organized, immature cartilage-like tissue a potentially necessary intermediate step. Indeed, it is possible that the fully organized structure of adult articular cartilage may only be attained after an extended period under *in vivo* biochemical and biomechanical conditions.

Pericellular matrix synthesis appears to be a "high priority" for chondrocytes synthesizing *de novo* matrix in agarose since it is the first matrix zone reconstructed and its assembly is associated with the highest metabolic rates (Figs 4b and 5). Combined with observations indicating that the chondrocyte response to compression changes with matrix assembly [2,20], these results suggest that the pericellular matrix plays an important role in mechanotransduction. This specialized matrix [29], adjacent to the cell membrane, is particularly rich in proteoglycans, poor in large-diameter collagen fibrils, and includes macromolecular constituents which differentiate it from the further-removed matrix. Its physical location implies that it must play a major role in cell–matrix interactions. Therefore, it seems likely that physical factors which may play a role in the chondrocyte response to compression, including solid deformations, physicochemical changes, alterations in transport properties of bioactive solutes, fluid flows, and electrical streaming potentials, are ultimately communicated to chondrocytes via the pericellular matrix. This matrix might therefore be usefully regarded as an extension of the cell itself which is essential to the expression of the chondrocyte phenotype, including matrix synthesis, remodelling, and the response to compression.

Concomitant with rapid, but steadily decreasing rates of matrix synthesis (Fig. 4b) during agarose culture, chondrocytes exhibited steadily increasing cell volumes (Fig. 3). This suggested alterations to cell volume regulatory processes secondary to changes in the cell microphysical environment due to matrix damage. This was also consistent with the interpretation that cellular hypertrophy can indicate rapid metabolic activity and an attempted repair response. At the same time, these observations emphasize that

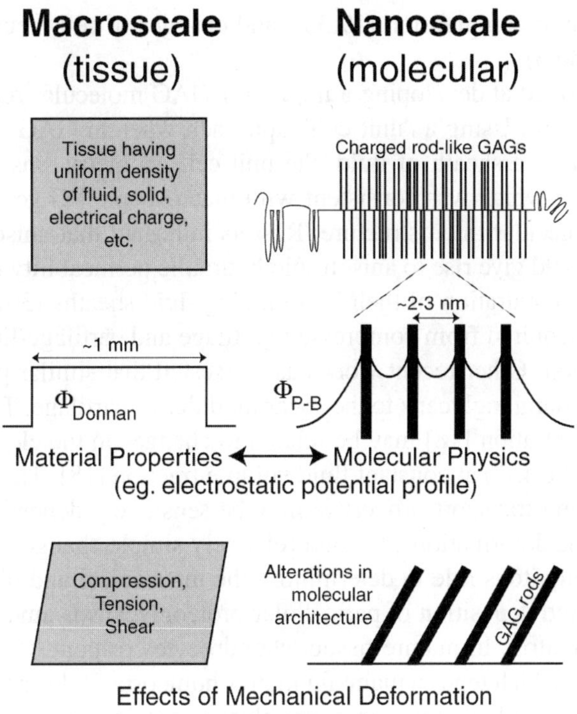

Fig. 6. Schematic representation of relationships between tissue-scale properties of cartilage and molecular-scale behavior of matrix proteoglycans (adapted from [4]). For example, the macroscopic Donnan electrical potential profile representing the charged GAG-filled tissue as a medium of uniform fixed charge density may be compared to the space-varying potential profile derived from the Poisson–Boltzmann model, pictured as surrounding each rod-like charged GAG chain of the aggrecan proteoglycan molecule. The Donnan model assumes a constant potential Φ_{Donnan} while the P–B model accounts for the space-varying potential between GAG chains. These models are each cast at different length scales (macroscopic versus molecular) and can thereby give significantly different predictions of the swelling stress and stiffness of the charged, poroelastic tissue. Consideration of effects of tissue mechanical deformations on proteoglycan architecture may further refine understanding of these relationships.

cell volume alone may be a relatively poor indicator of metabolic activity: the chondrocyte response to static compression exhibits opposite trends, whereby cell volume and metabolism correlate [5].

Proteoglycans play a central governing role in the poroelastic mechanics and biochemistry of the cartilage extracellular matrix. Double-layer mediated electrostatic interactions between glycosaminoglycan molecules [10] (Fig. 6) imbue the “proteoglycan gel” with swelling forces that contribute directly to the tissue compressive modulus, and indirectly to the tensile modulus due to pre-stress of the collagen network [25]. Due in part to their association with the Poisson–Boltzmann equation (Fig. 6), these electrically mediated swelling forces are highly nonlinear functions of PG gel fluid volume fraction [4], contributing to a marked stiffening of the matrix as it is compressed. Fluid flow in cartilage is also governed largely by PG gel density [24]. Classical models for flow through arrays of cylindrical fibers [13] have been shown to provide useful models for cartilage hydraulic permeability [7] when GAG molecules are treated as rod-like particles. Solute transport within the cartilage extracellular matrix can also be strongly dependent upon PG gel density [24], particularly for solutes which interact with PGs electrically, sterically (due to large physical size), or via biochemical binding. Since adult articular cartilage is avascular, transport limitations of bioactive solutes through the PG gel may play an important role in mediating the cell response to compression [19,33], and changes in cell metabolic activity with time in agarose culture (Figs 4b and 5).

Some recent work has aimed at developing a model for GAG molecular rearrangements during cartilage compression [30] (Fig. 6). Using a "unit cell" approach wherein GAG rods were approximated as solid cylinders surrounded by a sheath of fluid, the unit cell geometry was required to deform during tissue compression in a way which was consistent with macroscopic PG gel deformations and some of the basic constraints of molecular microstructure. Results indicated that anisotropic deformations of an initially isotropic PG gel could give rise to anisotropic hydraulic permeability due to selective orientation of GAGs and nonuniform dehydration of their surrounding fluid sheaths [30]. Predictions of the model were consistent with data acquired from compressed cartilage and cartilage-like materials under physiological levels of compression. Other recent work has suggested that similar phenomena may underlie a direct contribution by glycosaminoglycans to the shear modulus of cartilage. The dependence of the shear modulus on bath salt concentration [18] may be related to changes in the electrostatic energy density of GAGs (and associated "unit cells") at constant fluid volume fraction [18]. Therefore, the contributions of PGs to cartilage material and transport properties may be sensitively dependent upon the details of the molecular response to tissue deformations, beyond relatively simple changes in fluid volume fraction.

Proteoglycans play a ubiquitous role in determining the mechanical and transport properties of cartilage. In agarose culture, rapid deposition of pericellular proteoglycans is among the earliest events in the establishment of *de novo* matrix. In mature tissue, chondrocytes respond to physical stimuli from their microenvironment, in ways which tend to maintain matrix homeostasis during cartilage use, overuse, and even damage. It seems inescapable that matrix proteoglycans mediate the "flow of physical information" during mechanical compression, from the scale of the tissue to that of the cell. Proteoglycans therefore represent the source of a rich array of physicochemical information for chondrocytes, as well as one of their most important means of biological response to such stimuli.

Acknowledgements

This work was supported by NIH Grant AR45779, AO/ASIF Grant 96-Q76, and a fellowship from the Arthritis Society of Canada (TMQ). We thank Drs. S.T. Kung and M.D. Buschmann for helpful discussions, and V. Gaschen, A. Maurer, P. Perumbuli and M. Ponticello for technical assistance.

References

[1] P.D. Benya and J.D. Shaffer, Dedifferentiated chondrocytes reexpress the differentiated collagen phenotype when cultured in agarose gels, *Cell* **30** (1982), 215–224.

[2] M.D. Buschmann, Y.A. Gluzband, A.J. Grodzinsky and E.B. Hunziker, Mechanical compression modulates matrix biosynthesis in chondrocyte/agarose culture, *J. Cell Sci.* **108** (1995), 1497–1508.

[3] M.D. Buschmann, Y.A. Gluzband, A.J. Grodzinsky, J.H. Kimura and E.B. Hunziker, Chondrocytes in agarose culture synthesize a mechanically functional extracellular matrix, *J. Orthop. Res.* **10** (1992), 745–758.

[4] M.D. Buschmann and A.J. Grodzinsky, A molecular model of proteoglycan-associated electrostatic forces in cartilage mechanics, *J. Biomech. Eng.* **117** (1995), 179–192.

[5] M.D. Buschmann, E.B. Hunziker, Y.J. Kim and A.J. Grodzinsky, Altered aggrecan synthesis correlates with cell and nucleus structure in statically compressed cartilage, *J. Cell Sci.* **109** (1996), 499–508.

[6] M.A. Campbell, C.J. Handley, V.C. Hascall, R.A. Campbell and D.A. Lowther, Turnover of proteoglycans in cultures of bovine articular cartilage, *Arch. Biochem. Biophys.* **234**(1) (1984), 275–289.

[7] S.R. Eisenberg and A.J. Grodzinsky, Electrokinetic micromodel of extracellular matrix and other polyelectrolyte networks, *Physicochemical Hydrodynamics* **10** (1988), 517–539.

[8] L.D. Fischer and G. vanBelle, *Biostatistics: A Methodology for the Health Sciences*, John Wiley & Sons, New York, 1993.

[9] P.M. Freeman, R.N. Natarajan, J.H. Kimura and T.P. Andriacchi, Chondrocyte cells respond mechanically to compressive loads, *J. Orthop. Res.* **12**(3) (1994), 311–320.

[10] A.J. Grodzinsky, Electromechanical and physicochemical properties of connective tissues, *CRC Critical Reviews in Biomedical Engineering* **9**(2) (1983), 133–199.

[11] H.J.G. Gundersen and E.B. Jensen, The efficiency of systematic sampling in stereology and its prediction, *J. Microsc.* **147**(3) (1987), 229–263.

[12] H.J.G. Gunderson, The nucleator, *J. Microsc.* **151**(1) (1988), 3–21.

[13] J. Happel, Viscous flow relative to arrays of cylinders, *AIChE J.* **5** (1959), 174–177.

[14] T.E. Hardingham and A.J. Fosang, Proteoglycans: many forms and many functions, *FASEB J.* **6** (1992), 861–870.

[15] H.J. Häuselmann, M.B. Aydelotte, B.L. Schumacher, K.E. Kuettner, S.H. Gitelis and E.J.M.A. Thonar, Synthesis and turnover of proteoglycans by human and bovine articular chondrocytes cultured in alginate beads, *Matrix* **12** (1992), 116–129.

[16] D.K. Heinegard and E.R. Pimentel, *Cartilage Matrix Proteins*, Raven Press, New York, 1992.

[17] A.L. Horwitz and A. Dorfman, The growth of cartilage cells in soft agar and liquid suspension, *J. Cell. Biol.* **45**(2) (1970), 434–438.

[18] M. Jin and A. Grodzinsky, Effect of electrostatic interactions between GAGs on the shear stiffness of cartilage: a molecular model and experiments, *Macromolecules* **34** (2001), 8330–8339.

[19] Y.J. Kim, L.J. Bonassar and A.J. Grodzinsky, The role of cartilage: streaming potential, fluid flow and pressure in the stimulation of chondrocyte biosynthesis during dynamic compression, *J. Biomech.* **28**(9) (1995), 1055–1066.

[20] M.M. Knight, D.A. Lee and D.L. Bader, The influence of elaborated pericellular matrix on the deformation of isolated articular chondrocytes cultured in agarose, *Biochim. Biophys. Acta* **1405** (1998), 67–77.

[21] A.M. Kucharska, K.E. Kuettner and J.H. Kimura, Biochemical characterization of long-term culture of the swarm rat chondrosarcoma chondrocytes in agarose, *J. Orthop. Res.* **8** (1990), 781–792.

[22] D.A. Lee and D.L. Bader, Compressive strains at physiological frequencies influence the metabolism of chondrocytes seeded in agarose, *J. Orthop. Res.* **15** (1997), 181–188.

[23] S. Lohmander, Turnover of proteoglycans in guinea pig costal cartilage, *Arch. Biochem. Biophys.* **180** (1977), 93–101.

[24] A. Maroudas, Biophysical chemistry of cartilaginous tissues with special reference to solute and fluid transport, *Biorheology* **12** (1975), 233–248.

[25] A. Maroudas, Balance between swelling pressure and collagen tension in normal and degenerate cartilage, *Nature* **260** (1976), 808–809.

[26] A. Maroudas, G. Palla and E. Gilav, Racemization of aspartic acid in human articular cartilage, *Connect Tiss. Res.* **28**(3) (1992), 161–169.

[27] R.L. Mauck, M.A. Soltz, C.C. Wang, D.D. Wong, P.H. Chao, W.B. Valhmu, C.T. Hung and G.A. Ateshian, Functional tissue engineering of articular cartilage through dynamic loading of chondrocyte-seeded agarose gels, *J. Biomech. Eng.* **122**(3) (2000), 252–260.

[28] H. Muir, Heberden Oration, 1976. Molecular approach to the understanding of osteoarthrosis, *Ann. Rheum. Dis.* **36**(3) (1977), 199–208.

[29] A.R. Poole, I. Pidoux, A. Reiner and L. Rosenberg, An immunoelectron microscope study of the organization of proteoglycan monomer, link protein and collagen in the matrix of articular cartilage, *J. Cell. Biol.* **93** (1982), 921–937.

[30] T.M. Quinn, P. Dierickx and A.J. Grodzinsky, Glycosaminoglycan network geometry may contribute to anisotropic hydraulic permeability in cartilage under compression, *J. Biomech.* **34**(11) (2001), 1483–1490.

[31] T.M. Quinn, A.J. Grodzinsky, M.D. Buschmann, Y.J. Kim and E.B. Hunziker, Mechanical compression alters proteoglycan deposition and matrix deformation around individual cells in cartilage explants, *J. Cell Sci.* **111** (1998), 573–583.

[32] T.M. Quinn, A.J. Grodzinsky, E.B. Hunziker and J.D. Sandy, Effects of injurious compression on matrix turnover around individual cells in calf articular cartilage explants, *J. Orthop. Res.* **16**(4) (1998), 490–499.

[33] T.M. Quinn, V. Morel and J.J. Meister, Static compression of articular cartilage can reduce solute diffusivity and partitioning: implications for the chondrocyte biological response, *J. Biomech.* **34**(11) (2001), 1463–1469.

[34] B. Rahfoth, J. Weisser, F. Sternkopf, T. Aigner, K. von der Mark and R. Brauer, Transplantation of allograft chondrocytes embedded in agarose gel into cartilage defects of rabbits, *Osteoarthritis Cartilage* **6**(1) (1998), 50–65.

[35] R.L.Y. Sah, Y.J. Kim, J.Y.H. Doong, A.J. Grodzinsky, A.H.K. Plaas and J.D. Sandy, Biosynthetic response of cartilage explants to dynamic compression, *J. Orthop. Res.* **7** (1989), 619–636.

Biorheology 39 (2002) 39–45
IOS Press

Electrical signals for chondrocytes in cartilage

W.M. Lai, D.D. Sun, G.A. Ateshian, X.E. Guo and V.C. Mow
Departments of Mechanical Engineering, Orthopaedic Surgery and Biomedical Engineering, Columbia University, New York, NY 10027, USA
Tel.: +1 212 854 4236; Fax: +1 212 854 4404; E-mail: WML1@Columbia.edu

Abstract. An important step toward understanding signal transduction mechanisms modulating cellular activities is the accurate predictions of the mechanical and electro-chemical environment of the cells in well-defined experimental configurations. Although electro-kinetic phenomena in cartilage are well known, few studies have focused on the electric field *inside* the tissue. In this paper, we present some of our recent calculations of the electric field *inside* a layer of cartilage (with and without cells) in an open circuit one-dimensional (1D) stress relaxation experiment. The electric field inside the tissue derives from the streaming effects (streaming potential) and the diffusion effect (diffusion potential). Our results show that, for realistic cartilage material parameters, due to deformation-induced inhomogeneity of the fixed charge density, the two potentials compete against each other. For softer tissue, the diffusion potential may dominate over the streaming potential and vice versa for stiffer tissue. These results demonstrate that for proper interpretation of the mechano-electrochemical signal transduction mechanisms, one must not ignore the diffusion potential.

1. Introduction

The biologic activities of the chondrocyte population are regulated by genetic, and other biologic and biochemical factors, as well as environmental factors. It has often been noted that physical environmental factors, such as stress, flow, electric field, etc. are as strong as biologic factors in regulating cellular activities [10,21]. In recent years, there has been much research on the effects of mechanical and/or hydrostatic/osmotic pressure loading on cartilage explant metabolism (see [21] and [11] and the references therein). Such studies have been specifically aimed at elucidating possible “mechano-signal” transduction mechanism(s) that might govern the chondrocytes’ biosynthetic activities in maintaining and organizing the extracellular matrix (ECM) comprising the tissue (see the reviews by Comper [3], and Mow and Ratcliffe [22]). Although the electrical events in cartilage have been observed by many researchers for more than three decades [1,2,5–8,13,17–20], few studies however have focused on the details of the electrical potential *within* the ECM where the chondrocytes reside. The development of more elaborate constitutive models of cartilage in recent years has enabled us to calculate the various mechanical and electrical signals that the chondrocytes might see *in situ* [16]. In this paper, we present some of the results that we have obtained for the electric field in and around a cell inside a tissue under a 1D stress relaxation experimental configuration. The contribution of diffusion potential to the total potential field, ignored by most cartilage researcher, is also assessed.

0006-355X/02/$8.00

2. Streaming potential and diffusion potential in the absence of electric current

Using the triphasic theory [15], one can derive the following equation for the electric current density in cartilage (see also [9] and [12])

$$I_e = -g_o \nabla P - \frac{RT}{F_c}\Big[\sum_{\alpha=+,-} g_\alpha \nabla\big(\ln(\gamma_\alpha c^\alpha)\big)\Big] - \chi_o \nabla\psi, \tag{1}$$

where P is the pressure (hydraulic and osmotic), c^+ and c^- are cation and anion concentrations (per unit tissue water volume), γ_α are activity coefficients, F_c is Faraday constant, ψ is electric potential, R is universal gas constant, T is absolute temperature, g_o, g_α and χ_o are material parameters which are functions of the ion concentrations and the frictional coefficients. Equation (1) states that the electric current density at any point is the vector sum of three kinds of currents: the convection current (the first term on the right-hand side of Eq. (1)), the diffusion current (the second term) and the conduction current (the last term). The driving force for the convection current is the mechano-chemical force ∇P, generated by the gradient of the pressures. The driving force for the diffusion current is the electro-chemical force $\nabla(\gamma_\alpha c^\alpha)$, generated by the gradient of the concentration c^α for the ionic species; and the driving force for the conduction current is the electric force $\nabla\psi$, generated by the gradient of the electric potential ψ. The fluid flow, driven by the pressure gradient, convects the cations and anions (of unequal concentration), resulting in a *convection current*. The diffusion of cations and anions at different speeds or at different directions, driven by the gradient of concentrations, results in a *diffusion current*, and the movements of ions under a non-zero gradient of electric potential ψ give rise to a *conduction current*. When external circuits are not provided for the tissue to sustain a net flow of ions and electrons, the tissue is in a state of zero current. This condition is the most commonly used experimental configuration in the study of charged biological tissues for the determination of their electrical behaviors. For such cases, at every point in the tissue, the sum of the three currents must vanish. In this currentless experimental condition, there is no externally applied electric potential, the potential is entirely *induced* by the convection current and the diffusion current. This induced potential generates a conduction current to oppose the convection current and the diffusion current so as to achieve the zero current condition. The potential induced by convection in the presence of a pressure gradient is the *streaming potential*. The potential induced by the diffusion in the presence concentration gradient is the *diffusion potential*.

3. Confined compression stress relaxation experiment in and around a cell

The length scale for cells differs from that for the tissue by more than two orders of magnitude (~1 mm for the tissue and ~10 μm for the cell), therefore, it is necessary to use a two-scale finite-element model to analyze the events surrounding a cell. In the following, we first present our parametric study of the macro-scale problem so as to gain understanding on the relative importance of streaming and diffusion potential in this problem. The parameter that has an important effect on the diffusion potential is the intrinsic aggregate modulus H_a of the ECM.

3.1. Macroscale problem (ECM in the absence of cells)

A schematic diagram for a 1D ramped displacement stress relaxation experiment is shown in Fig. 1. The layer of tissue is confined within a rigid-cylindrical container. The side-wall and bottom are insulated

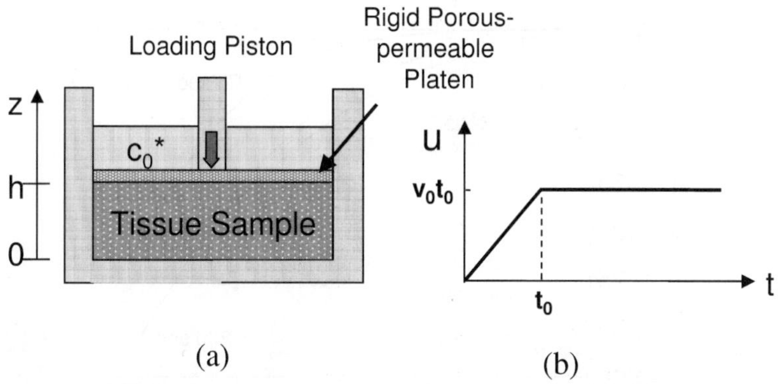

Fig. 1. (a) Schematic of an open-circuit, 1D ramped-displacement, stress-relaxation experiment. The bathing NaCl solution concentration c^* is kept fixed during the experiment, and the motion of the loading piston is prescribed in (b). The surface-to-surface compressive strain is 10%, $t_0 = 200$ s and $h = 1$ mm.

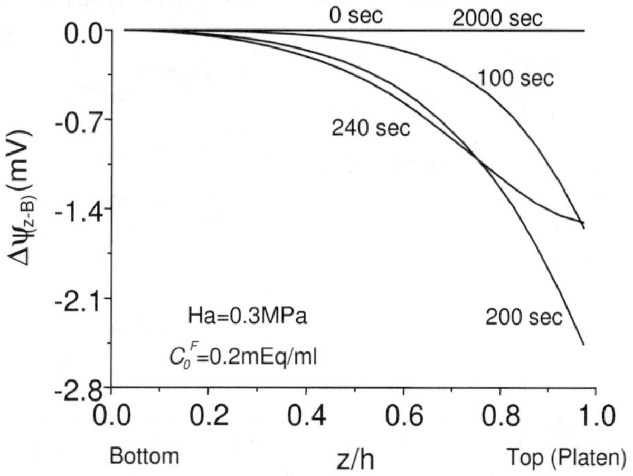

Fig. 2. Electric potential distribution *inside* the tissue at various times for $H_a = 0.3$ MPa. The potential increases in the direction toward the bottom indicating that it is dominated by the diffusion potential. Initial FCD $c_0^F = 0.2$ mEq/ml; diffusivities of Na^+ and Cl^- are 0.5×10^{-9} and 0.8×10^{-9} m^2/s, respectively, porosity $\phi_0^w = 0.8$, friction coefficient between water and solid matrix $K = 7 \times 10^{14}$ N s/m^4.

against flow of water and ions. Compressing the tissue from its top is a rigid porous-permeable loading platen. The ramped-displacement imposed on this loading platen is given in Fig. 1(b). The tissue is initially equilibrated in a NaCl solution of concentration $c^* = 0.15$ M NaCl. Prior to the application of the ramped displacement $U(t)$, the tissue is in equilibrium where the anions distribution is given by the Donnan equilibirium distribution law $c^+(z,0)c^-(z,0) = \gamma_\pm^2 c^{*2}$ [19,22]. With v^α ($\alpha = s, +$ and $-$) denoting the velocity of solid, cation and anion, respectively, the currentless condition (open circuit) is expressed as $c^+(v^+ - v^s) = c^-(v^- - v^s)$. The solution of this problem is obtained using the finite element formulation of Sun et al. [23].

Figure 2 shows the electric potential distribution $\Delta\psi_{z-B}[= \psi(z,t) - \psi(0,t)]$ at various times, for a tissue with aggregate modulus $H_a = 0.3$ MPa. It is seen here that $\Delta\psi_{z-B}$ is negative for all times, that is $\psi(0,t) \geqslant \psi(z,t)$ so that the electric potential increases from porous platen toward the bottom of the specimen, indicating that the diffusion potential dominates over the streaming potential. This large diffusion potential effect is due to the compaction of the charged solid matrix (see Fig. 3), caused by the

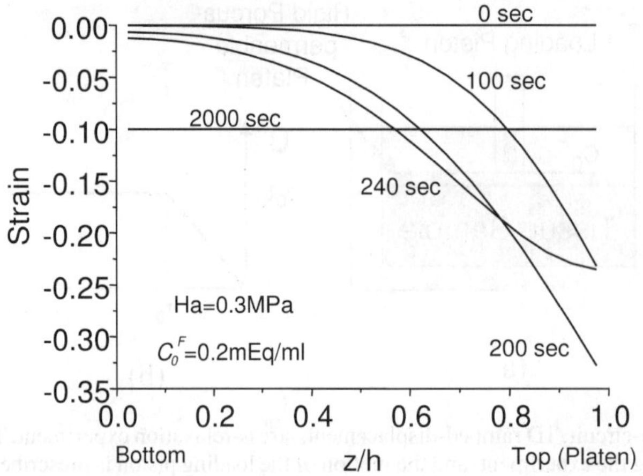

Fig. 3. Strain distribution *inside* the tissue at various times. The strain is caused by frictional drag force of permeation between water and solid matrix. The strain increases monotonically in the upward (flow) direction, so does the FCD (not shown).

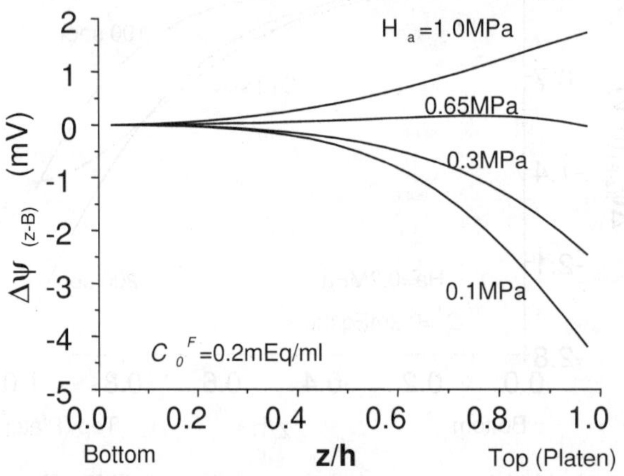

Fig. 4. Electric potential distribution *inside* the tissue at time $t = 200$ s (i.e., at the end of the compression-ramp phase) for four values of aggregate modulus. For more rigid tissue ($H_a > 0.65$ MPa), the streaming potential effect dominates whereas for softer tissues ($H_a < 0.65$ MPa), the diffusion potential effect dominates.

drag forces, exerted by the fluid on the solid matrix, during the ramped-phase of the experiment [14]. For tissues with larger compressive stiffness, the opposite may be true, i.e., the streaming potential dominates with more positive potential near the loading platen. The electric potential for four values of aggregate modulus H_a are shown in Fig. 4. The results show that the electric fields inside the tissue reverse its polarity at approximately $H_a = 0.65$ MPa.

3.2. Microscale problem – electric field in and around a cell

Once the macro-scale model is solved, its solid displacement, water and ion chemical/electrochemical potentials are then used to define the boundary conditions at the interface of the micro-scale cell–matrix model. In the following we discuss the results of our calculation using this two-scale approach. The chondrocyte is assumed to be spherical with a radius $= 5\ \mu$m. Due to the sparse population of chondrocytes in articular cartilage (10% in volume in mature tissue [24]), only one chondrocyte is assumed

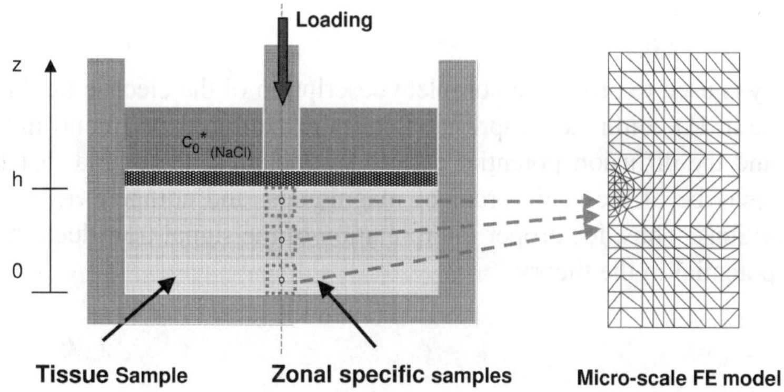

Fig. 5. The schematic representations of the two-scale triphasic finite element model for tissue and cell. The right figure is the micro-scale model.

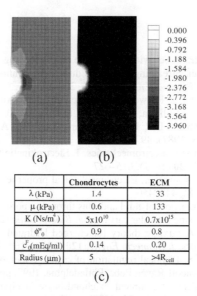

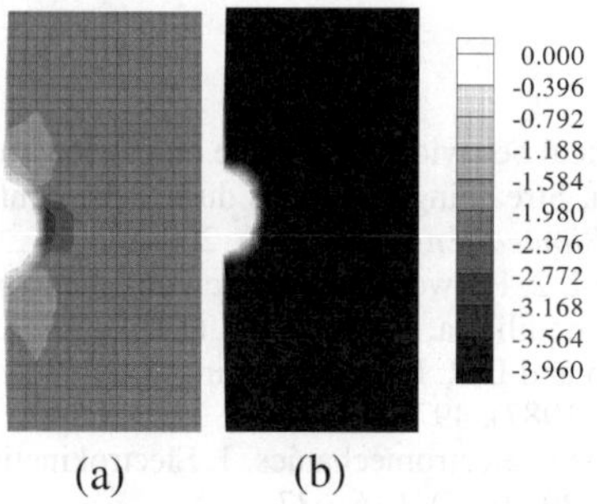

	Chondrocytes	ECM
λ (kPa)	1.4	33
μ (kPa)	0.6	133
K (Ns/m^4)	5×10^{10}	0.7×10^{15}
ϕ^w_0	0.9	0.8
c^F_0(mEq/ml)	0.14	0.20
Radius (μm)	5	$>4R_{cell}$

(c)

Fig. 6. The contour plots of the electric field in and around a cell at the ramp peak: (a) Cell near the tissue surface. (b) Cell at the middle zone depth. The electrical potential shown is relative to the center of the cell. (c) Properties of ECM and chondrocyte are listed in the table, where λ and μ are Lame constants for the solid phase.

to be embedded inside the micro-scale tissue domain (Fig. 5). That is, we assume that the interactions between and among chondrocytes are negligible. The cell is also modeled as a triphasic medium but with different material parameters from those of the ECM, as represented in the literature [11]. Further, it is assumed that the cell adheres to the ECM. The cell–matrix domain is chosen to be 5 times of the cell radius. The external bathing solution was chosen to be 0.15 M NaCl solution. The calculated results show that the electric field depends on the location of the cell in the tissue as well as time. For example, at the peak of the ramped displacement, the electrical potential difference between the territorial matrix and the center of the cell is −2.2 mV for a cell near the surface versus −4.2 mV for a cell at 1/2 depth from the surface. It should be noted that the cell near the surface was compressed more than the cell at mid-depth so that the difference in FCD between the cell and ECM is less for the cell near the surface. For a complete picture of the electric field surrounding a cell, we show in Fig. 6 a contour plot of the electric field (relative to the center of the cell) in and around a cell in the surface zone.

4. Conclusions

The triphasic theory is used to provide a complete description of the electric field in and around a cell inside a layer of tissue in 1D confined compression stress relaxation experiment. In the theory, both the streaming potential and the diffusion potential effects are included. We found that for cartilage ECM, these two potentials may be of the same order of magnitudes, and compete *against* each other under realistic loading conditions. Thus, for proper interpretation of the signal transduction data, one must not ignore the diffusion potential in the theory.

Acknowledgement

This work was supported in part by NIH Grant No.: AR41913, and No.: AR42850.

References

[1] C.A.L. Bassett and R.J. Pawluk, Electrical behavior of cartilage during loading, *Science* **178** (1972), 982–983.
[2] A.C. Chen, T.T. Nguyen and R.L. Sah, Streaming potentials during the confined compression creep test of normal and proteoglycan-depleted cartilage, *Ann. Biomed. Eng.* **25** (1997), 269–277.
[3] W.D. Comper, *Extracellular Matrix*, Vol. 2, Harwood Academic Publishers, Australia, 1996, pp. 1–386.
[4] F.G. Donnan, The theory of membrane equilibria, *Chem. Rev.* **1** (1924), 73–90.
[5] E.H. Frank, A.J. Grodzinsky, T.J. Koob and D.R. Eyre, Streaming potentials: A sensitive index of enzymatic degradation in articular cartilage, *J. Orthop. Res.* **5** (1987), 497–508.
[6] E.H. Frank and A.J. Grodzinsky, Cartilage electromechanics. I. Electrokinetic transduction and the effects of electrolyte pH and ionic strength, *J. Biomechanics* **20** (1987), 615–627.
[7] A.J. Grodzinsky, H. Lipshitz and M.J. Glimcher, Electromechanical properties of articular cartilage during compression and stress relaxation, *Nature* **275** (1978), 448–450.
[8] W.Y. Gu, W.M. Lai and V.C. Mow, Transport of fluid and ions through a porous-permeable charged-hydrated tissue, and streaming potential data on normal bovine articular cartilage, *J. Biomechanics* **26** (1993), 709–723.
[9] W.Y. Gu, W.M. Lai and V.C. Mow, A mixture theory for charged hydrated soft tissues containing multi-electrolytes: passive transport and swelling behaviors, *J. Biomech. Engng.* **120** (1998), 169–180.
[10] F.A. Guilak, R.L. Sah and L.A. Setton, Physical regulation of cartilage metabolism, *Basic Orthopaedic Biomechanics*, V.C. Mow and W.C. Hayes, eds, Lippincott-Raven Pubs, Philadelphia, 1997, pp. 179–207.
[11] F. Guilak and V. Mow, The mechanical environment of chondrocyte: a biphasic finite element model of cell–matrix interactions in articular cartilage, *J. Biomechanics* **33** (2000), 1663–1673.
[12] J.M. Huyghe, J.D. Janssen, Quadriphasic mechanics of swelling incompressible porous media, *Int. J. Engng. Sci.* **35** (1997), 793–802.
[13] Y.J. Kim, L.J. Bonassar and A.J. Grodzinsky, The role of cartilage streaming potential, fluid flow and pressure in the stimulation of chondrocyte biosynthesis during dynamic compression, *J. Biomechanics* **28** (1995), 1055–1066.
[14] W.M. Lai and V.C. Mow, Drag-induced compression of articular cartilage during permeation experiment, *Biorheology* **17** (1980), 111–123.
[15] W.M. Lai, J.S. Hou and V.C. Mow, A triphasic theory for swelling and deformation behavior of articular cartilage, *J. Biomech. Engng.* **113** (1991), 245–258.
[16] W.M. Lai, V.C. Mow, D.N. Sun and G.A. Ateshian, On the electric potentials inside a charged soft hydrated biological tissue: Streaming potential vs. diffusion potential, *J. Biomech. Engng.* **122** (2000), 336–346.
[17] R.C. Lee, E.H. Frank, A.J. Grodzinsky and D.K. Roylance, Oscillatory compressional behavior of articular cartilage and its associated electromechanical properties, *J. Biomech. Engng.* **103** (1981), 280–292.
[18] P.A. Lotke, J. Black and S.J. Richardson, Electromechanical properties in human articular cartilage, *J. Bone Joint Surgery* **56A** (1974), 1040–1046.
[19] A. Maroudas, Physicochemical properties of cartilage in the light of ion exchange theory, *Biophys. J.* **8** (1968), 575–595.
[20] A. Maroudas, H. Muir and J. Wingham, The correlation of fixed negative charge with glycosaminoglycan content of human articular cartilage, *Biochim. Biophys. Acta* **177** (1969), 492–500.
[21] V.C. Mow, C.B. Wang and C.T. Hung, The extracellular matrix, interstitial fluid and ions as a mechanical signal transducer in articular cartilage, *Osteoarthritis Cart.* **7** (1999), 41–58.

[22] V.C. Mow and A. Ratcliffe, Structure and function of articular cartilage and meniscus, in: *Basic Orthopaedic Biomechanics*, V.C. Mow and W.C. Hayes, eds, Lippincott-Raven Pubs, Philadelphia, 1997, pp. 113–177.
[23] D.N. Sun, W.Y. Gu, X.E. Guo, W.M. Lai and V.C. Mow, A mixed finite element formulation of triphasic mechano-electrochemical theory for charged, hydrated biological soft tissues, *Int. J. Num. Meth. Eng.* **45** (1999), 1375–1402.
[24] R.A. Stockwell, *Biology of Cartilage Cells*, Cambridge University Press, Cambridge, 1979.

Biorheology 39 (2002) 47–53
IOS Press

Measuring principles of frictional coefficients in cartilaginous tissues and its substitutes

J.M. Huyghe [a,*], C.F. Janssen [a], C.C. van Donkelaar [a] and Y. Lanir [b]
[a] *Department of Biomedical Engineering, Engineering Mechanics Institute, Eindhoven University of Technology, Eindhoven, The Netherlands*
[b] *Julius Silver Institute for Biomedical Engineering, Technion, Haifa, Israel*

Abstract. The frictional properties of cartilaginous tissues, such as the hydraulic permeability, the electro-osmotic permeability, the diffusion coefficients of various ions and solutes, and the electrical conductance, are vital data to characterise the extracellular environment in which chondrocytes reside. This paper analyses one-dimensional measurement principles of these coefficients. Particular attention is given to the deformation dependence of them and the highly deformable nature of the tissues. A suggested strategy is the combination of a diffusion experiment using radiotracer methods, an electro-osmotic flow experiment and an electro-osmotic pressure experiment at low electric current.

1. Introduction

Cartilaginous tissue and its substitutes are very deformable molecular mixtures of an ionized solid and an ionised fluid. For a correct description of their time-dependent behaviour we need to measure frictional properties, including the hydraulic permeability, the electro-osmotic permeability, diffusion coefficients of various ions and electrical conductivity. The highly deformable nature of these materials complicate these measurements significantly. The very flow of substances through the medium causes deformation of the material which in turn affects the frictional coefficients governing their flow [6]. The aim of the present analysis is to develop an experimental strategy which allows correct quantification of frictional coefficients of charged media under a given state of deformation. Given the complexity of this undertaking, we restrict our attention to the one-dimensional case.

2. The frictional coefficients

Hydraulic permeability of cartilaginous tissues is in the order of 10^{-16} $m^4 N^{-1} s^{-1}$ [3]. As a result of this, a classical Darcy-type set-up, consisting of a measurement of flow through a sample subjected to a pressure gradient, yields very low flows for relatively large pressure gradients. Typically, a sample of 4 mm diameter and 2 mm thickness, subject to a 0.1 MPa pressure drop, yields a flow of the order of a few microliters per day. Accurate measurement of such low flow is already a challenge, yet one would prefer – if at all possible – to use thicker samples to increase the accuracy for measurement of the pressure gradient. However, this requires measurement of even lower flows. Under the given conditions a

*Address for correspondence: Dr. J.M. Huyghe, Department of Biomedical Engineering, Eindhoven University of Technology, P.O. Box 513, 5600MB Eindhoven, The Netherlands. Tel.: +31 40 2473137; Fax: +31 40 2447355; E-mail: jacques@wfw.wtb.tue.nl.

flow dependent, non-homogenous strain field has been shown to significantly affect the hydraulic permeability itself while in turn the strain affects the frictional properties of the sample in a non-homogenous way. Lanir [6] showed that diffusion coefficients of sodium and chloride are significantly affected by compressive strain, again underwriting the need for experimental methods of quantification of frictional coefficients under near zero deformation. Permeability and ion diffusion-convection in incompressible electrochemomechanics is described using the following constitutive equation for the velocities v^γ of constituents γ [5]:

$$-C^\beta \nabla \mu^\beta = \sum_{\gamma=f,+,-} B^{\beta\gamma}(v^\gamma - v^s), \quad \beta = f,+,- \tag{1}$$

in which C^β is the molecular concentration of phase β per unit mixture volume, μ^β its molecular electrochemical potential, $B^{\beta\gamma}$ a symmetric matrix of frictional coefficients, index f is the fluid and index s is the solid phase. Considering that under physiological salt concentrations, the probability of two ions interacting with one another is low [2,5],

$$B^{+-} = 0. \tag{2}$$

If we substitute Eq. (2) in Eqs (1), we find:

$$-C^f \nabla \mu^f = B^{ff}(v^f - v^s) + B^{f+}(v^+ - v^s) + B^{f-}(v^- - v^s), \tag{3}$$

$$-C^+ \nabla \mu^+ = B^{+f}(v^f - v^s) + B^{++}(v^+ - v^s), \tag{4}$$

$$-C^- \nabla \mu^- = B^{-f}(v^f - v^s) + B^{--}(v^- - v^s). \tag{5}$$

The frictional coefficients are related to the two diffusion coefficients D^+ and D^- measured through a radiotracer method [6] by:

$$D^\beta = \frac{RTC^\beta}{B^{\beta\beta}}. \tag{6}$$

They are obtained at a prescribed state of deformation, shortcircuiting conditions, no fluid flow relative to the solid and a given ionic concentration. In Eq. (6), we assume the following expression to hold for the electrochemical potentials:

$$\mu^\beta = RT \ln \frac{C^\beta}{C^f} + \overline{V}^\beta p + z^\beta F\xi + \mu_0^\beta, \quad \beta = +,- \tag{7}$$

in which R is the gas constant, T the absolute temperature, F Faraday's constant, $\overline{V}^\beta$ the molar volume of ion β, z^β the valence of component β, ξ the electrical potential, p the pressure and μ_0^β the reference potentials. Given the diffusion coefficients D^+ and D^-, we are left with three unknown frictional coefficients in Eqs (3), (4) and (5), namely. Preferably, we would like to measure these coefficient in the same well-defined state in which the two diffusion coefficients were measured. Excluding gradients in concentration in the frictional relationship (1) yields the electrokinetic relationships:

$$j = -L^p \nabla p - L^{pe} \nabla \xi, \tag{8}$$

$$i = -L^{ep} \nabla p - L^e \nabla \xi, \tag{9}$$

where j represents the volume flux

$$j = \sum_{\gamma=f+,-} \overline{V}^{\gamma} C^{\gamma}(v^{\gamma} - v^{s}), \tag{10}$$

i the electric flux

$$i = \sum_{\gamma=f+,-} F z^{\gamma} C^{\gamma}(v^{\gamma} - v^{s}), \tag{11}$$

L^p the short-circuit hydraulic permeability, L^{pe} and L^{ep} the electro-osmotic permeability and L^e the electrical conductivity. L^{pe} and L^{ep} are equal by Onsager reciprocity. L^p, L^e and L^{pe} are related to the values of $B^{\beta\gamma}$ in Eq. (1) by:

$$L^p = \sum_{\gamma=f,+,-} \sum_{\beta=f,+,-} \overline{V}^{\gamma}\overline{V}^{\beta} L^{\beta\gamma}, \tag{12}$$

$$L^e = F^2 \sum_{\gamma=f,+,-} \sum_{\beta=f,+,-} z^{\gamma} z^{\beta} L^{\beta\gamma}, \tag{13}$$

$$L^{pe} = F \sum_{\gamma=f,+,-} \sum_{\beta=f,+,-} \overline{V}^{\gamma} z^{\beta} L^{\beta\gamma} \tag{14}$$

in which $L^{\beta\gamma}$ are the conductances:

$$L^{\beta\gamma} = C^{\gamma} C^{\beta} (B^{-1})^{\beta\gamma}. \tag{15}$$

Therefore, one can reasonably assume that if there is an experimental procedure to measure the three electrokinetic coefficients L^p, L^{ep} and L^e, the three remaining frictional coefficients B^{ff}, B^{f+} and B^{f-} can be solved from Eqs (12), (13) and (14).

3. Electrokinetic coefficients

Two types of driving forces can be applied to measure the electrokinetic coefficients L^p, L^{ep} and L^e in a sample: a pressure gradient and an electrical potential gradient. In view of the problems associated with pressure application, we investigate the option of the electrical potential. One can either apply an electric field under no flow conditions $j = 0$ or under no pressure gradient conditions $\nabla p = 0$. The latter is termed electro-osmotic flow experiment while the former is an electro-osmotic pressure experiment. The Eqs (8) and (9) reshape into

$$j = -L^{pe}\nabla\xi, \tag{16}$$

$$i = -L^{e}\nabla\xi \tag{17}$$

for the electro-osmotic flow experiment and into

$$j = -L^{p}\nabla p - L^{pe}\nabla\xi = 0, \tag{18}$$

$$i = -(L^{e} - L^{pe}L^{ep}/L^{p})\nabla\xi \tag{19}$$

for the electro-osmotic pressure experiment. Hence, if we measure the flow, the current and the electrical potential gradient in the electro-osmotic flow experiment, we can calculate the electro-osmotic permeability L^{pe} and the electrical conductance L^{e}. If we measure the pressure gradient, the current and the electrical potential gradient in the electro-osmotic pressure experiment we can calculate the ratio L^{pe} over L^{p} and the no-flow electrical conductance $L^{e} - L^{pe}L^{ep}/L^{p}$. Hence, we can compute all four electrokinetic coefficients from an electro-osmotic flow and pressure experiments. This includes a check of Onsager's symmetry.

4. Deformation and electro-osmotic flow

In the foregoing, no deformation has been taken into account. We use the electrochemomechanical theory [5] under infinitesimal strain to evaluate deformational behaviour. We consider a homogenous sample sandwiched between two solutions of equal composition and in thermodynamic equilibrium with the solution. The free energy of the sample is described by a function W of the strain and composition of the sample. The equilibrium is assumed stable. The function W includes the elastic energy, the mixing energy and possible cross terms of mixing and elasticity. It may include phenomena like Donnan osmosis, chemical expansion stress, excluded volume effects or hydration forces. The boundary conditions of an electro-osmotic flow experiment are on the left-hand side of the sample:

$$u^s = 0, \tag{20}$$

$$\frac{\partial W}{\partial C^f} + \overline{V}^f p = \mu^f_{\text{left}}, \tag{21}$$

$$\frac{\partial W}{\partial C^+} + \overline{V}^+ p + F\xi = \mu_{\text{left}}, \tag{22}$$

$$\frac{\partial W}{\partial C^-} + \overline{V}^- p - F\xi = \mu_{\text{left}} \tag{23}$$

with μ^f_{left} the chemical potential of the water, μ_{left} the electrochemical potential of the cations and anions in the solution contacting the left-hand side of the sample and u^s the displacement of the solid. The electrochemical potential of the solution on the left-hand side of the sample is the same for anions and cations because we set the electrical potential to vanish at that side. The solution contacting the right-hand side of the sample has the same composition, only it is subject to an electrical potential ξ_e higher than the solution on the left. On the right the boundary conditions are:

$$\frac{\partial W}{\partial \epsilon} - p = 0, \tag{24}$$

$$\frac{\partial W}{\partial C^f} + \overline{V}^f p = \mu^f_{\text{left}}, \tag{25}$$

$$\frac{\partial W}{\partial C^+} + \overline{V}^+ p + F\xi = \mu_{\text{left}} + F\xi_e, \tag{26}$$

$$\frac{\partial W}{\partial C^-} + \overline{V}^- p - F\xi = \mu_{\text{left}} - F\xi_e. \tag{27}$$

Substitution of the frictional Eqs (1) into the mass balances of the respective constituents demonstrates that the driving forces μ^f, μ^+ and μ^- should vary linearly over the thickness of the sample – i.e., between the values prescribed by both solutions. I.e., Eq. (25) should be valid across the entire sample and the value of the electrochemical potentials of the ions inside the sample take the form:

$$\frac{\partial W}{\partial C^+} + \overline{V}^+ p + F\xi = \mu_{\text{left}} + \frac{Fx}{d}\xi_e, \tag{28}$$

$$\frac{\partial W}{\partial C^-} + \overline{V}^- p - F\xi = \mu_{\text{left}} - \frac{Fx}{d}\xi_e \tag{29}$$

in which x is the distance from the left edge and d the thickness of the sample. Because of momentum balance, Eq. (24) is valid over entire thickness of the sample. Equations (28), (29) are summed to eliminate the electrical potential:

$$\frac{\partial W}{\partial C^+} + \overline{V}^+ p + \frac{\partial W}{\partial C^-} + \overline{V}^- p = 2\mu_{\text{left}}. \tag{30}$$

Electroneutrality requires:

$$C^- - C^+ = \frac{C^{fc}}{1+\epsilon} \tag{31}$$

in which C^{cf} is the initial fixed charge density per unit volume of mixture. Saturation requires:

$$\epsilon = \overline{V}^f (C^f - C_0^f) + \overline{V}^+ (C^+ - C_0^+) + \overline{V}^- (C^- - C_0^-). \tag{32}$$

Equations (24), (25), (30)–(32) are 5 equation with 5 unknowns, namely the concentrations C^f, C^+, C^-, the strain ϵ and the pressure p. None of the coefficients in the equations depend on the position in the sample, nor on the imposed electrical potential, nor on the ensuing volume flow. So, as we solve these non-linear equations for these five unknown, we find for each of them a single constant value which does not depend on the applied electric potential. Unlike in the Darcy experiment, in the electro-osmotic flow experiment the viscous drag does not result in deformation. The viscous drag is compensated by the electric field acting on the charged solid.

5. Deformation and electro-osmotic pressure

We consider the same homogenous sample subjected to the same electric field while disallowing any volume flow across the sample. The boundary conditions on the left-hand side of the sample are given by Eqs (20)–(23). On the right-hand side they are:

$$\frac{\partial W}{\partial \epsilon} - p = -p_e, \tag{33}$$

$$\frac{\partial W}{\partial C^f} + \overline{V}^f p = \mu^f_{\text{left}} + \overline{V}^f p_e, \tag{34}$$

$$\frac{\partial W}{\partial C^+} + \overline{V}^+ p + F\xi = \mu_{\text{left}} + F\xi_e + \overline{V}^+ p_e, \tag{35}$$

$$\frac{\partial W}{\partial C^-} + \overline{V}^- p - F\xi = \mu_{\text{left}} - F\xi_e + \overline{V}^- p_e. \tag{36}$$

The solution contacting the right-hand side of the sample has the same composition, only it is subject to an electrical potential ξ_e relative to the left solution and a pressure p_e relative to the left solution. ξ_e and p_e are related by Eq. (18):

$$p_e = -\frac{L^{pe}}{L^p}\xi_e. \tag{37}$$

Substitution of the frictional Eqs (1) into the mass balances of the respective constituents demonstrates that the driving forces μ^f, μ^+ and μ^- should vary linearly over the thickness of the sample – i.e., between the values prescribed by both solutions. Therefore the expressions for the three electrochemical potentials inside the sample are:

$$\frac{\partial W}{\partial C^f} + \overline{V}^f p = \mu^f_{\text{left}} + \frac{\overline{V}^f p_e x}{d}, \tag{38}$$

$$\frac{\partial W}{\partial C^+} + \overline{V}^+ p + F\xi = \mu_{\text{left}} + \frac{(F\xi_e + \overline{V}^+ p_e)x}{d}, \tag{39}$$

$$\frac{\partial W}{\partial C^-} + \overline{V}^- p - F\xi = \mu_{\text{left}} + \frac{(-F\xi_e + \overline{V}^- p_e)x}{d}. \tag{40}$$

Because of momentum balance, Eq. (33) is valid over the entire sample. Equations (31)–(33) and (38)–(40) are a set of 6 equations with 6 unknowns: p, ξ, ϵ, C^f, C^+ and C^-. Summing the Eqs (39) and (40) results in elimination of the electrical potential from the set of equations. In general, any of the unknowns may depend on position in this case. Provided simplifying assumptions are handled, one can demonstrate a linear variation of the pressure across the sample leading to a varying strain across the sample. Although here no viscous drag is present, the electrostatic forces result in a non-uniform strain across the sample. However, because no volume flow is to be measured in a electro-osmotic pressure experiment – typically one would measure the pressure, the electric potential gradient and the electric current – one can reduce the values of those quantities to such an extent that the strain is negligible.

6. Discussion

A procedure has been conceptualised for the evaluation of 5 frictional coefficients of cartilaginous tissues or substitutes. The procedure should be feasible under near zero deformation. It includes a radio-tracer method for the diffusion coefficients [6] and a combination of electro-osmotic flow and pressure experiments for the electrokinetic coefficients. It is shown that provided electrochemomechanical theory applies, unlike the classical Darcy experiment, the electro-osmotic flow experiments does not induce deformation for a wide class of ionised incompressible media, while for the electro-osmotic pressure experiments, deformation inducing non-linearities can be avoided through lowering the electric field.

Experimental data from the literature [1] and from our lab [4] find a linear relationship between volume flow and electric current and electric current and electric potential gradient in an electro-osmotic flow experiment, indicating that the applied electric potential does not affect the frictional coefficients. The absence of deformation during electro-osmotic flow is a possible explanation of the experimental phenomenon.

Acknowledgements

The authors thank the Technology Foundation STW, the technological branch of The Netherlands Organisation for Scientific Research and the Schuurman Schimmel–van Outeren Foundation for their financial support. The research of J.M. Huyghe has been made possible through a fellowship of the Royal Netherlands Academy for Arts and Sciences.

References

[1] A.J. Grodzinsky, Mechanical and electrical properties and their relevance to physiological processes: Overview, in: *Methods for Cartilage Research*, A. Maroudas and K.E. Kuettner, eds, Academic Press, New York, USA, 1990, pp. 275–281.
[2] W.Y. Gu, W.M. Lai and V.C. Mow, A mixture theory for charged-hydrated soft tissues containing multi-electrolytes: passive transport and swelling behaviors, *J. Biomech. Eng.* **120** (1998), 169–180.
[3] G.B. Houben, M.R. Drost, J.M. Huyghe, J.D. Janssen and A. Huson, Non-homogenous permeability of canine anulus fibrosus, *Spine* **22** (1997), 7–16.
[4] J.M. Huyghe, C.F. Janssen, Y. Lanir, C.C. van Donkelaar, A. Maroudas and D.H. van Campen, Experimental measurement of electrical conductivity and electro-osmotic permeability of ionised porous media, in: *Porous Media: Theoretical, Experimental and Numerical Applications*, W. Ehlers and J. Bluhm, eds, Springer-Verlag, Heidelberg, Germany, submitted.
[5] J.M. Huyghe and J.D. Janssen, Quadriphasic mechanics of swelling incompressible porous media, *Int. J. Eng. Sci.* **35** (1997), 793–803.
[6] Y. Lanir, J. Seybold, R. Schneiderman and J.M. Huyghe, Partition and diffusion of sodium and chloride ions in soft charged foam: the effect of external salt concentration and mechanical deformation, *Tiss. Eng.* **4** (1998), 365–378.

Biorheology 39 (2002) 55–61
IOS Press

Influence of ion channels on the proliferation of human chondrocytes

David Wohlrab *, Susanne Lebek, Thomas Krüger and Heiko Reichel
Department of Orthopaedics, Martin Luther University, Halle-Wittenberg, Germany

Abstract. The goal of the study was to examine connections between ion channel activity and the proliferation of human chondrocytes. Chondrocytes were isolated form human osteoarthritic knee joint cartilage. In this study the concentration-dependent influence of the ion channel modulators tetraethylammonium (TEA), 4-aminopyridine (4-AP), 4′,4′ diisothiocyanato-stilbene-2,2′-disulfonic acid (DIDS), 4-acetamido-4′-isothiocyano-2,2′-disulfonic acid stilbene (SITS), verapamil (vp) and lidocaine (lido) on the membrane potential and the proliferation of human chondrocytes was investigated using flow cytometry and the measurement of ^{3}H-thymidine incorporation as measure for the cell proliferation. The results show an effect of the used ion channel modulators causing a change of the membrane potential of human chondrocytes. The maximal measurable effects of the membrane potential were listed with 0.25 mmol/l verapamil (−18%) and 0.1 mmol/l lidocaine (+20%). When measuring DNA distribution, it became apparent that the human chondrocytes are diploid cells with a very low proliferation tendency. After 12 days culture duration, lidocaine and 4-AP cause an increase of the DNA synthesis rate being a limited effect. These results allow the conclusion of an influence of ion channel modulators on chondrocyte proliferation. To gain knowledge of the regulation of chondrocyte proliferation via ion channel modulators could serve the research of new osteoarthritis treatment concepts.

Keywords: Chondrocytes, ion channel modulators, membrane potential, proliferation

1. Introduction

As all living cell systems, human chondrocytes are provided with a membrane potential. For its origin the existence of ion channels at the cell membrane is an essential prerequisite. For this development, different active and passive transportation systems are responsible, especially ion channels in the cell membrane [6]. In non-human chondrocytes, different ion channels could already be identified [9,11,15]. A connection between the potassium channel activity and the proliferation has already been detected in different human cell systems [12]. Whereas, the proof of a connection between ion channel activity of human chondrocytes and the proliferation has yet to be established.

The objective of the performed experiments was to gain the membrane potential of human chondrocytes by flow cytometry.

Furthermore, it should be determined the concentration dependent influence of the ion channel modulators tetraethylammonium (TEA), 4-aminopyridine (4-AP), 4′,4′diisothiocyanato-stilbene-2,2′-disulfonic acid (DIDS), 4-acetamido-4′-isothiocyano-2,2′-disulfonic acid (SITS), verapamil (vp) and lidocaine (lido) on the DNA distribution of human chondrocytes and the ^{3}H-thymidine incorporation as measure for the DNA synthesis capacity.

*Address for correspondence: David Wohlrab, MD, Department of Orthopaedics, Martin Luther University, Halle-Wittenberg, 06097 Halle, Germany. Tel.: +49 345 557 4802/05; Fax: +49 345 557 4809; E-mail: david.wohlrab@medizin.uni-halle.de.

2. Material and methods

Electrophysiological investigations were performed on human chondrocytes isolated from the osteoarthitic knee joint cartilage of 11 patients (5 female, 6 male) suffering from gonarthritis between the ages of 40 to 79 years (mean 63.5 ± 12.8 years).

2.1. Chondrocyte preparation

The cartilage was isolated from cartilage- bone- fragments resected during the insertion of knee prostheses. Immediately after the resection, the cartilage- bone- fragments were potted in sterile L15 medium (Seromed, Berlin, Germany). Thereafter, the cartilage was separated from the bone, reduced to 1 mm^3 pieces and handled as described previously [17].

2.2. Electrophysiological investigations

To determine and characterize the membrane potential of the chondrocytes, these were incubated in solutions with known ionic concentrations (10 mmol/l hepes, 10 mmol/l glucose, 0.5 mmol/l $MgCl_2$, 0.3 mmol/l $CaCl_2$, pH = 7.4), distinguished only by the concentration of potassium. Additionally, these solutions contained the ionophor valinomycin (Molecular Probes, Leiden, Holland) and the fluorescent pigment oxonol VI (Molecular Probes). By valinomycin, the cell membrane becomes permeable for potassium and the fluorescent intensity of oxonol VI is altered depending on the membrane potential. First, the chondrocytes were incubated in a 150 mmol/l potassium solution and then, dissolved in a solution with less potassium concentration. Immediately after, a flow cytometric analysis of approximately 70,000 cells was performed by the determination of the fluorescence intensity of oxonol VI. Using the Nernst equation and knowing the exact ionic concentration of the used solutions it is possible to calculate the according membrane potentials. With the obtained curve and its mathematical function all measured fluorescence intensities of oxonol VI could be calculated.

The influence of ion channel modulators on the membrane potential has been measured in human chondrocytes after incubation in different solutions. These solutions (10 mmol/l Hepes, 10 mmol/l glucose, 0.5 mmol/l $MgCl_2$, 0.3 mmol/l $CaCl_2$) contained an ion channel modulator of different concentration each.

For the measurement of the influence of different ion channel modulators on the membrane potential and the proliferation of human chondrocytes we used TEA, 4-AP, DIDS, SITS, verapamil and lidocaine. TEA (Fluka GmbH, Buchs, Switzerland) and 4-AP (Sigma GmbH, Steinheim, Germany) are known unspecific potassium channel blockers [4,7,14]. DIDS (Sigma GmbH) is an anion channel blocker with an inhibiting effect of chloride exchange simultaneously; ion channels with conducting capacity for Cl^- and some other anions (e.g., aspartate) are mainly blocked [14]. SITS (Sigma GmbH) is known as a strong inhibitor of the anion transport. Verapamil (ICN-Biomedicals Inc., Aurora/Ohio, USA) blocks Ca^{2+}-channels in smooth and heart muscle cells. Additionally, it has vasodilatative and antiarrhythmic properties. Lidocaine (Sigma) is known as a Na^+-channel blocker.

Different concentrations of the ion channel modulations were used (20 and 40 mmol/l TEA, 1 and 2 mmol/l 4-AP, 0,1 and 0.3 mmol/l DIDS, 0.25 and 0.5 mmol/l SITS, 0.25 and 0.5 mmol/l verapamil, lidocaine 0.1 and 0.2 mmol/l).

2.3. Determination of the DNA distribution

The determination of the DNA distribution of human chondrocytes denoting the proliferation were performed using the Cycle Test™ PLUS DNA Reagent Kit (Becton Dickinson, San Jose, USA) with the fluorescent pigment propidium iodide. The fluorescence intensity being proportional of the DNA distribution has been measured by flow cytometry. For data evaluation the average as well as the standard deviation were calculated. The testing of the values for statistical significance ($p < 0.05$) was carried out with the aid of the t test and the one-way variance analysis, respectively.

2.4. ^{3}H-thymidine incorporation

For proliferation measurement, chondrocytes were cultivated from 6, 12 or 18 days at 37.0°C with 5% CO_2. Thereafter measuring of the ^{3}H-thymidine incorporation as measure for the cell proliferation was carried out.

Subsequently, this was followed by the addition of 20 μl ^{3}H-methyl-thymidine (specific activity 60.3 Ci/mmol; American Radio Labeled Chemicals Inc., St. Louis, USA).

Two hours after administering ^{3}H-thymidine, the medium was siphoned off from the cell chamber with the aid of a cell harvester (Berthold Inc., Bad Wildbad, Germany). Every culture chamber was fed with 200 μl trypsin and after 20 minutes, the cell suspension was siphoned off through a filter. Subsequently, this was followed by measuring the radioactivity of the cells in the filter paper with the aid of a Tricarb (Berthold Inc., Bad Wildbad, Germany).

The ion channel modulators dissolved in PBS were added on the second day of cell culture. The control investigations done at the same time were mixed with the same amount of PBS.

To gain a start value, uncultivated cells were exposed to the influence of ^{3}H-thymidine.

3. Results

3.1. Membrane potential calibration curve

The chondrocytes incubated in different extracellular potassium concentrations $[K^+]_e$ were measured by flow cytometry to gain a calibration curve of the membrane potential. According to the measured fluorescence values the actual membrane potentials were calculated with the Nernst equation. For the calibration curve a mathematical function allowing the calculation of the membrane potential for each fluorescence value was determined [6]. The results as well as their mathematical function are shown in Fig. 1.

3.2. The influence of ion channel modulators on the membrane potential

A considerable decrease of the membrane potential of the chondrocytes has been registered with 4-AP, verapamil and SITS. There were no differences in the different concentrations, however. The influence of TEA leads to a more positive membrane potential compared with a control group. With a concentration of 0.1 mmol/l, lidocaine causes an increase of the membrane potential of 20% compared to the control. This is a significant increase compared to TEA (Fig. 2).

An influence of DIDS on the membrane potential of the chondrocytes could not be found (Fig. 2).

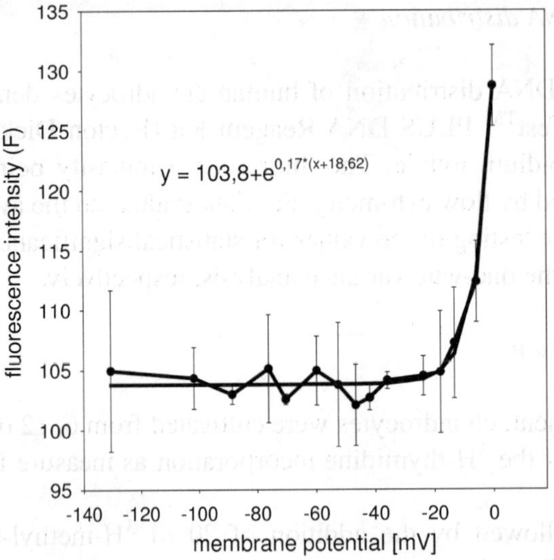

Fig. 1. Results of the flow cytometric characterization of the membrane potential of human chondrocytes using the fluorescent pigment oxonol VI and the ionophor valinomycin ($n = 10$).

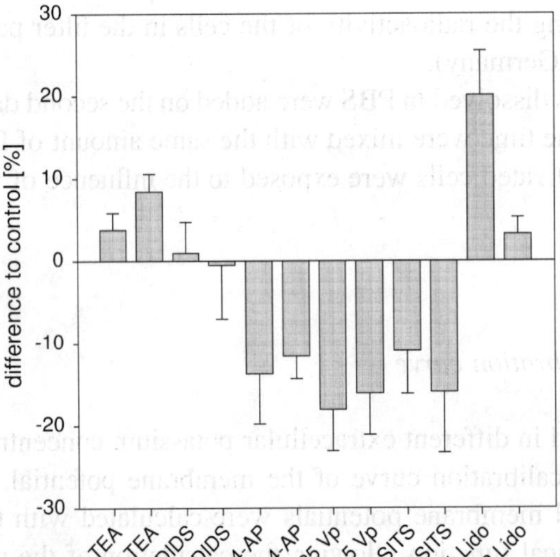

Fig. 2. Behaviour of the membrane potential of human chondrocytes under the concentration dependent influence of different ion channel modulators ($n = 6$) (mmol).

3.3. *DNA distribution*

The analysis of the results of the DNA distribution of human chondrocytes denoting the proliferation showed the following pattern: only approximately 95% of the cells are in G0/G1 phase of the cell cycle, 1% in the S phase and 4% in the G2/M phase. The subsumption of the results of the evaluation (Fig. 3) shows even after the analysis of the distribution in the several cell cycle phases a small drift of the populations in different donators. In Fig. 4 the results of all investigations are represented in a diagram.

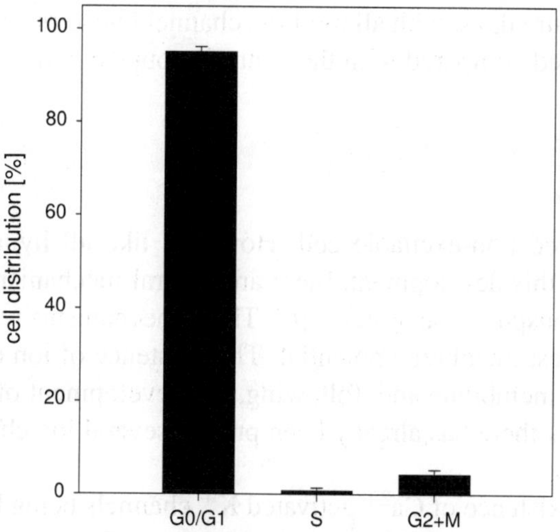

Fig. 3. Cell cycle distribution of human chondrocytes ($n = 11$).

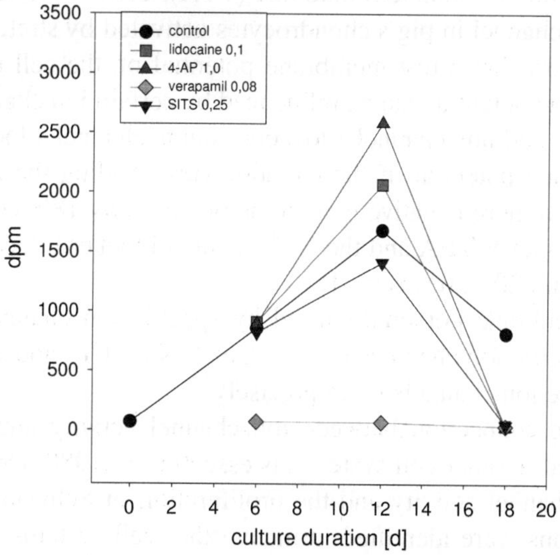

Fig. 4. Example of the DNA synthesis rate of human chondrocytes depending on the cultivation period. ^{3}H-thymidine was measured as a unit of proliferation. Each assessment consisting of eight single measurements provided the mean value.

These results show that the investigated human chondrocytes are diploid cells with an extremely low proliferative tendency.

3.4. ^{3}H-thymidine incorporation

The different ion channel modulators were added to the human chondrocytes on the 2nd culture day. ^{3}H-thymidine incorporation was measured on the 6th, 12th and 18th culture day. By the 6th culture day (4 days after adding ion channel modulators) there were no changes detected in the ^{3}H-thymidine incorporation rate compared to the control group. On the 12th culture day, an increase of the ^{3}H-thymidine incorporation was verified with 0.1 mmol/l lidocaine as well as with 1.0 mmol/l 4-AP. However, these effects are limited. After 18 culture days, with all used ion channel blockers a decrease of the ^{3}H-thymidine incorporation rate was observed compared with the control group (Fig. 4).

4. Discussion

Chondrocytes are considered non-excitable cell. However, like all living cell systems, they have a rest membrane potential. For this development, there are several mechanisms responsible, in particular, different active and passive transportation systems [6]. These mechanisms result in the development of an equilibrium potential called rest membrane potential. The existence of ion channels is a prerequisite for ion transport through the cell membrane and, following, the development of a rest membrane potential.

In non-human chondrocytes there has already been proven several ion channels regulated by different ion channel modulators.

Long et al. [9] proved the existence of Ca^{2+} activated K^+ channels being blocked by TEA or Ba^{2+} and activated by a catalytic subunit of protein kinase A. Furthermore, Sugimoto et al. [16] described specific ion channels for Na^+ as well as K^+ and Cl^- in rabbit's chondrocytes. These ion channels could be inhibited by ion channel specific blockers tetrodotoxine (TTX), TEA, 4-AP and SITS. Martina et al. [11] successfully identified a K^+ channel in pig's chondrocytes activated by stretching of the channel.

The findings presented show that a rest membrane potential on the cell membrane of chondrocytes exists, too. This rest membrane potential can be influenced by certain ion channel modulators. By adding potassium channel blockers, sodium channel blockers, anion channel blockers and calcium channel blockers change the membrane potential of the chondrocytes. Adding the ion channel blockers 4-AP, verapamil and SITS result in a more negative membrane potential by 18% compared to a control group. The unspecific K^+ channel blocker TEA and the Na^+ channel blocker lidocaine lead to a more positive membrane potential by 8% and 20%, respectively.

These changes of the membrane potential caused by specific ion channel modulators are first evidence of the existence of specific ion channels for K^+, Ca^{2+}, Na^+, Cl^- and anions, respectively. Further investigations will specify the ion channels more precisely.

To gain knowledge of the connection between ion channel activity and cell proliferation, already proven in some and presumed in other cell systems, is essential. In 1988, Deutsch et al. [2] proved such a connection between K^+ channel activity and the proliferation of Schwann cells and B lymphocytes. Thereafter, similar connections were identified in many other cell systems, such as tumor cell populations, human [12] and mouse keratinocytes [5], melanoma cells [13], neuroblastoma and astrocytoma cells [8]. A connection between free intracellular Ca^{2+} concentration and cell proliferation has already been known [1–3]. Whether the proliferation of chondrocytes can be influenced by ion channel activity is not known yet. First results show changed proliferation behaviour by modifying the ion channel activity.

Up to now, it has been established that an increased proliferation by modification of ion channel activity can be traced to a conversion of cells from the G0 phase into the G1 phase of the cell cycle [10, 16]. Furthermore, it appears likely that the retention of cells in the G1 phase is partly abrogated [10,16]. These mechanisms lead to an increase of cells into the cell cycle resulting in a numerical shifting of cells within the phases by changing the regulation mechanisms in the course of cell proliferation. Accordingly, as approximately 90% of human osteoarthritic chondrocytes are in the G0/G1 phase, many cells can be stimulated to proliferate by ion channel modulation.

So far connections of this kind are unknown in human as well as in non-human chondrocytes. Our results show, that by inhibition of different ion channels on the cell membrane of human chondrocytes the proliferation behaviour can be influenced. The inhibition of Na^+ and K^+ channels, respectively, lead temporarily to a distinct increase of the proliferation. The inhibition of ion channels over a longer period lead presumably to a decrease of proliferation.

As 95% of human chondrocytes exist in G0/G1 phase (Fig. 3) an increased proliferation by increased conversion of cells from G0 into G1 or G1 into S is suppositive. Certainly, further investigations are necessary to characterize possible connections between ion channel activity and proliferation behaviour as well as cell cycle distribution of human chondrocytes.

References

[1] A.M. Al Ani, A.G. Messenger, J. Lawry, S.S. Bleehen and S. MacNeil, Calcium/calmodulin regulation of the proliferation of human epidermal keratinocytes, dermal fibroblasts and mouse B16 melanoma cells in culture, *Br. J. Dermatol.* **119** (1988), 295–306.

[2] S. Amigorena, D. Choquet, J.L. Teillaud, H. Korn and W.H. Fridman, Ion channels and B cell mitogenesis, *Mol. Immunol.* **27** (1990), 1259–1268.

[3] D. Breitkreutz, H.J. Stark, P. Plein, M. Baur and N.E. Fusenig, Differential modulation of epidermal keratinization in immortalized (HaCaT) and tumorigenic human skin keratinocytes (HaCaT-ras) by retinoic acid and extracellular Ca^{2+}, *Differentiation* **54** (1993), 201–217.

[4] C. Deutsch, K^+ channels and mitogenesis, *Prog. Clin. Biol. Res.* **334** (1990), 2–21.

[5] S.C. Harmon, D. Lutz and J. Ducote, Potassium channel openers stimulate DNA synthesis in mouse epidermal keratinocyte and whole hair follicle cultures, *Pharmacol. Rev.* **6** (1993), 170–178.

[6] B. Hille, *Ionic Channels of Excitable Membranes*, Sinauer Associates Inc., Sunderland, 1992, pp. 504–524.

[7] S. Inohara, Studies and perspectives of signal transduction in the skin, *Exp. Dermatol.* **1** (1992), 207–220.

[8] Y.S. Lee, M.M. Sayeed and R.D. Wurster, Inhibition of cell growth by K^+ channel modulators is due to interference with agonist-induced Ca^{2+} release, *Cellular Signaling* **5** (1993), 803–809.

[9] K.J. Long and K.B. Walsh, A calcium-activated potassium channel in growth plate chondrocytes: regulation by protein kinase A, *Biochem. Biophys. Res. Commun.* **201** (1994), 776–781.

[10] J. Lübbe, P. Kleihues and G. Burg, Das Tumorsupressor-Gen p53 und seine Bedeutung für die Dermatologie, *Hautarzt* **45** (1994), 741–745.

[11] M. Martina, J.W. Mozrzymas and F. Vittur, Membrane stretch activates a potassium channel in pig articular chondrocytes, *Biochim. Biophys. Acta* **1329** (1997), 205–210.

[12] M.T. Mauro, R.R. Isseroff, R. Lasarow and A.P. Pappone, Ion channels are linked to differentiation in keratinocytes, *J. Membrane Biol.* **132** (1993), 201–209.

[13] B. Nilius and G. Droogmans, A role for potassium channels in cell proliferation?, *News in Physiological Sciences* **16** (1994), 1–12.

[14] B. Nilius, G. Schwarz and G. Droogmans, Control of intracellular calcium by membrane potential in human melanoma cells, *Am. J. Physiol.* **265** (1993), 1501–1510.

[15] T. Sugimoto, M. Yoshino, M. Nagao, S. Ishii and H. Yabu, Voltage-gated ionic channels in cultured rabbit articular chondrocytes, *Comp. Biochem. Physiol.* **115C** (1996), 223–232.

[16] D. Wohlrab, *Der Einfluß elektrophysiologischer Membraneigenschaften auf die Proliferation humaner Keratinozyten*, Tectum, Marburg, 1998, pp. 1–89.

[17] D. Wohlrab, J. Wohlrab, H. Reichel and W. Hein, Is the proliferation of human chondrocytes regulated by ionic channels?, *J. Orthop. Sci.* **6** (2001), 155–159.

Biorheology 39 (2002) 63–67
IOS Press

Cell cytoskeleton and tensegrity

K.Yu. Volokh, O. Vilnay and M. Belsky
Faculty of Civil Engineering, Technion – I.I.T., Haifa 32000, Israel
E-mail: cvolokh@aluf.technion.ac.il

Abstract. The role of tensegrity architecture of the cytoskeleton in the mechanical behavior of living cells is examined by computational studies. Plane and spatial tensegrity models of the cytoskeleton are considered as well as their non-tensegrity counterparts. Local buckling including deep postbuckling response of the compressed microtubules of the cytoskeleton is considered. The tensioned microfilaments cannot sustain compression. Large deformation of the whole model is accounted and fully nonlinear analysis is performed. It is shown that in the case of local buckling of the microtubules non-tensegrity models exhibit qualitatively the same linear stiffening as their tensegrity counterparts. This result raises the question of experimental validation of the local buckling of microtubules. If the microtubules of real cells are not straight, then tensegrity (in a narrow sense) is not a necessary attribute of the cytoskeleton architecture. If the microtubules are straight then tensegrity is more likely to be the cytoskeletal architecture.

1. Introduction

The cell mechanism of transition of mechanical signals (deformations) into biochemical output is not well understood and various scenarios have been proposed for the explanation of the aforementioned mechanotransduction. Among other are [7]: modification of electrical potentials; variations in the chemical environment of cells; G-protein-linked receptors; mechanically activated ion channels; and cytoskeleton. The latter one [1] is a result of recent experiments refuting the traditional fluid balloon model of a cell [3,5]. It was discovered that a microstructural framework comprising microtubules and microfilaments is responsible for the cell contractility. This cytoskeletal framework is spread over the cell and it seems to be the main load-bearing element of the cell. Ingber [1,4] proposed that the cytoskeletal framework enjoys a specific architecture called *tensegrity*. By using tensegrity model shown in Fig. 3 [2,6,9] it was possible to explain the experimentally observed linear stiffening of living cells [8,10,11].

Tensegrity architecture of the cytoskeleton implies that isolated compressed microtubules are attached at every node of the tensioned network of the microfilaments. An important feature of tensegrity assemblies (in a narrow sense) is that they are statically and kinematically indeterminate. Static indeterminacy means possibility of pre-stressing and stabilization of the whole assembly. This stabilization is necessary because kinematic indeterminacy means existence of small displacements that do not produce elongations of microtubules and microfilaments. As a result of the kinematic indeterminacy the mechanical response of cytoskeletal frameworks is nonlinear and linear stiffening of the cell may be observed. However, the lack of constraints of tensegrity structures is not the only possible reason of the nonlinear cell response and its linear stiffening. Also, local buckling of microtubules of a fully constrained cytoskeleton can be the main source of cell non-linearity. A question suggests itself: what is a mechanical response of a non-tensegrity and fully constrained cytoskeletal model? Is it possible to observe the linear stiffening as a result of the buckling of microtubules only without involvement of geometrical degeneracy of tensegrity? The goal of this note is to answer this question. Computer simulations examine plane and spatial under-constrained tensegrity models as well as their fully constrained "counterparts".

0006-355X/02/$8.00

2. Computer simulation

Four cytoskeletal models shown in Figs 1–4 are considered. First two models are plane (2D) and latter ones are spatial (3D). Models shown in Figs 1(a) and 3(a) are kinematically indeterminate underconstrained tensegrity assemblies while their "counterparts" shown in Figs 2(a) and 4(a) are kinematically determinate fullyconstrained and non-tensegrity. All models are statically indeterminate and allowed for pre-stressing. Double line in figures marks microtubules and regular line marks microfila-

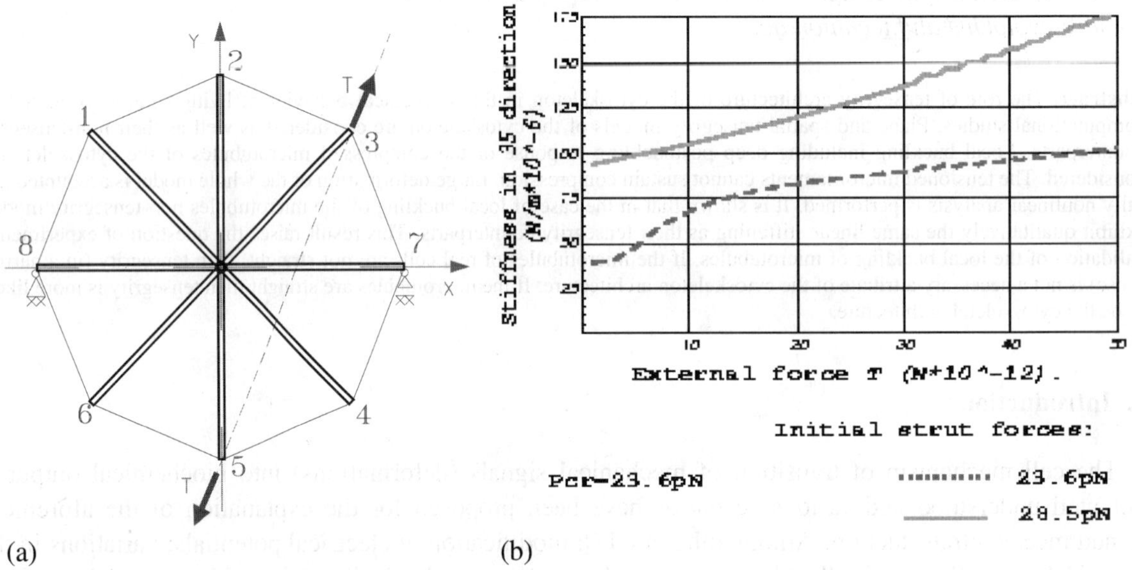

Fig. 1. Plane kinematically indeterminate underconstrained tensegrity cell (a); stiffness versus force (b).

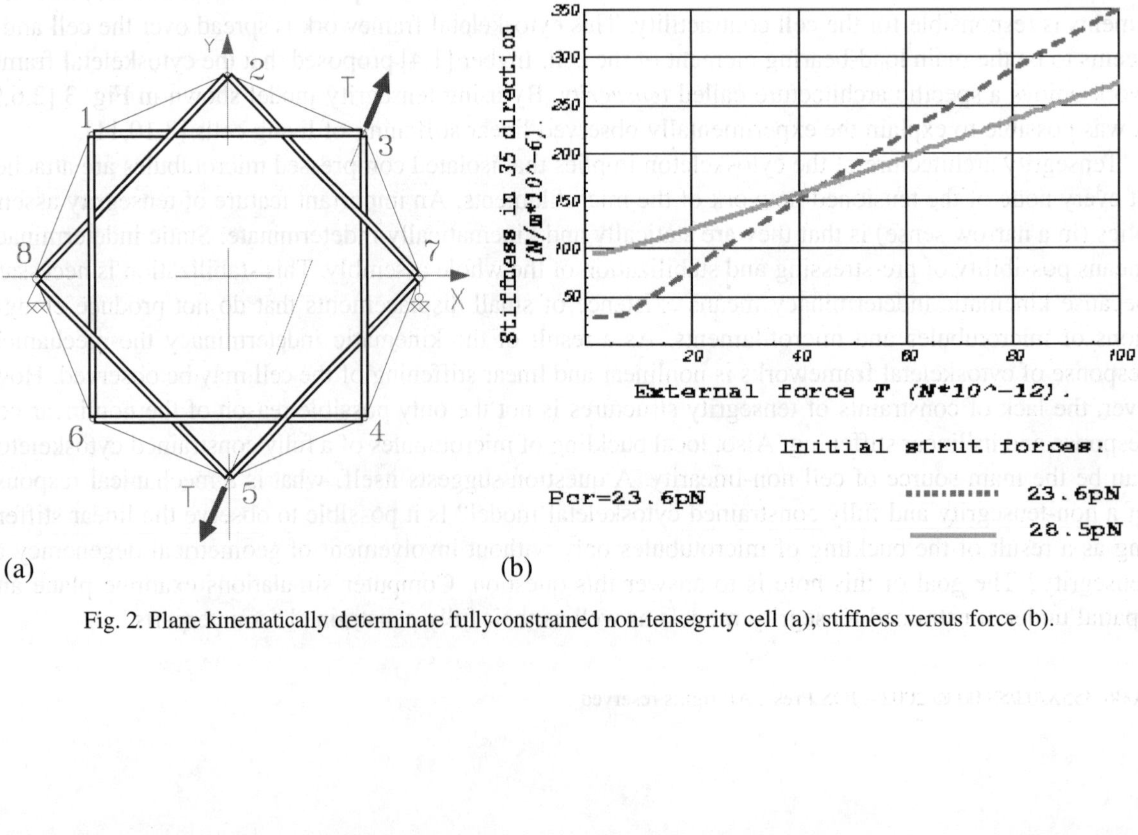

Fig. 2. Plane kinematically determinate fullyconstrained non-tensegrity cell (a); stiffness versus force (b).

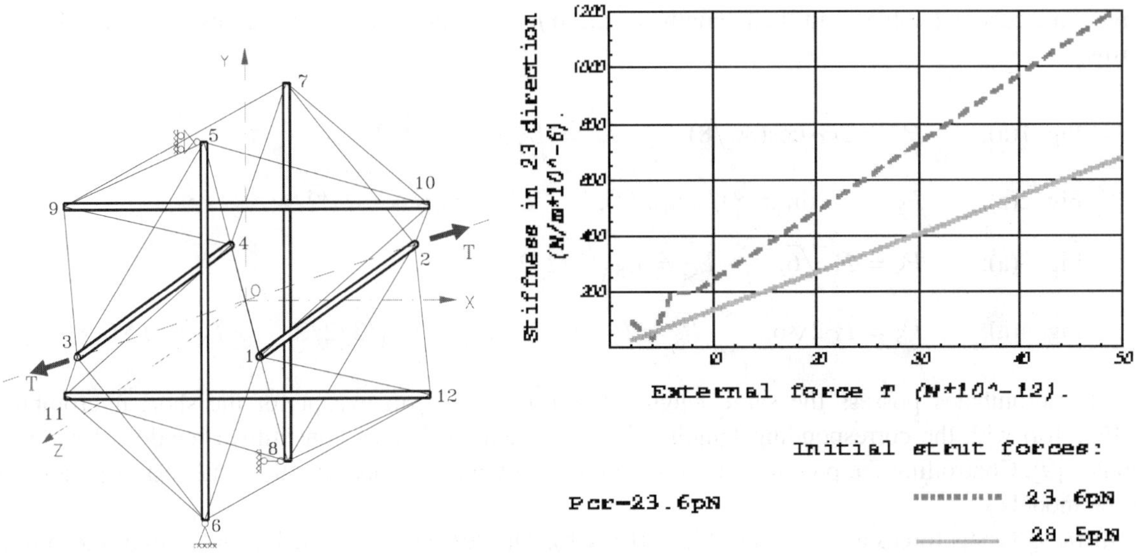

(a) (b)

Fig. 3. Spatial kinematically indeterminate underconstrained tensegrity cell (a); stiffness versus force (b).

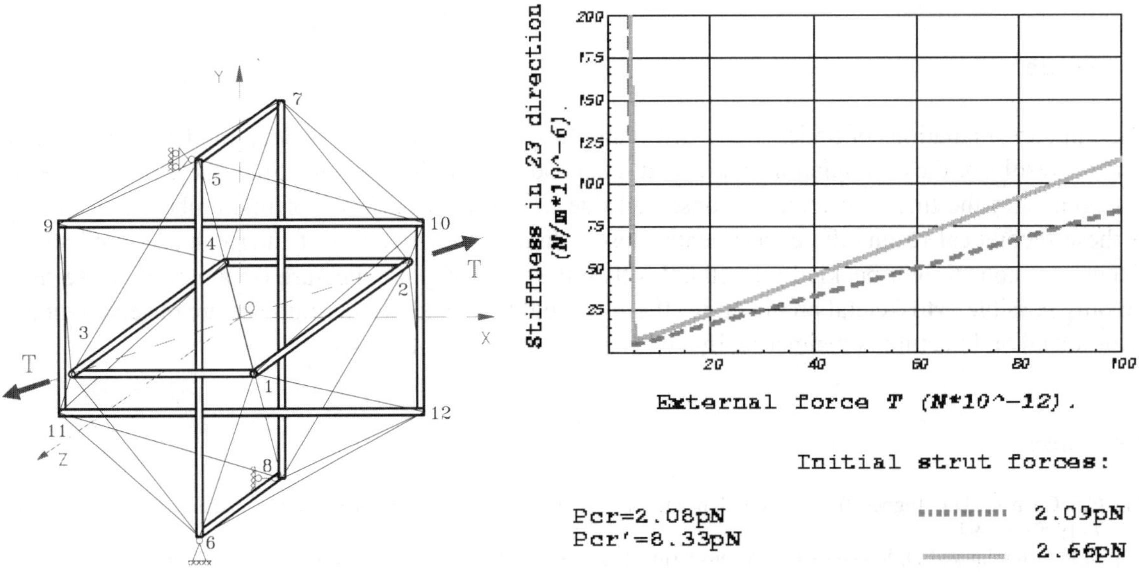

(a) (b)

Fig. 4. Spatial kinematically determinate fullyconstrained non-tensegrity cell (a); stiffness versus force (b).

ments. Three degrees of freedom are prohibited for 2D models and six degrees of freedom are prohibited for 3D models in order to exclude rigid body motions. Elasticity modulus is $E_S = 1.2$ GPa for struts-microtubules; and $E_C = 2.6$ GPa for cables-microfilaments. The bending stiffness of struts and cross-sectional areas of struts and cables are accordingly $(EI)_S = 2.15 \cdot 10^{-23}$ N m^2, $A_S = 190$ nm^2, $A_C = 18$ nm^2 for three first models (Figs 1–3). The corresponding magnitudes of the fourth model (Fig. 4) are $(EI)_S = 1.9 \cdot 10^{-24}$ N m^2, $A_S = 40$ nm^2, $A_C = 2.6$ nm^2. The relations between pre-stressing forces in struts and cables and their lengths *at the reference state* shown in figures take the following forms:

Fig. 1(a): $P_S = 2P_C \cos(3\pi/8), \quad L_C = L_S\sqrt{2-\sqrt{2}}/2;$

Fig. 2(a): $P_S = P_C \sin(\pi/8)/\sin(\pi/4), \quad L_C = L_S\sqrt{4-2\sqrt{2}}/4;$

Fig. 3(a): $P_S = P_C\sqrt{6}, \quad L_C = L_S\sqrt{3/2}/2;$

Fig. 4(a): $P_S = P_C 2\sqrt{6}, \quad P'_S = P_C\sqrt{6}, \quad L_C = L_S\sqrt{3/2}/2, \quad L'_S = 0.5L_S.$

All microtubules possess the same length *at rest* $L_0 = 3\ \mu$m except for the short microtubules in Fig. 4(a) with the corresponding length $0.5L_0$. This length defines constitutive relations for microtubules [9]. Controlling the pre-stressing force in microtubules it is possible to fit the rest of parameters of the models.

Stiffness versus forces are shown in Figs 1(b)–4(b). The stiffness is defined as the ratio of the applied force to the elongation in its direction. Two cases of pre-stressing are accounted. First, microtubules are compressed up to the critical buckling load. Second, microtubules are buckled at the reference state. It is readily seen that all models exhibit linear stiffening response independently of their architecture.

3. Closure

Computer simulations of under- (tensegrity) and fully-constrained cytoskeletal models have been presented. Results of these simulations show that independently of the specific structural geometry (regular or degenerate) the linear stiffening is observed due to the local buckling of microtubules. In the light of these theoretical results the experimental observation of the buckling of microtubules is of interest. If there are non-straight buckled microtubules then the tensegrity architecture (in a narrow sense) is not the only possible cytoskeletal architecture. If only straight unbuckled microtubules are observed then the cytoskeletal architecture is geometrically degenerate and it may be tensegrity.

References

[1] C.S. Chen and D.E. Ingber, Tensegrity and mechanoregulation: from skeleton to cytoskeleton, *Osteoarthritis and Cartilage* **7** (1999), 81–94.

[2] M.F. Coughlin and D. Stamenovich, A tensegrity structure with buckling compression elements: application to cell mechanics, *J. Appl. Mech.* **64** (1997), 480–486.

[3] A.K. Harris, P. Wild and D. Stopak, Silicone rubber substrata: a new wrinkle in the study of cell locomotion, *Science* **208** (1980), 177–180.

[4] D.E. Ingber, Tensegrity: the architectural basis of cellular mechanotransduction, *Ann. Rev. Physiol.* **59** (1997), 575–599.

[5] A.J. Maniotis, C.S. Chen and D.E. Ingber, Demonstration of mechanical connections between integrins, cytoskeletal filaments, and nucleoplasm that stabilize nuclear structure, *Proc. Nat. Acad. Sci. USA* **94** (1997), 849–854.

[6] C. Oddou, S. Wendling, H. Petite and A. Meunier, Cell mechanotransduction and interactions with biological tissues, *Biorheology* **37** (2000), 17–25.

[7] J.F. Stoltz, D. Dumas, X. Wang, E. Payan, D. Mainard, F. Paulus, G. Maurice, P. Netter and S. Muller, Influence of mechanical forces on cells and tissues, *Biorheology* **37** (2000), 3–14.

[8] O. Thoumine, T. Ziegler, P.R. Girard and R.M. Nerem, Elongation of confluent endothelial cells in culture: The importance of fields of force in the associated alterations of their cytosceletal structure, *Exp. Cell Res.* **219** (1995), 427–441.

[9] K.Yu. Volokh, O. Vilnay and M. Belsky, Tensegrity architecture explains linear stiffening and predicts softening of living cells, *J. Biomech.* **33** (2000), 1543–1549.
[10] N. Wang and D.E. Ingber, Control of cytosceletal mechanics by extracellular matrix, cell shape, and mechanical tension, *Biophys. J.* **66** (1994), 2181–2189.
[11] N. Wang, J.P. Butler and D.E. Ingber, Mechanotransduction across the cell surface and through the cytosceleton, *Science* **260** (1993), 1124–1127.

Biorheology 39 (2002) 69–78
IOS Press

Deformation properties of articular chondrocytes: A critique of three separate techniques

D.L. Bader [a,*], T. Ohashi [b], M.M. Knight [a], D.A. Lee [a] and M. Sato [b]
[a] *IRC in Biomedical Materials and Medical Engineering Division, Queen Mary, University of London, UK*
[b] *Graduate School of Mechanical Engineering, Tohoku University, Sendai, Japan*

Abstract. This paper presents a series of techniques, which examine the deformation characteristics of bovine articular chondrocytes. The direct contact approach employs well established methodology, involving AFM and micropipette aspiration, to yield structural properties of local regions of isolated chondrocytes. The former technique yields a non-linear response with increased structural stiffness in a central location on a projected image of the chondrocyte. A simple viscoelastic model can be used with data from the micropipette aspiration technique to yield a mean value of Young's modulus, which is similar to that recently reported (Jones et al., 1999). An indirect approach is also described, involving the response of chondrocytes seeded within compressed agarose constructs. For 1% agarose constructs, the resulting cell strain, yields a gross cell modulus of 2.7 kPa. The study highlights the difficulties in establishing unique mechanical parameters, which reflect the deformation behaviour of articular chondrocytes.

1. Introduction

Under normal physiological conditions, articular cartilage provides a load-bearing and low-friction covering for the surface of joints. The tissue may be considered as a fibre reinforced polymer gel containing cells, termed chondrocytes, at a relatively low density [1]. The properties of the extracellular components and their interactions determine the physical and mechanical properties of the tissue. This has led to numerous studies examining the mechanical properties of the extracellular matrix of articular cartilage [1]. By contrast, the chondrocytes play a minor role in the mechanical properties of cartilage. However their anabolic and catabolic activities in producing and maintaining the extracellular matrix, mainly the proteoglycan, aggrecan, and type II collagen, are crucial. The mechanical environment to which articular cartilage is normally subjected is known to influence the metabolism of its chondrocytes. The deformation of chondrocytes is believed to be important in mechanotransduction with the possible involvement of stretch activated ion channels within the cell membrane and the cytoskeleton [8,14].

Guilak and Mow [4] have developed a finite element model of the chondrocyte within the extracellular matrix and have shown that the stress–strain and fluid flow environment around the chondrocytes is dependent on the mechanical properties and morphology of the cells. The deformation of a structural unit, such as the chondrocyte, is determined by its response to mechanical forces. Several techniques exist for exerting small forces on individual cells, including cell poking, micropipette aspiration, cell

[*]Address for correspondence: IRC in Biomedical Materials and Department of Engineering, Queen Mary, University of London, Mile End Road, London, UK. Tel.: +1 44 207 882 5274; Fax: +1 44 208 983 1799; E-mail: D.L.Bader@qmw.ac.uk.

0006-355X/02/$8.00

indentation [19] and the atomic force microscope. In reporting both structural and material properties of a range of cells, proponents of individual techniques have also highlighted limitations which still need to be resolved [2,6,7]. With regard to the mechanical behaviour of primary human chondrocytes, Jones et al. [9] employed the micropipette aspiration technique. These authors reported that the human chondrocyte behaved as a viscoelastic solid, with an estimated Young's modulus of 0.65 ± 0.63 kPa. An alternative approach involved studying the deformation of chondrocytes seeded within a compressed homogeneous gel [3,10–14]. The former study [3] seeded isolated cells from a rat chondrosarcoma into 2% agarose gel, which was subjected to a series of static compressive strains. The finite element model used to estimate the experimental set-up yielded a value of chondrocyte elasticity of 4 kPa, very similar to that of the agarose gel.

The present study evaluates a series of techniques to examine the deformational behaviour of bovine articular chondrocytes, both in isolation and seeded in a well-established biomaterial construct.

2. Materials and methods

2.1. Chondrocyte isolation and culture

Bovine articular chondrocytes were isolated using a well established sequential enzyme digestion procedure. Full depth cartilage was removed from the proximal articular surfaces of the metacarpal-phalangeal joints of mature steers, finely diced using a scalpel blade and cultured overnight at 37°C and 5% CO_2 in Dulbecco's modified Eagle medium supplemented with 20% (v/v) heat-inactivated fetal bovine serum (DMEM + 20% FBS, Gibco, MD, USA). The cartilage slices were chopped into small pieces using a scalpel, centrifuged at 1000 *g*, and incubated in DMEM + 20% FBS + 700 unit/ml pronase (Wako Life Science, Osaka, Japan) at 37°C for 1 h. The supernatant was removed, replaced with DMEM + 20% FBS + 100 unit/ml collagenase (Sigma, MO, USA), and incubated at 37°C for 16 h. Subsequently, the supernatant containing released chondrocytes was passed through a 70 μm pore size sieve (Falcon, NJ, USA) to remove undigested tissue. Where appropriate, cells were maintained in monolayer culture for several days prior to testing.

2.2. Atomic Force Microscopy (AFM) experiments

The AFM system, recently described in detail [18], was used in one set of experiments (Fig. 1). To review briefly, a triangular silicon nitride cantilever (Olympus Optical Co. Ltd., Tokyo, Japan) was used with a length of 200 μm and a spring constant of 0.02 N/m. The tip is pyramidial in shape 0.6 μm in height and 4 μm in diagonal length of base. The radius of curvature of the tip is less than 20 nm. The cantilever tip was carefully approached to the surface of individual rounded chondrocytes in solution, imaged through a conventional light microscope (IX70, Olympus), a CCD camera (TM1650, Toshiba, Tokyo, Japan), and a TV monitor (12M310, Tokyo Electric Industry, Tokyo, Japan). The tip of the cantilever interacted with the cell surface and the force was indirectly recorded under indentation control of the tip. Various locations were tested on a series of individual chondrocytes. These locations were defined by eye to represent both edge regions and the centre of the projected image of the chondrocyte. Structural properties, in the form of the force–deformation relationship, were obtained.

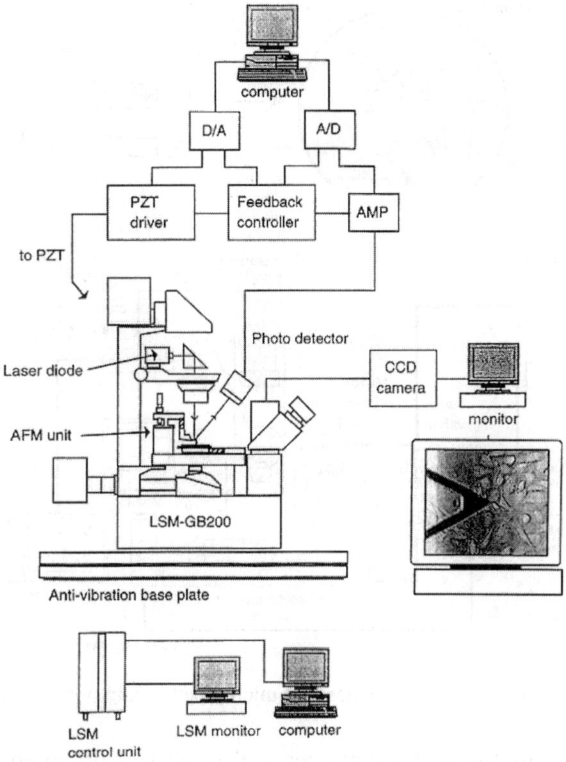

Fig. 1. Schematic of components of atomic force microscope.

2.3. Micropipette aspiration experiments

The micropipette aspiration technique, reported previously [9,17], was employed to deform the cell membrane. The schematic of the experimental apparatus is shown in Fig. 2. Micropipettes were made by drawing capillary tubes (1 mm in O.D.) with a micropipette puller (PP-83, Narishige, Tokyo, Japan) and were then fractured on a microphorge (MF-90, Narishige, Tokyo, Japan) to an inner diameter of between 5 and 15 μm. They were coated with a silicone solution (Sigmacote, Sigma, MO, USA) to prevent cell adhesion. The chondrocytes were detached from the substrate by treatment with 0.05% (w/v) trypsin-EDTA in buffer solution. Approximately 1 ml of cell suspension was placed in a cell chamber (2 mm in height and 20 mm in width) designed to allow for the entry of the micropipette from the side. The micropipette was filled with PBS using a syringe, fixed to the stage of an inverted microscope (IMT-2, Olympus, Tokyo, Japan) and then connected to a reservoir and and a micro-injector via an in-line pressure transducer. The pressure control system was filled with PBS. After approaching the micropipette tip to the cell surface by controlling a hydraulic micro-manipulator (MO-203, Narishige, Tokyo, Japan), a series of 5–10 step increments in pressure, P, were applied into the micropipette ranging from 0 to -1.5 kPa by operating the micro-injector to aspirate the cell surface and allowing to equilibrate for about 60 s. The transducer was calibrated with a mercury manometer before each experiment. During the application of pressure, the deformation process of the cell into the micropipette was observed on a TV monitor screen (TM1550, Ikegami, Tokyo, Japan) through a CCD camera (TM1650, Toshiba, Tokyo, Japan). The aspirated length, L, was measured with a video dimension analyzer (E-1000, Elcow, Hakodate, Japan) and continuously recorded on an X–Y recorder (WX2400, Graphtech, Yokohama, Japan) with

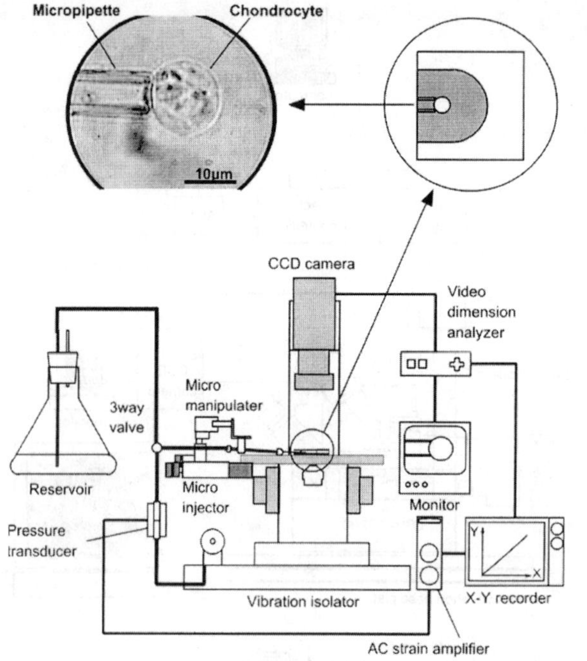

Fig. 2. Schematic of components of micropipette aspiration technique.

increasing P. Prior to aspiration, the initial diameter of the cell was obtained by averaging the measured values of the vertical and horizontal diameter with the video dimension analyzer, as well as the pipette inner diameter. All the lengths on the TV monitor were calibrated with an appropriate scale.

On the basis of the experimental data, the Young's modulus was determined using a theoretical model previously developed by Theret et al. [20] to analyze the mechanical properties of endothelial cells. In this model, the cell is assumed to be a homogeneous, elastic half-space material and the Young's modulus, E, is given as:

$$E = \frac{3a\Delta P\Phi(\eta)}{2\pi L}, \tag{1}$$

where a and b are the inner and outer radii of the micropipette and $\Phi(\eta)$ is defined as the wall function with $\eta = (b - a)/a$. In the boundary conditions of this model, the normal displacement of the cell surface at the micropipette–cell contact region is equal to zero, corresponding to $\Phi(\eta) = 2.1$ for the practical range used in this study. The Young's modulus was determined from the slope of the linear regression of the normalized length L/a versus the negative pressure P.

2.4. Compression of cell-seeded constructs

The chondrocyte suspension, prepared as above, was added to an equal volume of two agarose concentration (2% and 6%, Type IX, Sigma, Poole, UK) in Earle's Balanced Salt Solution to give final concentrations of 4×10^6 cells/ml in 1% and 3% agarose. The chondrocyte/agarose suspension was plated in a perspex mould and allowed to gel at 4°C for 20 min. Cylindrical constructs, 5 mm in diameter and 5 mm in height, were cut from the gel.

Individual cell-agarose constructs were mounted in a Universal testing machine (Model 4464, Instron UK) and hydrated in EBSS. Constructs were compressed between impermeable plattens to 20% strain at a strain rate of 20%.min^{-1}. The strain was maintained for a period of 60 minutes, while the applied load was recorded on a 2.5 N load cell at a sampling frequency of 1 Hz.

For analysis of cell deformation, cylindrical constructs were cut longitudinally into half-cylinder constructs. The agarose constructs were labelled for 1 hour at 37°C with the viable stain Calcein-AM (5 μM, Cambridge Bio-Science, UK) prepared in DMEM + 20% FCS. Individual half-cylinder constructs were placed within a test rig [11] mounted on the stage of an inverted microscope (Eclipse TE200, Nikon) associated with a confocal system (Noran, Thermo Instruments, Bicester, UK).

A 20% uniaxial unconfined compressive strain was applied to the construct at a strain rate of 20%.min^{-1}. Following a 10 minute stress relaxation period, individual cells were visualised using the confocal laser scanning microscopy. Cell deformation was quantified in terms of a deformation index ($I = X/Y$), representing the ratio of cell diameters parallel (X) and perpendicular (Y) to the axis of compression [11,14]. Based on the following assumptions, previously validated [14] namely,

Cells in unstrained agarose are spherical ($X = Y = Z$);
Cells in compressed agarose deform to an oblate ellipsoid ($X < Y = Z$);
Cell deformation occurs at constant cell volume.

The percentage reduction in cell X diameter, or cell strain (ε) was calculated from the deformation index (I)

$$\varepsilon = (1 - (I)^{2/3}) \times 100\%. \qquad (2)$$

3. Results

3.1. Atomic Force Microscopy (AFM) experiments

Calibration of the cantilever system yielded a highly significant linear relationship ($r = 1.0$) between cantilever deflection and force. The slope of the model was 0.02 N/m. The AFM technique yielded a characteristic non linear force–displacement behaviour (Fig. 3) for each cell tested ($n = 6$). Each response was reproducible, with a degree of hysteresis evident when the indentation was removed (data not shown). The non-linearity was also evident when different locations were measured across the same cell (Fig. 4). There was considerable variation in the response across the cell with the central region of the cell exhibiting increased resistance to indentation.

3.2. Micropipette aspiration experiments

The results from the micropipette aspiration techniques, yielded a linear relationship (Fig. 5), with the individual values of Young's modulus, estimated from Eq. (1), indicated for 3 individual cells. The modulus values were highly dependent on the inner diameter of the micropipette. For example, micropipettes with inner diameter of 9.3 μm and 14.6 μm, yielded mean values of 0.81 $\pm$ 0.42 and 0.30 $\pm$ 0.10 kPa, respectively (for $n = 6$ in both cases).

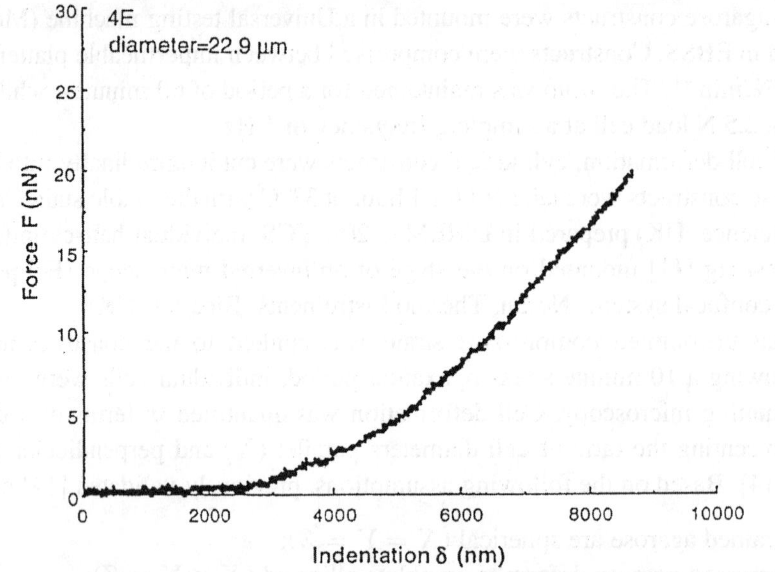

Fig. 3. Force displacement response of single chondrocyte measured using the AFM system.

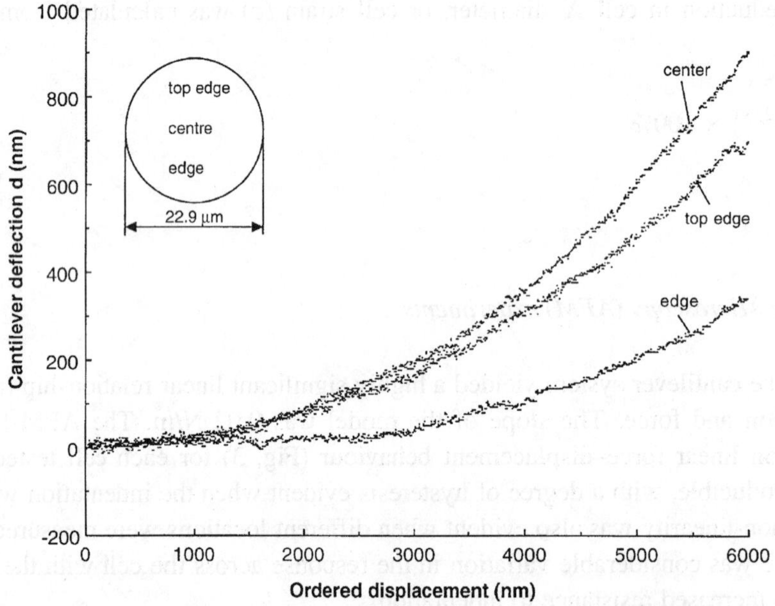

Fig. 4. Structural response curves across a singleisolated chondrocyte measured using the AFM system. The inset indicates the location of the response with respect to the projected image.

3.3. Compression of cell-seeded constructs

Following the application of the 20% compressive strain, the stress in both 1% and 3% agarose decreased rapidly at first and then more slowly towards an equilibrium value (Fig. 6). Constant levels of stress were attained after approximately 5 minutes in 1% agarose and 15 minutes in 3% agarose. The relaxation modulus measured 30 minutes after the start of compression 1% and 3% agarose was 1.9 kPa

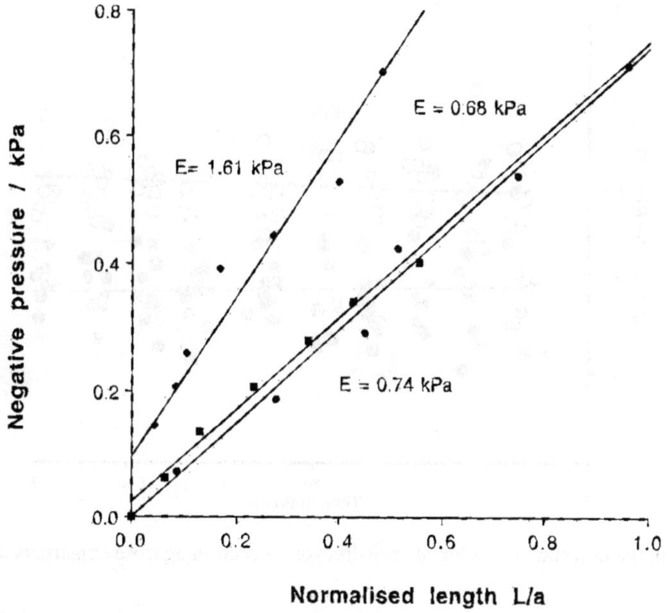

Fig. 5. Change in normalised length with negative pressure for 3 isolated chondrocytes examined using micropipette aspiration technique.

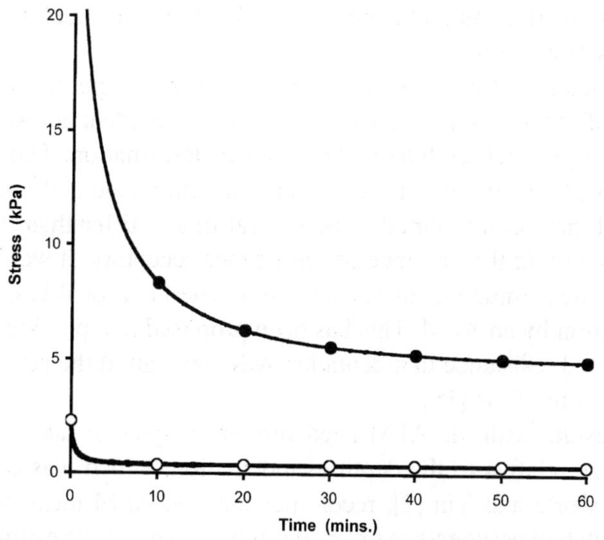

Fig. 6. Stress relaxation at 20% compressive strain of agarose constructs. Open circles represent 1% agarose, closed circles represent 3% agarose.

and 25.0 kPa, respectively. Throughout the entire measurement period, i.e., 10–60 minutes following applied strain, there was no statistically significant change in the values of deformation index for both 1% and 3% agarose (Fig. 7). The mean deformation indices for samples of 100 cells were 0.80 in 1% agarose and 0.71 in 3% agarose. This difference was found to be statistically significant ($p < 0.001$). The equivalent strain values were 13.8% and 20.4% for 1% and 3% agarose, respectively.

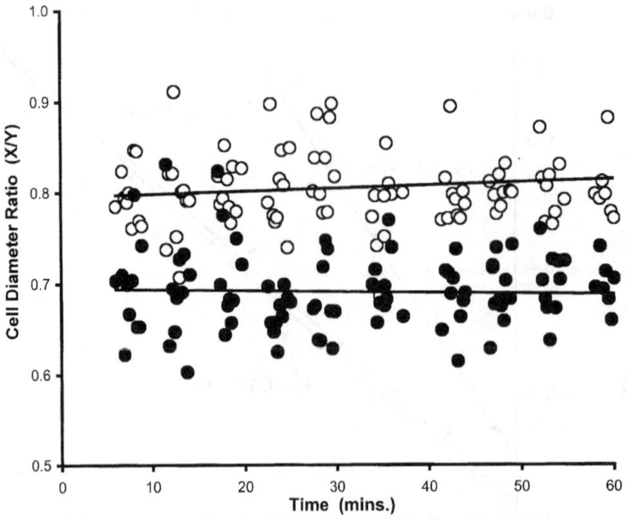

Fig. 7. Temporal changes in the deformation index of chondrocytes seeded in agarose constructs compressed to 20% strain.

4. Discussion

This paper examined the deformation behaviour of bovine articular chondrocytes using three distinct techniques. Two direct contact techniques were associated with the local deformation of the surface of isolated cells, while the alternative indirect approach involved the gross deformation of cells seeded in a viscoelastic homogeneous biomaterial.

The AFM system provided repeatable measures of the structural properties of localised regions of individual cells. Indeed, the differences in resistance to deformation evident across the cell (Fig. 4) confirms the importance of the underlying sub-cellular material under deformation. Thus it may be postulated that the increased resistance exhibited by the central region of the projected cell image may be a result of the underlying nucleus, which has been reported to be several times stiffer than the cytoplasm [5,9,14,15], local cytoskeleton organisation or the presence of cell surface receptors. It would be possible to visualise some of these cellular features simultaneously, using an inverted confocal laser scanning microscope, in conjunction with deformation by an AFM. This has been proposed in a previous study [18]. Interestingly in a previous study, increased resistance to indentation was apparent at the periphery of endothelial cells, subjected to unidirectional fluid flow [18].

It is evident that the results with the AFM need further analysis to take into account the geometry of the tip and its small size relative to the depth of indentation, which was considerable in the present study. A recent paper by Costa and Yin [2], recommended that AFM measurements would have to be combined with biaxial stretch experiments in order to provide quantitative estimates of material constants for samples.

The results of the micropipette aspiration method are in close agreement to those recently reported [9]. The nature of the technique, however, involves local deformation of the cell membrane. Thus the extent of the aspired length will be influenced by both the inner diameter of the micropipette and the proximity of cellular features, including both cytoskeletal elements and cell surface receptors. Further studies are underway investigating the response of chondrocyte originating from different tissue regions [16].

The model used to yield material parameters is based on the cell as an infinite half-space in order to develop an analytical solution for a viscoelastic solid cell. A recent publication [6] has attempted to provide more realistic treatment of the micropipette contact problem in the form of a computational model. It is anticipated that this approach will be able to account for the inhomogeneities in the cellular properties.

When using both AFM and micropipette aspiration techniques, only rounded cells were selected for mechanical perturbation. It is, however, well established that chondrocytes are not phenotypically stable in prolonged 2D culture and hence this "pre-selection" may have led to the investigation of cells which were, indeed, a sub-population in terms of differentiation potential or state of cell cycle. By contrast chondrocytes are phenotypically stable in 3D constructs made of agarose or alginate gels. In compressed 3% agarose constructs, the cells deformed to approximately 20%, equivalent to the overall level of the applied strain. This suggests that the mean compressive modulus of the chondrocytes was less than the equilibrium modulus of the 3% agarose construct measured as 25 kPa. This confirms previous work by the authors [11]. However, in 1% agarose, the cell strain was only 13.8% in the X direction parallel to the axis of compression. This strain was a result of a stress (σ) of 0.38 kPa within the surrounding agarose. This enabled an approximate values of cell modulus ($E = \sigma/\varepsilon$) to be calculated at 2.7 kPa. This was similar to the results published using a similar approach [3] and considerably higher than those estimated by local techniques, such as micropipette aspiration.

The study highlights the difficulties in establishing unique mechanical parameters, which represent the deformation behaviour of articular chondrocytes. A knowledge of the material properties of chondrocytes is of value in understanding the regulation of metabolism in both intact and diseased tissue. In addition, mechanical characterisation of both cells and scaffold materials is of value in predicting the effectiveness of mechanical conditioning of cell-seeded scaffolds and the production of tissue engineered cartilage with mechanical functionality.

Acknowledgements

Financial support is acknowledged from both Monbusho and the British Council for a sabbatical stay (DLB) at Tohoku University and for long term funding from the EPSRC (UK). The authors wish to express appreciation to Mr. Naoya Sakamoto and Mr. Junpei Sato for their assistance in performing the AFM and the pipette aspiration experiments.

References

[1] D.L. Bader and D.A. Lee, *Structure-Properties of Soft Tissues. Articular Cartilage in Structural Biological Materials*, M. Elices, ed., Pergamon Press, Oxford, 2000, pp. 73–103.

[2] K.D. Costa and F.C.P. Yin, Analysis of indentation: Implications for measuring mechanical properties with atomic force microscopy, *ASME J. Biomech. Eng.* **121** (1999), 462–471.

[3] P.M. Freeman, R.N. Natarajan, J.H. Kimura and T.P. Andriacchi, Chondrocyte cells respond mechanically to compressive loads, *J. Orthop. Res.* **12** (1994), 311–320.

[4] F. Guilak and V.C. Mow, Determination of the mechanical response of the chondrocyte in situ using finite element modeling and confocal microscopy, *ASME Adv. Bioeng.* **22** (1992), 21–24.

[5] F. Guilak, Compression-induced changes in the shape and volume of the chondrocyte nucleus, *J. Biomech.* **28** (1995), 1529–1541.

[6] M.A. Haider and F. Guilak, An axisymmetric boundary integral model for incompressible linear viscoelasticity: application to the micropipette aspiration contact problem, *ASME J. Biomech. Eng.*, **122** (2000), 236–244.

[7] R.M. Hochmuth, Micropipette aspiration of living cells, *J. Biomech.* **33** (2000), 15–22.

[8] B.D. Idowu, M.M. Knight, D.L. Bader and D.A. Lee, Confocal analysis of cytoskeletal organization within isolated chondrocytes sub-populations seeded in agarose, *Histochemistry* **32** (2000), 165–174.

[9] W.R. Jones, H. Ping Ting-Beall, G.M. Lee, S.S. Kelley, R.M. Hochmuth and F. Guilak, Alterations in the Young's modulus and volumetric properties of phondrocytes isolated from normal and osteoarthritic human cartilage, *J. Biomech.* **32** (1999), 119–127.
[10] M.M. Knight, D.A. Lee and D.L. Bader, Distribution of chondrocyte deformation in compressed agarose gel using confocal microscopy, *Cell Eng.* **1** (1996), 97–102.
[11] M.M. Knight, D.A. Lee and D.L. Bader, The influence of elaborated pericellular matrix on the deformation of isolated articular chondrocytes cultured in agarose, *Biochem. Biophys. Acta* **251** (1998), 580–585.
[12] D.A. Lee and D.L. Bader, The development and characterisation of an *in vitro* system to study strain induced cell deformation in isolated chondrocytes, *In Vitro Cell. Devel. Biol.* **31** (1995), 828–835.
[13] D.A. Lee and D.L. Bader, Compressive strains at physiological frequencies influence the metabolism of chondrocytes seeded in agarose, *J. Orthop. Res.* **15** (1997), 181–188
[14] D.A. Lee, M.M. Knight, J.F. Bolton, B.D. Idowu, M.V. Kayser and D.L. Bader, Chondrocyte deformation within compressed agarose constructs at the cellular and sub-cellular levels, *J. Biomech.* **33** (2000), 81–95.
[15] A.J. Maniotis, C.S. Chen and D.E. Ingber, Demonstration of mechanical connections between integrins, cytoskeletal filaments and nucleoplasm that stabilizes nuclear structure, in: *Proceedings of the National Academy of Sciences of the USA*, 1994, pp. 849–854.
[16] T. Ohashi, M. Fujita, D.L. Bader and M. Sato, Changes in mechanical properties of isolated articular chondrocytes with time in culture, in: *Proceedings of the 10th ICBME*, Singapore, December 2000.
[17] M. Sato, M.J. Levesque and R.M. Nerem, Micropipette aspiration of cultured bovine aortic endothelial cells exposed to shear stress, *Arteriosclerosis* **7** (1987), 276–286.
[18] M. Sato, K. Nagayama, N. Kataoka, M. Sasaki and K. Hane, Local mechanical properties measured by atomic force microscopy for cultured bovine endothelial cells exposed to shear stress, *J. Biomech.* **33** (2000), 127–135.
[19] D. Shin and K. Athanasiou, Cytoindentation for obtaining cell biomechanical properties, *J. Orthop. Res.* **17** (1999), 880–890.
[20] D.P. Theret, M.J. Levesque, M. Sato, R.M. Nerem and L.T. Wheeler, The application of a homogeneous half-space model in the analysis of endothelial cell micropipette measurements, *ASME J. Biomech. Engng.* **110** (1988), 190–199.

Biorheology 39 (2002) 79–88
IOS Press

The influence of repair tissue maturation on the response to oscillatory compression in a cartilage defect repair model

Christopher J. Hunter [a] and Marc E. Levenston [a,b,*]

[a] *Georgia Institute of Technology, Wallace H. Coulter Department of Biomedical Engineering, Atlanta, GA, USA*

[b] *Georgia Institute of Technology, George W. Woodruff School of Mechanical Engineering, Atlanta, GA, USA*

Abstract. This study examined the effects of mechanical compression on engineered cartilage in a novel hybrid culture system. Cylindrical holes were cut in discs of bovine articular cartilage and filled with agarose gels containing chondrocytes. These constructs were compressed in radiolabeled medium under static or oscillatory unconfined compression. Oscillatory compression at 1 Hz significantly stimulated synthesis above static control levels. Control experiments indicate that oscillatory compression does not stimulate freshly cast gels (without annuli), but does so after several weeks. This may be because physiologic fluid flow levels do not occur until sufficient extracellular matrix has accumulated. Finite element models predict minimal fluid flow in the gel core, and minimal differences in flow patterns between free and constrained gels. However, the models predict fluid pressures in constrained gels to be substantially higher than those in free gels. Our results suggest that pressure variations may influence synthesis of engineered cartilage matrices, with implications for construct development and post-implantation survival.

1. Introduction

Moderate to severe injuries to articular cartilage often require surgical interventions to reduce pain and improve joint mobility. However, current treatments fail to restore long-term functionality and do little to prevent further degradation of the joint. Tissue engineering may provide new therapies that are capable of repairing the existing damage and reversing the disease process, therefore offering long-term solutions to joint repair. Many different studies have demonstrated the potential use of tissue engineered cartilage to treat focal joint defects *in vitro* [2,19,26]. While some tissues appear to grow rather easily *in vivo*, it has been suggested that additional exogenous stimuli may be required for formation of engineered cartilage.

Previous studies have demonstrated that both cartilage explants and isolated chondrocytes cultured in polymer carriers will alter matrix synthesis in response to mechanical stimulation. Static compression inhibits cell proliferation and matrix synthesis of cells grown in agarose, alginate, and collagen as well as tissue explants [1,3,6,10,13,21,23]. Static loading also inhibits collagen types I and II gene expression in explant cultures, but may not affect long-term expression of aggrecan core protein mRNA [1,25]. In contrast, oscillatory compression stimulates matrix synthesis and gene expression in tissue, agarose, and alginate cultures [3,5,13,21,22,24].

*Address for correspondence: Marc E. Levenston, School of Mechanical Engineering, Atlanta, GA 30332-0405, USA. Tel.: +1 404 894 4219; Fax: +1 404 385 1397; E-mail: marc.levenston@me.gatech.edu.

Compression of articular cartilage induces many different cell-scale physicochemical phenomena, including matrix strain, fluid flow, electrical currents, and pH changes. Comparisons between spatial distributions of cellular responses and predicted changes in physical phenomena suggest that chondrocytes in tissue explants detect macroscopic mechanical stimuli via induced fluid flow [5,9,12]. Hence the cells' ability to detect external loads should be dependent upon the mechanical properties of the extracellular matrix, as the magnitudes of the induced fluid velocities will vary with tissue stiffness and permeability.

While there have been many attempts, no engineered cartilage to date has completely reproduced the mechanical properties of native tissue [4,15–18,20]. The engineered tissues are generally softer, more permeable, and more homogeneous than the native tissue. As the engineered tissue matures, the material properties may change, with decreased permeability, increased stiffness, and decreased water content. The resulting differences in local biomechanical stimuli may cause both spatial and temporal changes in chondrocyte metabolism [27].

In order to study the influence of altered tissue mechanics in a controlled environment, we have employed an *in vitro* cartilage defect repair model [7,8]. Engineered cartilage was placed into an annulus of articular cartilage explant tissue, allowing for study of the effects of biochemical and mechanical stimuli on the repair process. This hybrid construct captured many aspects of *in vivo* repair, including cell necrosis adjacent to the cut surface, and biochemical interactions between the native and engineered tissues [8]. The local biochemical signals detected by cells in the this repair model were significantly different from those encountered by an engineered tissue grown in isolated culture. In the current study, hybrid constructs and free gels were subjected to static and oscillatory compression, and matrix synthesis rates were monitored. A finite element model was constructed to estimate the local biomechanical stimuli experienced by the cells, and the results were compared to the experimental data. Lastly, a series of models were constructed to estimate how those local biomechanical stimuli might change as the engineered tissue matures *in situ*.

2. Methods

2.1. Materials

Bovine stifle joints were from Research 87, Marlborough, MA. Papain was from Worthington Biochemical. Collagenase type II, low melting point (LMP) agarose, phosphate buffered saline (PBS), high-glucose DMEM, non-essential amino acids, sodium pyruvate, L-glutamine, and gentamicin were from Gibco BRL, and fetal bovine serum was from Cellgro. Ascorbate, neomycin, kanamycin, β-agarase, and Hoechst 33258 dye were from Sigma. DMMB dye was from Aldrich Chemicals. Radiolabelled L-5-^{3}H-proline and sodium ^{35}S-sodium sulfate were from American Radiolabelled Chemicals.

2.2. Construct assembly and cell culture

Articular cartilage was isolated from one month old bovine stifle joints within 24 hours of slaughter. Full-thickness cores (8 mm diameter) were removed from the femoral condyles and patellofemoral grooves. The cartilage cores were trimmed to 3 mm height by removing both the superficial and deep zone cartilage, and circular holes (4 mm diameter) were cut to form annuli of tissue with 40 μl defects (Fig. 1). The annuli were then soaked in DMEM supplemented with 0.05 mg/ml neomycin and 0.1 mg/ml kanamycin for 24 hours in a 37°C, 5% CO_2 incubator. Chondrocytes were isolated out of the remaining cartilage from the joint surface by dicing the tissue into 1 mm^3 pieces and transferring the pieces to 0.2%

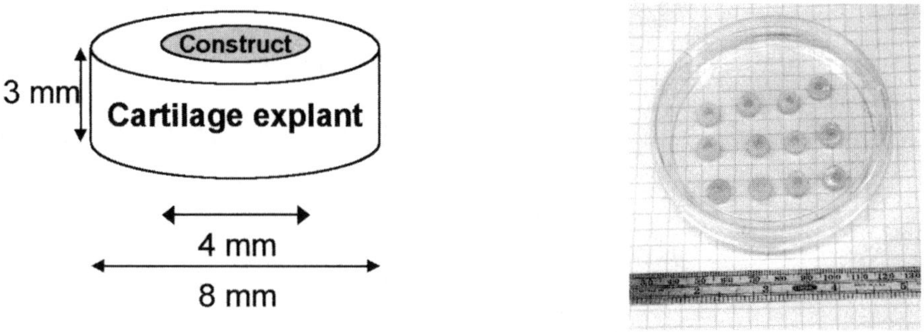

Fig. 1. Defect repair model. An engineered cartilage is implanted into a simulated defect in an explant of articular cartilage.

type 2 collagenase in DMEM (10 ml per gram of tissue) supplemented with 0.05 mg/ml neomycin and 0.1 mg/ml kanamycin for 24 hours in a 37°C, 5% CO_2 incubator. The digest solution was centrifuged at $160 \times g$ for 5 minutes to pellet the cells, which were then rinsed twice with DMEM.

Agarose gels were assembled by suspending cells at 10^7 cells/ml in a solution of LMP agarose, FBS, and 5× DMEM (final concentrations 2%, 10%, and 1×, respectively), and allowing the agarose to gel *in situ* in the cartilage defects for 10 minutes (Fig. 1). In the case of free gels, the cell/agarose mixture was cast in cartilage defects, allowed to gel, then pushed out of the defect and cultured in isolation. Once the agarose had completely gelled, 2 ml of culture medium was added over each hybrid construct. Samples were cultured in DMEM containing 10% FBS, 0.1 mM NEAA, 4 mM L-glutamine, 1 mM sodium pyruvate, 5 μg/ml gentamicin sulfate, 4 mM L-proline, and 50 μg/ml ascorbic acid. Constructs were precultured for 3 days (both hybrids and free gels) or for 15 days (free gels only) prior to compression, with medium changes every two days.

2.3. Mechanical compression

Constructs were compressed in a custom-designed bioreactor system, consisting of removable bioreactor cassettes loaded onto a digitally-controlled electromechanical load frame operating under displacement control (Fig. 2). The upper half of the bioreactor cassettes were attached to the mobile plate of the load frame, and the lower half to the stationary plate. The body of a linear stepper motor was attached to the stationary plate, and the mobile plate rode on the motor's lead screw, guided by three stainless steel guide-rods. The stepper motor was driven by a custom-designed digital controller, and lead screw rotation was confirmed with a rotary encoder. The entire load frame was housed inside a 37°C, 5% CO_2 incubator for experiments. Each bioreactor cassette held eight samples in independent culture wells containing 1 ml of medium, and two cassettes were mounted onto each load frame. Constructs were cultured in either free-swelling conditions, statically compressed by 10% of original thickness, or dynamically compressed by $10 \pm 4\%$ of original thickness at either 0.1 or 1 Hz.

2.4. Radiolabel incorporation

Hybrid constructs ($n = 8$) were compressed for 24 hours in 1 ml medium supplemented with 10 μCi/ml of L-5-^{3}H-proline and 5 μCi/ml of ^{35}S-sodium sulfate to measure protein and proteoglycan synthesis rates, respectively. Constructs were removed and separated into core and annulus regions, then rinsed four times for 30 minutes each at 4°C with PBS containing 0.8 mM sodium sulfate and 1 mM L-proline to remove unincorporated radiolabel. Gels were lyophilized and immersed in 485 μl PBS

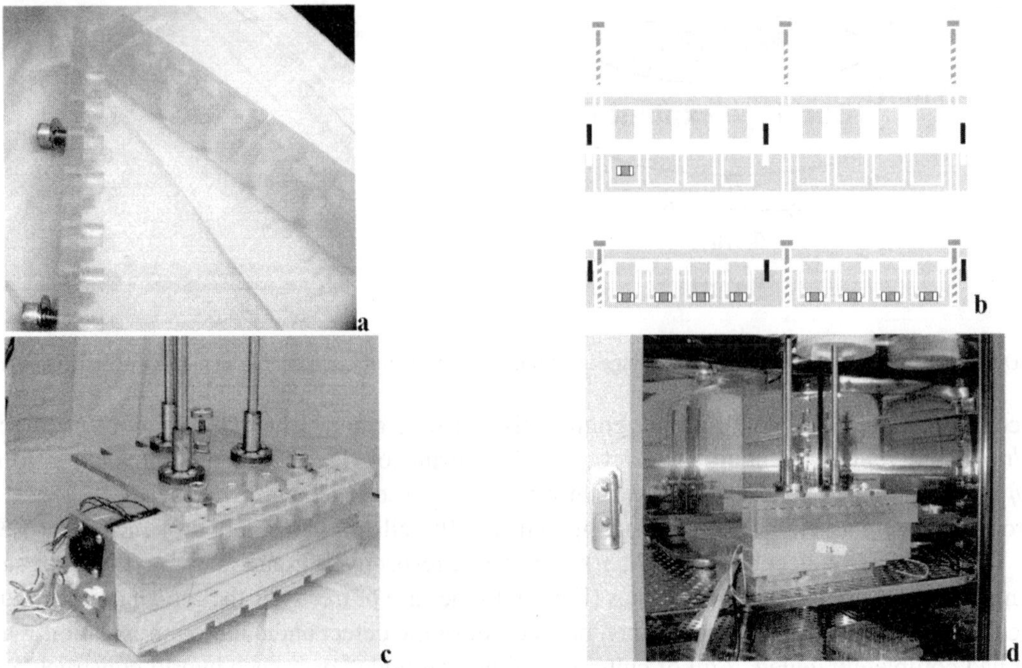

Fig. 2. Dynamic uniaxial compression system. Bioreactor cassettes (a, b) hold eight samples each in separate culture wells, and can be configured for static compression with spacers (b) or rigged to the loading frame (c) for dynamic compression (two cassettes per frame). Two independent load frames and all related controls fit into a standard tissue culture incubator (d).

containing 10 mM cysteine and 10 mM EDTA, heated at 70°C until molten and then cooled to 40°C. The mixture was supplemented with 10 μl of 500 U/ml β-agarase and incubated at 40°C for one hour, then supplemented with 5 μl of papain at 280 U/ml and further incubated at 55°C until completely digested. Digests were read on a scintillation counter to assess radioactive content. Radioactive content was normalized to DNA content, which was quantified using the fluorescent Hoechst 33258 dye-binding assay [11,28].

2.5. *Finite element models*

Hybrid constructs and free gels were modeled using a linearized form of a previously described three-field mixed poroelastic finite element formulation [14]. Quadratic 9-node elements with three internal pressure degrees-of-freedom were used to build an axisymmetric quarter-space model of the sample, taking advantage of the horizontal plane and vertical axis of symmetry (Fig. 3). The mesh contained 40 elements each for the core and annulus regions (40 elements for free gels), and was refined towards expected regions of large gradients, e.g., the interface between the core and annulus regions of hybrid constructs. Material properties for articular cartilage and freshly cast agarose gels were chosen to be consistent with values measured in our laboratory (Table 1). Maturation of agarose gels was simulated by simultaneously increasing the equilibrium modulus and decreasing the permeability, with the most "mature" gels having a modulus and permeability identical to those of articular cartilage. The platen was modeled as rigid and impermeable, and no slip was allowed between the platen and the sample. The lateral edge of the sample was free-draining. Vertical displacement of the platen was specified to match

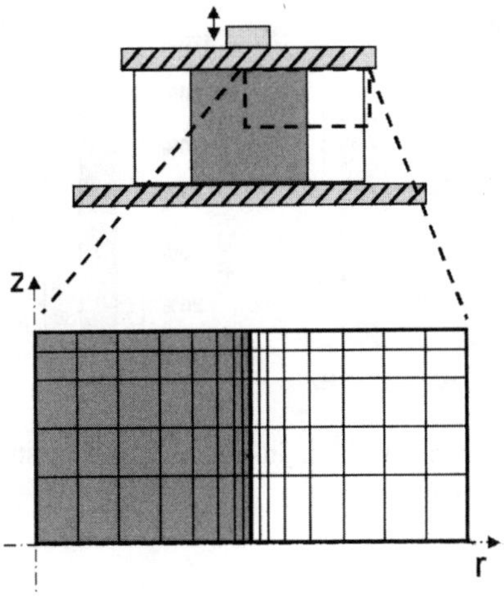

Fig. 3. Finite element models of loaded constructs were generated, taking advantage of the vertical axis and horizontal plane of symmetry. The platens were modeled as rigid, impermeable boundaries, and no slip was allowed between the construct and platen.

Table 1

Material properties used in finite element models

Maturation state	E (MPa)	v	k (m^4/N s)	ϕ^f (%)
0	0.015	0.1	2e-13	95
1	0.025	0.1	1e-13	90
2	0.040	0.1	5e-14	85
3	0.1	0.1	2e-14	85
4 (native cartilage)	1.0	0.1	2e-15	80

the oscillatory compression applied to experimental samples. The root-mean-square (RMS) values of the fluid pressure and fluid velocity over a cycle were calculated at sinusoidal steady state.

2.6. Statistical analysis

Data were analyzed using one-way ANOVA and Tukey's test for post-hoc analysis. Differences were considered significant at or below the $p = 0.05$ level.

3. Results

Radiolabel incorporation studies demonstrated a significant effect of mechanical compression on synthesis rates in hybrid constructs. Static compression inhibited both ^{3}H-proline and ^{35}S-sulfate incorporation by approximately 60% in the gel core and the tissue annulus of hybrid constructs ($p < 0.001$ and $p < 0.001$, respectively). Superposition of oscillatory compression had no effect on either ^{3}H or ^{35}S incorporation at 0.1 Hz ($p = 0.29$ and $p = 0.56$), but 1 Hz compression significantly increased both incorporation rates ($p = 0.009$ and $p = 0.02$) (Fig. 5). Interestingly, baseline synthesis per cell was substantially higher (approx. 4 times) in the gel core than in the tissue annulus (data not shown).

Static compression inhibited both ^{3}H-proline and ^{35}S-sulfate incorporation rates in free gels after three days of preculture ($p < 0.001$ and $p < 0.001$, respectively) (Fig. 6a). Superposition of oscillatory

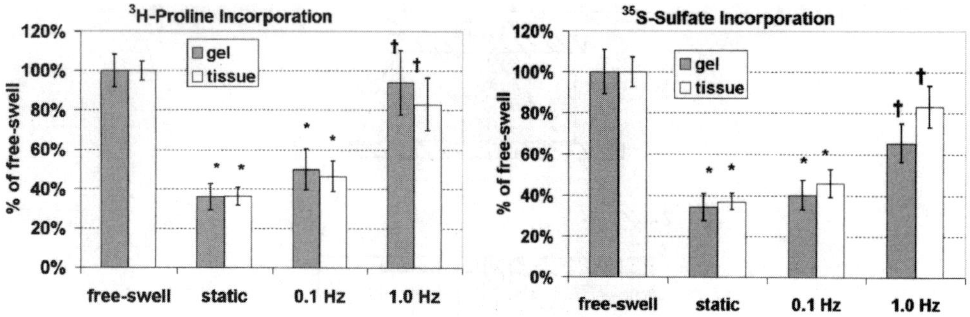

Fig. 4. Oscillatory compression stimulates matrix synthesis in loaded hybrid construct. Stars: $p < 0.05$ vs. "free-swell"; daggers: $p < 0.05$ vs. "static" ($n = 8$, means ± sem).

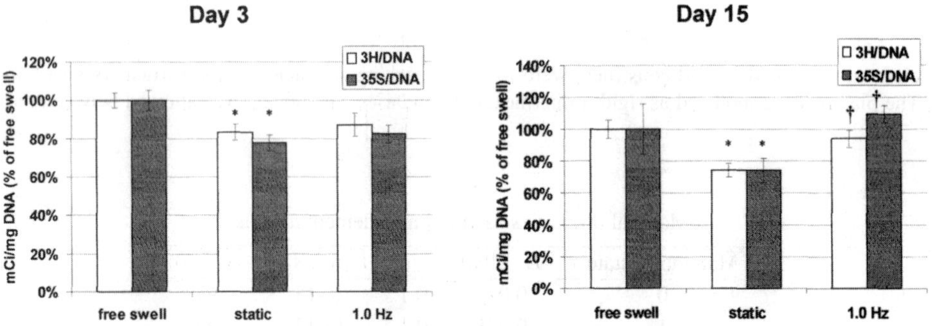

Fig. 5. Oscillatory compression has no effect upon matrix synthesis in gels compressed after 3 days of preculture, but significantly stimulates synthesis in gels compressed after 15 days of preculture. Stars: $p < 0.05$ vs. "free-swell"; daggers: $p < 0.05$ vs. "static" ($n = 8$, means ± sem).

compression had no effect on either ^{3}H or ^{35}S incorporation ($p = 0.16$ and $p = 0.17$). After 15 days of preculture, static compression inhibited ^{3}H incorporation by approximately 20% ($p = 0.005$), while there was no change in ^{35}S incorporation ($p = 0.19$). Superposition of oscillatory compression at 1 Hz significantly increased both ^{3}H and ^{35}S incorporation ($p = 0.010$ and $p = 0.0012$) (Fig. 6b) in free gels precultured for 15 days.

The finite element analyses indicate that the local mechanical environment within a constrained gel is substantially different from that in a free gel. Regardless of gel maturation state, the RMS fluid pressures within constrained (hybrid) gels were several orders of magnitude higher than those within free gels (Fig. 6). The RMS pressure in free gels increased substantially with increased gel maturation, but the RMS pressure in hybrid gels was virtually unchanged (Figs 6, 7). In contrast, the RMS fluid velocities in hybrid and free gels were comparable (and fairly low), with little change due to maturation of gel material properties (Fig. 8).

4. Conclusion

Both native and tissue engineered cartilages have been shown to respond to mechanical compression with changes in matrix metabolism and gene expression. This response may play an important role in modulating tissue growth and maintenance in the joint, and may also be a useful tool for supporting tissue engineered joint repair. In this study, we examined the effects of mechanical compression on

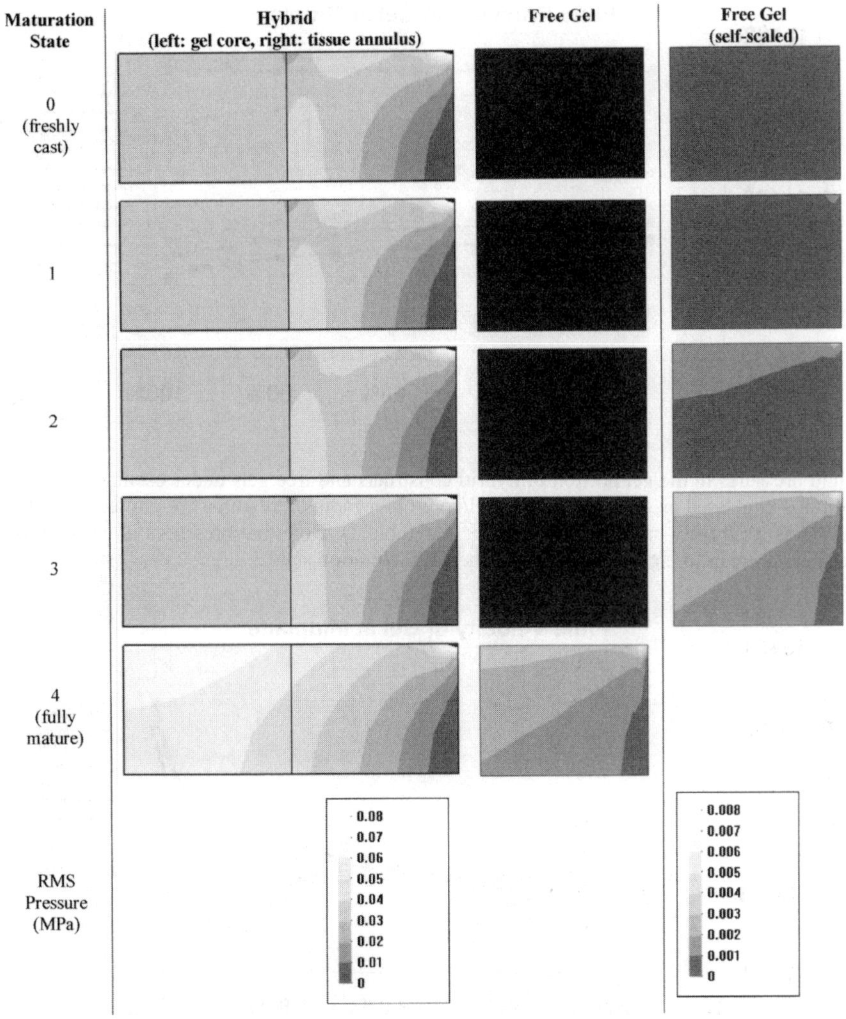

Fig. 6. Predicted RMS fluid pressure profiles in hybrid constructs and free gels under oscillatory compression ($10 \pm 4\%$, 1 Hz). Column 1 shows the pressure throughout the hybrid construct, and column 2 shows the pressure throughout a comparable free gel. Column 3 shows the pressure in the free gels rescaled.

biosynthesis by chondrocytes in an *in vitro* defect repair model. Compared to "isolated" construct culture models, this hybrid culture model may be more representative of the actual *in vivo* cartilage defect repair environment. For example, the presence of living tissue in the joint may directly or indirectly influence cells in the repair tissue. We have previously shown that biochemical interactions with the living tissue influence both proliferation and matrix synthesis in this repair model [8]. The mechanical environment of *in vivo* repair tissue may also be substantially different from that in isolated culture models. In an isolated construct, fluid is free to flow through the lateral surfaces. In a construct placed in a defect *in vivo* or in our hybrid culture model, however, fluid flow is restricted laterally. Our experimental and theoretical results suggest that consequent changes in the local biomechanical environment may substantially affect the biosynthetic activity of chondrocytes in the repair tissue.

Repair tissue in the hybrid culture model demonstrated an elevated response to both static and oscillatory mechanical compression as compared to free gels. Static compression significantly inhibited

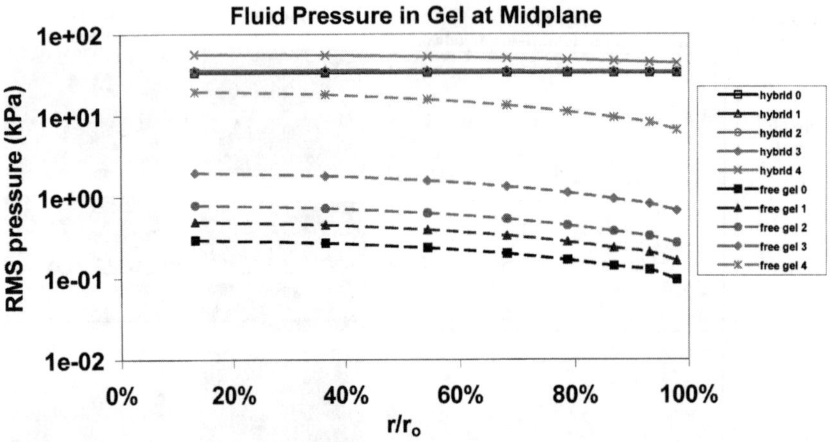

Fig. 7. Predicted fluid pressures in the gel portion of hybrid constructs and free gels under oscillatory compression ($10 \pm 4\%$, 1 Hz). Radial position is expressed as a percent of the gel/core outer radius. Pressures are plotted for hybrid and free gels of varying maturation states (0: freshly cast; 4: fully mature; see Table 1). Note that pressures in constrained (hybrid) gels are substantially higher than those in free gels, regardless of the gel maturation state.

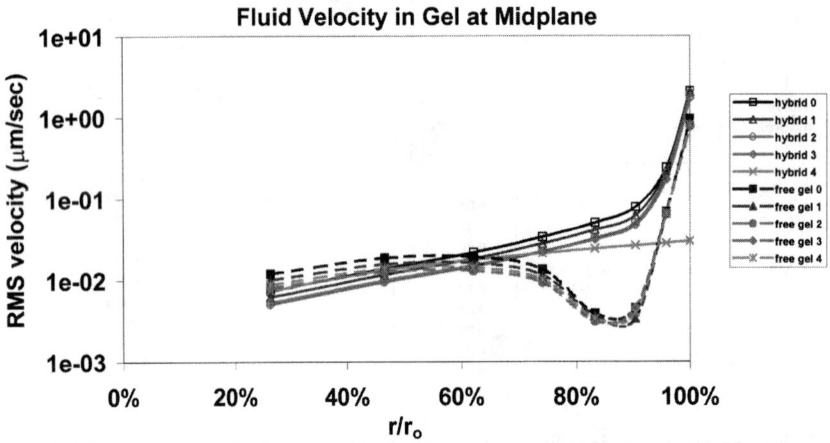

Fig. 8. Predicted fluid velocities in the gel portion of hybrid constructs and free gels under oscillatory compression ($10 \pm 4\%$, 1 Hz). Fluid velocities are plotted for hybrid and free gels of varying maturation states (0: freshly cast; 4: fully mature; see Table 1). Radial position is expressed as a percent of the gel/core outer radius. Note that fluid velocities are comparable in constrained (hybrid) and free gels.

biosynthesis in both freshly cast hybrid and free gels, but the level of inhibition was significantly greater in the hybrid gels. This suggests that other factors not directly related to compression (e.g., biochemical interactions or diffusional restrictions) also play a role in this system. Oscillatory compression stimulated biosynthesis in freshly cast hybrid gels, but not in freshly cast free gels. Consistent with previously reported results [3], however, oscillatory compression did stimulate biosynthesis in free gels that matured through 15 days of preculture. Taken together, these experimental results indicate that the local mechanical environment can substantially alter the responsiveness of repair tissue to macroscopic mechanical stimuli.

The finite element analyses provide a potential explanation for the differences in mechanical stimulation observed in our experiments. Interstitial fluid flow induced by oscillatory compression has been proposed as a mechanism for mechanical stimulation in cartilage explant cultures [5,12]. Interestingly, our theoretical analysis predicts little difference in fluid flow patterns between free gels and gels in the hybrid culture model. In contrast, interstitial fluid pressure was approximately two orders of magnitude higher in freshly cast hybrid gels than in similar free gels. Likewise, while predicted fluid flow was similar for free gels at all maturation states, predicted fluid pressure increased substantially with simulated gel maturation. These results suggest that fluid pressure may be as important as fluid flow in influencing chondrocyte behavior in agarose gel cultures, and may be a factor governing the behavior of tissue engineered cartilage *in vivo*.

Acknowledgements

This work was supported in part by the Georgia Tech/Emory Center for the Engineering of Living Tissues, a National Science Foundation Engineering Research Center funded by award number EEC-9731643, by the NIH under the Cellular Engineering Training Program, grant number GM08433, and by a grant from the Medtronic Foundation.

References

[1] A. Baer, L. Setton, J. Wang, F. Nickisch and F. Guilak, Static compression of chondrocytes in alginate culture does not alter aggrecan gene expression as measured by competitive PCR, in: *Proceedings, 45th Annual Meeting of the Orthopaedic Research Society*, Anaheim, CA, 1999.

[2] B.D. Boyan, C.H. Lohmann, J. Romero and Z. Schwartz, Bone and cartilage tissue engineering, *Clin. Plast. Surg.* **26** (1999), 629–645.

[3] M.D. Buschmann, Y.A. Gluzband, A.J. Grodzinsky and E.B. Hunziker, Mechanical compression modulates matrix biosynthesis in chondrocyte/agarose culture, *J. Cell Sci.* **108** (1995), 1497–1508.

[4] M.D. Buschmann, Y.A. Gluzband, A.J. Grodzinsky, J.H. Kimura and E.B. Hunziker, Chondrocytes in agarose culture synthesize a mechanically functional extracellular matrix, *J. Orthop. Res.* **10** (1992), 745–758.

[5] M.D. Buschmann, Y.J. Kim, M. Wong, E. Frank, E.B. Hunziker and A.J. Grodzinsky, Stimulation of aggrecan synthesis in cartilage explants by cyclic loading is localized to regions of high interstitial fluid flow, *Arch. Biochem. Biophys.* **366** (1999), 1–7.

[6] C.J. Hunter, S.M. Imler, P. Malaviya, R.M. Nerem and M.E. Levenston, Mechanical compression alters gene expression and extracellular matrix synthesis by chondrocytes cultured in collagen-I gels, *Biomaterials* **23** (2002), 1249–1259.

[7] C.J. Hunter, M.E. Levenston and R.M. Nerem, Engineered/native cartilage adhesion varies with scaffold selection, in: *Proceedings, 2001 BMES Annual Fall Meeting*, Durham, NC, 2001.

[8] C.J. Hunter, R.M. Nerem and M.E. Levenston, Interactions between native and engineered cartilage in a hybrid culture system, in: *Proceedings, Tissue Engineering Society International*, Orlando, FL, 2000.

[9] Y.J. Kim, L.J. Bonassar and A.J. Grodzinsky, The role of cartilage streaming potential, fluid flow and pressure in the stimulation of chondrocyte biosynthesis during dynamic compression, *J. Biomech.* **28** (1995), 1055–1066.

[10] Y.J. Kim, A.J. Grodzinsky and A.H. Plaas, Compression of cartilage results in differential effects on biosynthetic pathways for aggrecan, link protein, and hyaluronan, *Arch. Biochem. Biophys.* **328** (1996), 331–340.

[11] Y.J. Kim, R.L. Sah, J.Y. Doong and A.J. Grodzinsky, Fluorometric assay of DNA in cartilage explants using Hoechst 33258, *Analyt. Biochem.* **174** (1988), 168–176.

[12] Y.J. Kim, R.L. Sah, A.J. Grodzinsky, A.H. Plaas and J.D. Sandy, Mechanical regulation of cartilage biosynthetic behavior: physical stimuli, *Arch. Biochem. Biophys.* **311** (1994), 1–12.

[13] D.A. Lee and D.L. Bader, Compressive strains at physiological frequencies influence the metabolism of chondrocytes seeded in agarose, *J. Orthop. Res.* **15** (1997), 181–188.

[14] M.E. Levenston, E.H. Frank and A.J. Grodzinsky, Variationally derived 3-field finite element formulations for quasistatic poroelastic analysis of hydrated biological tissues, *Computat. Meth. Appl. Mechan. Eng.* **156** (1997), 231–246.

[15] P.X. Ma, B. Schloo, D. Mooney and R. Langer, Development of biomechanical properties and morphogenesis of in vitro tissue engineered cartilage, *J. Biomed. Mater. Res.* **29** (1995), 1587–1595.

[16] S.M. Malmonge, C.A. Zavaglia and W.D. Belangero, Biomechanical and histological evaluation of hydrogel implants in articular cartilage, *Brazil. J. Med. Biolog. Res.* **33** (2000), 307–312.

[17] R.L. Mauck, M.A. Soltz, C.C. Wang, D.D. Wong, P.H. Chao, W.B. Valhmu, C.T. Hung and G.A. Ateshian, Functional tissue engineering of articular cartilage through dynamic loading of chondrocyte-seeded agarose gels, *J. Biomech. Eng.* **122** (2000), 252–260.
[18] V.C. Mow and C.C. Wang, Some bioengineering considerations for tissue engineering of articular cartilage, *Clin. Orthop.* (1999), 204–223.
[19] S. Nehrer, H.A. Breinan, A. Ramappa, S. Shortkroff, G. Young, T. Minas, C.B. Sledge, I.V. Yannas and M. Spector, Canine chondrocytes seeded in type I and type II collagen implants investigated in vitro, *J. Biomed. Mater. Res.* **38** (1997), 95–104.
[20] G.M. Peretti, M.A. Randolph, V. Zaporojan, L.J. Bonassar, J.W. Xu, J.C. Fellers and M.J. Yaremchuk, A biomechanical analysis of an engineered cell-scaffold implant for cartilage repair, *Ann. Plastic Surg.* **46** (2001), 533–537.
[21] T.M. Quinn, A.J. Grodzinsky, M.D. Buschmann, Y.J. Kim and E.B. Hunziker, Mechanical compression alters proteoglycan deposition and matrix deformation around individual cells in cartilage explants, *J. Cell Sci.* **111** (1998), 573–583.
[22] P.M. Ragan, A.M. Badger, M. Cook and A.J. Grodzinsky, Chondrocyte gene expression of aggrecan and type IIA collagen is upregulated by dynamic compression and the response is related to the surrounding extracellular matrix density, in: *Proceedings, 46th Annual Meeting of the Orthopaedic Research Society*, Orlando, FL, 2000.
[23] P.M. Ragan, A.K. Staples, H.K. Hung, V. Chin, F. Binette and A.J. Grodzinsky, Mechanical compression influences chondrocyte metabolism in a new alginate disk culture system, in: *Proceedings, 44th Annual Meeting of the Orthopaedic Research Society*, New Orleans, LA, 1998.
[24] R.L. Sah, Y.J. Kim, J.Y. Doong, A.J. Grodzinsky, A.H. Plaas and J.D. Sandy, Biosynthetic response of cartilage explants to dynamic compression, *J. Orthop. Res.* **7** (1989), 619–636.
[25] W.B. Valhmu, E.J. Stazzone, N.M. Bachrach, F. Saed-Nejad, S.G. Fischer, V.C. Mow and A. Ratcliffe, Load-controlled compression of articular cartilage induces a transient stimulation of aggrecan gene expression, *Arch. Biochem. Biophys.* **353** (1998), 29–36.
[26] S. Wakitani, T. Goto, R.G. Young, J.M. Mansour, V.M. Goldberg and A.I. Caplan, Repair of large full-thickness articular cartilage defects with allograft articular chondrocytes embedded in a collagen gel, *Tiss. Eng.* **4** (1998), 429–444.
[27] J.S. Wayne, S.L. Woo and M.K. Kwan, Finite element analyses of repaired articular surfaces, *Proceedings of the Institution of Mechanical Engineers. Part H, J. Eng. Med.* **205** (1991), 155–162.
[28] J.J. Woessner, The determination of hydroxyproline in tissue and protein samples containing small proportions of this imino acid, *Arch. Biochem. Biophys.* **93** (1961), 440–447.

Biorheology 39 (2002) 89–96
IOS Press

The role of fibril reinforcement in the mechanical behavior of cartilage

LePing Li [a,*], M.D. Buschmann [b] and A. Shirazi-Adl [b]
[a] *BioSyntech Inc., Laval, Quebec, Canada*
[b] *Institute of Biomedical Engineering, Departments of Chemical & Mechanical Engineering, Ecole Polytechnique, Montreal, Quebec, Canada*

Abstract. Collagen fibril reinforcement was incorporated into a nonlinear poroelastic model for articular cartilage in unconfined compression. It was found that the radial fibrils play a predominant role in the transient mechanical behavior but a less important role in the equilibrium response of cartilage. The radial fibrils are in tension and can be highly stressed during compression, in contrast to low compressive stresses in all directions for the proteoglycan matrix after a small initial compression. The strain dependent fibril stiffening produces strong nonlinear transient response; the fibrils provide extra stiffness to balance a rising fluid pressure and to restrain stress increase in the proteoglycans. The fibril reinforcement, induced by the fluid pressure and flow, also accounts for a complex pattern of strain-magnitude and strain-rate dependence of cartilage stiffness.

Keywords: Cartilage mechanics, fibril reinforcement, finite element, poroelasticity, strain rate

1. Introduction

Articular cartilage consists of three major components, a proteoglycan matrix, a collagen fibril network and the interstitial water. It has been observed in experiments that the proteoglycan and collagen matrices are very different in biochemical compositions and thus different in mechanical properties. The negatively charged proteoglycans are ideal for supporting compressive stresses and are usually considered to be linear in constitutive behavior (up to 22% strain [1]). On the contrary, the collagen fibrils are much stiffer in tension and stiffen further when the tensile strain increases [3,10,11,14]; the fibrils have almost no resistance in compression, as compared to their ability in bearing tension [12]. However, these differences have not been well recognized in model studies; the proteoglycan and collagen matrices are often assumed as one material phase in formulation. As such, the different roles of the proteoglycan and collagen matrices in the mechanical behavior of cartilage cannot be determined, even if such models are able to partly describe the overall response of cartilage to external loadings (e.g., time varying integrated load in relaxation tests). On the other hand, the fibril reinforced or fiber-reinforced models [2,4,6–9] consider the fibrils as a distinct material phase and thus offer the potential to represent the actual tissue more precisely.

The fibril reinforced poroelastic models proposed by the present authors [6–9] assume cartilage as a three-phase composite, the fibrillar matrix (collagen), nonfibrillar matrix (the solid excluding collagen) and water. In these latter models, the fibrils are idealized as slender structural elements that support tension only. While in reality the collagen fibrils have certain resistance in shear and compression, in the

*Address for correspondence: LePing Li, Faculty of Kinesiology, University of Calgary, Calgary, Alberta, Canada T2N 1N4. Tel.: +1 403 220 8525; Fax: +1 403 284 3553; E-mail: poroelasticity@yahoo.com.

model analysis such attributes are included in the nonfibrillar matrix. The present paper demonstrates some features associated with the fibril reinforcement.

2. Methods

In order to focus on the behavior induced by fibril reinforcement, the material inhomogeneity is not considered here, and hence the homogeneous model [9] is adopted. The nonfibrillar matrix is modeled as linearly elastic, defined by the Young's modulus $E_{\rm m}$ and Poisson's ratio $\nu_{\rm m}$, either in compression or in tension. The fibrillar matrix is described by a tensile strain ($\varepsilon_{\rm f} > 0$) dependent modulus

$$E_{\rm f} = E_{\rm f}^0 + E_{\rm f}^{\varepsilon}\varepsilon_{\rm f}, \tag{1}$$

where the second term represents fibril stiffening and $E_{\rm f} = 0$ when $\varepsilon_{\rm f} < 0$. The permeability k varies with the disk dilatation ε per [5]

$$k = k_0 \exp(M\varepsilon). \tag{2}$$

The fibril volume fraction is not explicitly involved in this material model since $E_{\rm f}$ is defined as the modulus of the fibrillar matrix, rather than that of individual fibrils. In other words, $E_{\rm f} = E_{\rm t} - E_{\rm m}$ where $E_{\rm t}$ is the modulus of cartilage samples measured in tensile tests, and where $E_{\rm m}$ is the modulus determined in confined tests when the fibrils are in compression. According to this definition $E_{\rm m}$ includes the compressive stiffness of the real collagen network, if it does support a small compressive stress.

We first consider small deformations with unconfined geometry when the analytical solution is feasible for elastic behavior including the instantaneous response. When compression is applied suddenly, the interstitial water is trapped. Since incompressibility is assumed for the material phases, the total axial stress and the pore fluid pressure are determined by the requirement of volume conservation as follows:

$$\sigma_z^{\rm inst} = \left[\frac{3E_{\rm m}}{2(1+\nu_{\rm m})} + \frac{1}{2}E_{\rm f}^0 - \frac{1}{8}E_{\rm f}^{\varepsilon}\varepsilon_z\right]\varepsilon_z, \tag{3}$$

$$p_{\rm f}^{\rm inst} = -\left[\frac{E_{\rm m}}{2(1+\nu_{\rm m})} + \frac{1}{2}E_{\rm f}^0 - \frac{1}{8}E_{\rm f}^{\varepsilon}\varepsilon_z\right]\varepsilon_z. \tag{4}$$

On the other hand, if the compression is imposed at a sufficiently low speed, the fluid has little impact on the stresses and strains. The axial stress and radial strain for the drained cartilage are given as

$$\sigma_z^{\rm equil} = E_{\rm m}\varepsilon_z - \nu_{\rm m}(2E_{\rm f}^0 + E_{\rm f}^{\varepsilon}\varepsilon_r^{\rm equil})\varepsilon_r^{\rm equil}, \tag{5}$$

$$\varepsilon_r^{\rm equil} = \frac{\sqrt{(E_{\rm m} + E_{\rm f}^0\chi)^2 - 2E_{\rm f}^{\varepsilon}E_{\rm m}\nu_{\rm m}\chi\varepsilon_z} - E_{\rm m} - E_{\rm f}^0\chi}{E_{\rm f}^{\varepsilon}\chi} \quad \text{if } E_{\rm f}^{\varepsilon} > 0 \tag{6}$$

or

$$\varepsilon_r^{\rm equil} = -\frac{E_{\rm m}\nu_{\rm m}}{E_{\rm m} + E_{\rm f}^0\chi}\varepsilon_z \quad \text{if } E_{\rm f}^{\varepsilon} = 0 \text{ or no fibril stiffening}, \tag{7}$$

where $\chi = (1+\nu_{\rm m})(1-2\nu_{\rm m})$.

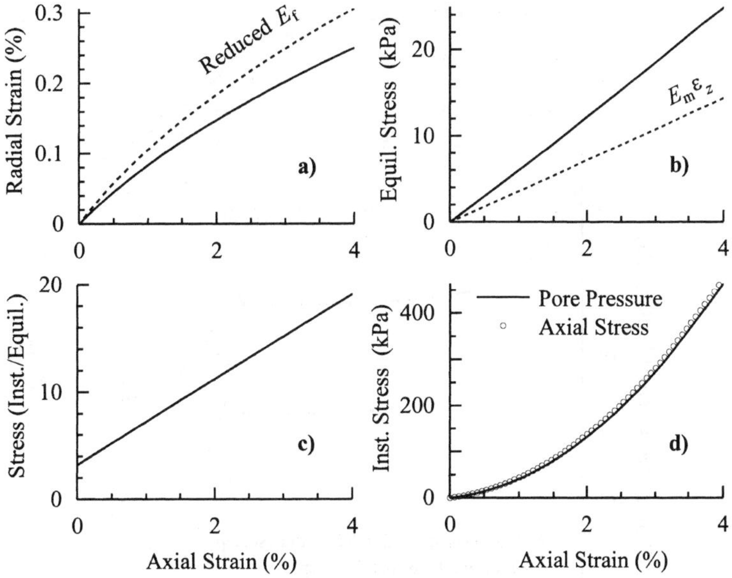

Fig. 1. Strain-dependent elastic response with small deformation for the case with $E_m = 0.36$ MPa, $\nu_m = 0.38$, $E_f^0 = 3$ MPa and $E_f^\varepsilon = 2000$ MPa. (a) Radial strain at equilibrium (the broken line shows the strain for $E_f^0 = 2$ MPa and $E_f^\varepsilon = 1500$ MPa); (b) compressive axial stress at equilibrium (the broken line shows the stress in absence of the fibrils); (c) the maximum possible ratio of the transient versus equilibrium axial stress (instantaneous versus equilibrium response); (d) the maximum possible fluid pressure and the total compressive axial stress (instantaneous response).

In general, when considering the transient behavior with large deformation, the finite element method is needed to extract solutions. In such cases, the fibril reinforced homogeneous and nonhomogeneous models have been employed to simulate measured data from unconfined compression tests, including the time varying load, the depth dependent radial strain and the strain–magnitude and strain–rate dependent cartilage stiffness [6–9]. In order to further demonstrate the role of fibril reinforcement in the mechanical behavior of cartilage in unconfined compression, a numerical example is considered here by employing the homogeneous model [9]. A sequence of two ramp compression steps is involved, allowing full relaxation after each ramp. Four loading protocols are adopted, as described in the caption of Fig. 2.

3. Results

Typical elastic responses undergoing small deformation are shown in Fig. 1, determined by Eqs (3)–(6). The instantaneous radial strain is half of the axial strain (not shown), regardless of the presence of the fibrils, due to incompressibility of the tissue. The equilibrium radial strain, however, is much smaller and is dependent on the fibril property (Fig. 1(a)). The curvature of the radial strain is completely produced by the fibril stiffening ($E_f^\varepsilon > 0$), comparing Eq. (6) with (7). The contribution of fibril reinforcement to the axial stress at equilibrium is demonstrated in Fig. 1(b) (the difference between the solid and broken lines, or the second term on the right hand side of (5)). The total axial stress (compressive) for instantaneous response (Fig. 1(d)), mainly contributed by the fluid pressure, is large and strongly nonlinear, in contrast to the equilibrium stress (Fig. 1(b)) that is low and almost linear with the axial strain. It is observed from (4) that the pressure is mainly induced by the fibrils due to small E_m. The difference between the total compressive axial stress ($-\sigma_z^{\text{inst}}$) and the pressure (p_f^{inst}) is actually the compressive axial stress of the

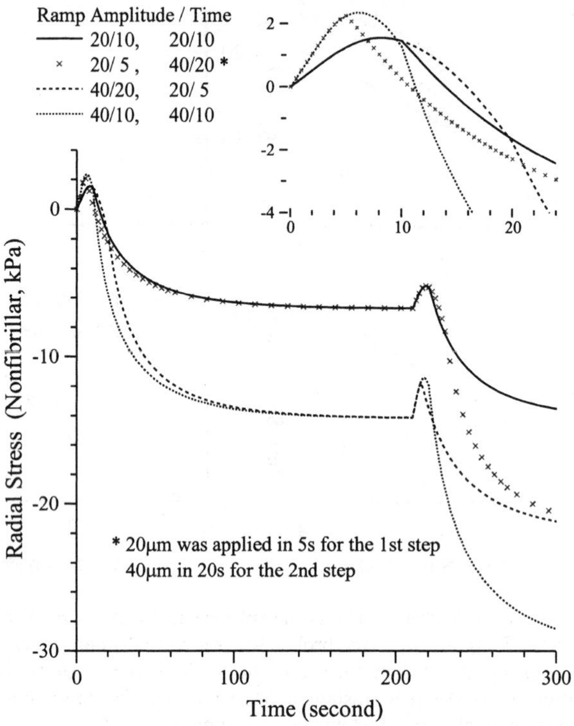

Fig. 2. Radial stress of the nonfibrillar matrix at the disk center for four different two-step ramp compression protocols: (i) 20 μm (2% nominal axial strain) was imposed in 10 s for both steps, respectively; (ii) 20 μm was imposed in 5 s for the first step and 40 μm in 20 s for the second step; (iii) 40 μm was imposed in 20 s for the first step and 20 μm in 5 s for the second step; (iv) 40 μm was imposed in 10 s for both steps, respectively. The material properties adopted are $E_{\mathrm{m}} = 0.36$ MPa, $\nu_{\mathrm{m}} = 0.38$, $E_{\mathrm{f}}^{0} = 3$ MPa, $E_{\mathrm{f}}^{\varepsilon} = 2000$ MPa, $k_0 = 0.003$ mm^4/N s and $M = 15$. The initial void ratio is 3.5. The radius and thickness of disk are 1.5 and 1.0 mm, respectively.

nonfibrillar matrix, determined as $E_{\mathrm{m}}(-\varepsilon_z)/(1 + \nu_{\mathrm{m}})$ which is very small. On the other hand, the ratio of instantaneous versus equilibrium axial stresses ($\sigma_z^{\mathrm{inst}}/\sigma_z^{\mathrm{equil}}$) increases approximately linearly with the axial strain (Fig. 1(c)).

Now we present the results obtained by employing the large deformation theory. The radial stress of the nonfibrillar matrix (referred to as *matrix stress* hereafter) at the disk center, which is initially tensile during the first ramp loading, soon becomes compressive as relaxation proceeds (Fig. 2). This stress remains compressive thereafter, exhibiting a small rebound at the following ramp compression. (The matrix stress would be tensile were the fibrils absent.) The rebound is similar to the initial rise, whose height is more dependent on the ramp speed than the compression amplitude for the cases considered (axial strain increment $\leqslant$ 4%). At the speed of 2 μm/s (20 μm in 10 s or 40 μm in 20 s), doubling the compression amplitude does *not* increase the peak stress (the rebound magnitude) at all. However, for the same compression amplitude, doubling the compression speed always increases the peak stress. On the other hand, at a higher speed (4 μm/s, 20 μm in 5 s or 40 μm in 10 s), doubling the compression amplitude *does* slightly increase the peak stress (the inset of Fig. 2).

For the radial stress of the fibrillar matrix (Fig. 3, referred to as *fibril stress* hereafter), the dependence on compression amplitude is more evident due to fibril stiffening on its strain. The fibril stiffening is also observed when comparing the fibril stress for the first step to that for the second step under the same compression amplitude and speed (the first step starts from an undeformed state but the second starts from

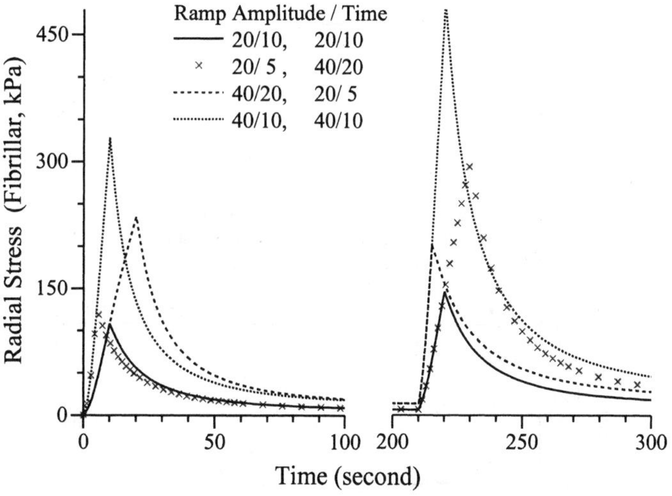

Fig. 3. Radial stress of the fibrillar matrix at the disk center for the cases considered in Fig. 2. For clarity of the figure, the stress is not shown for the time between 100 and 200 s.

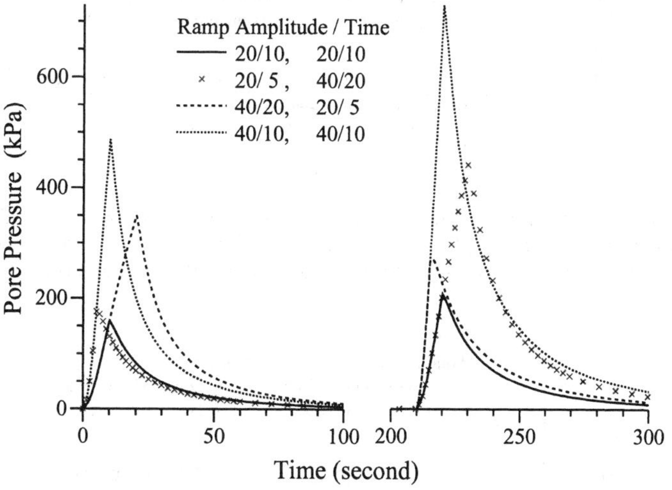

Fig. 4. Pore fluid pressure at the disk center for the cases considered in Fig. 2. For clarity of the figure, the pressure is not shown for the time between 100 and 200 s.

a deformed state). The fibrils are highly stressed at the peak transient and greatly relaxed at equilibrium. Apart from the large difference in magnitude, the pattern of the fibril stress is quite different from that of the matrix stress. When the compression continues, the matrix stress may decrease even before relaxation begins (the inset of Fig. 2, 40 μm in 20 s), while the fibril stress keeps growing (Fig. 3). The pore fluid pressure pattern (Fig. 4) is similar to that of the fibril stress (Fig. 3), indicating the overwhelming impact of fibril reinforcement on the pressurization, which in turn affects the total axial stress (not shown).

The stress distributions along the disk radius further demonstrate the mechanism of fibril reinforcement induced by the fluid flow and pressure (Figs 5 and 6). In this case, a sequence of two identical ramp compression/full relaxation steps is involved, each with 2% nominal axial strain increment. It is observed that while the matrix stress will be finally compressive at all positions, it becomes compressive first at the outer part of the disk (Fig. 5(a)). This is so because the fluid pressure at the outer part is lower and the

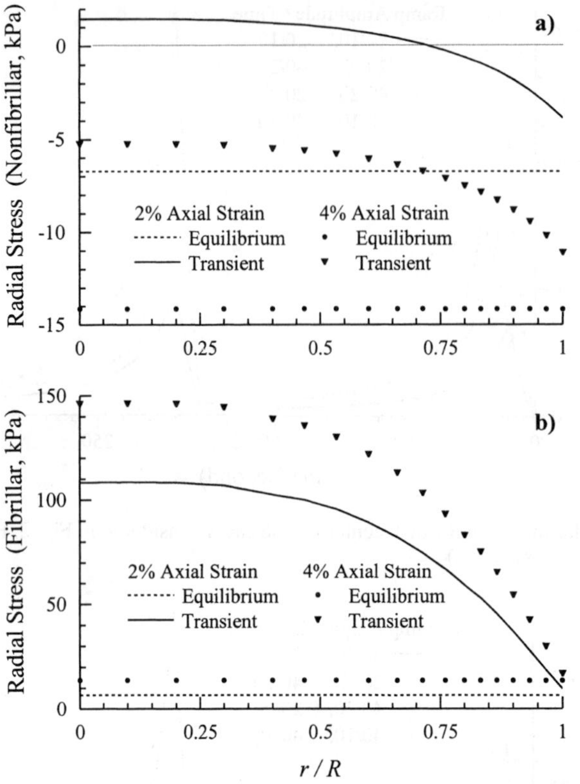

Fig. 5. Radial stress distribution along the disk radius for the case (i) considered in Fig. 2. (a) Radial stress of the nonfibrillar matrix; (b) radial stress of the fibrillar matrix. The stresses are shown for equilibrium and for the transient time just before relaxation proceeds in each ramp step, at 2% or 4% total nominal axial strain. At equilibrium, both stresses are identical in quantity but with opposite signs.

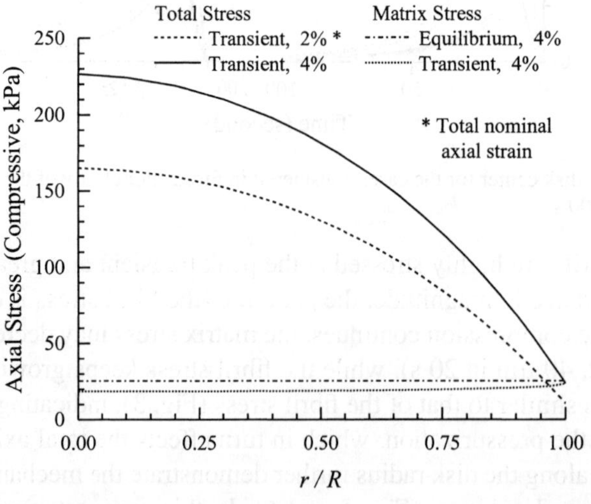

Fig. 6. Distribution of the total axial stress and the axial stress of the nonfibrillar matrix along the disk radius for the case (i) considered in Fig. 2. The total stress is shown for the transient time just before relaxation proceeds in each ramp step, at 2% or 4% total nominal axial strain. The matrix stress is shown for the same transient time and for equilibrium at 4% total nominal axial strain only. Both stresses are compressive and identical at equilibrium.

fibril stress induced by the radial strain there takes less time to surpass the pressure (note that the fibril stress is constant and balances the matrix stress at equilibrium, Fig. 5(b)). Now it is clear that the initial rise and rebound in the matrix stress (as shown in Fig. 2 for the disk center) decrease with the distance from the disk center (r) and completely disappear at the periphery (not shown). Furthermore, the matrix stress in both directions is very low and relatively invariable with position (r) and time (Figs 5 and 6). The fibril stress (Fig. 5(b)) and the total axial stress (Fig. 6) are several times higher at the center than that at the periphery of the disk, showing their strong dependence on the fluid pressure.

4. Discussion

The particular features of the present model attribute to the inclusion of fibril reinforcement. The advantages of modeling fibril reinforcement have also been demonstrated in other cases [7,8]. Fibril reinforcement was found to be essential to produce the strong and nonlinear transient behavior and to support the high pressurization of cartilage during the transient period. Modeling two distinct solid phases might also be helpful in simulating a pathologic process: the effects of degradation in individual phases on the cartilage response can be investigated independently. In preliminary studies, the models have been used to correlate the alterations of material properties to degeneration of cartilage disks.

The interactions among the three phases are influenced by strain-magnitude, strain-rate and loading history, in addition to the material properties. The equilibrium behavior is approximately linear with the axial strain, since the radial strain at equilibrium is very small so that fibril stiffening is not significantly exploited. At a low strain rate, the transient response of cartilage is closer to equilibrium behavior and hence E_m and ν_m have relatively larger impact on the behavior. At high strain rates, the response of cartilage is closer to the instantaneous behavior experiencing considerable radial strain. The fibril modulus, E_f, therefore predominantly influences the transient response where strong nonlinearity is present. In general, the transient stiffness of cartilage increases significantly with the strain when compression begins, witnessed by approximate volume conservation. A significant volume change follows that diminishes this trend sooner or later depending on the strain rate. The stiffness then turns to decrease if the strain rate is low and continues to increase if the strain rate is sufficiently high (not shown).

This study indicates the importance of introducing distinct phases for the proteoglycans and collagen fibrils into cartilage mechanical models. The representation of cartilage as a fibril reinforced tissue not only yields results in agreement with measurements but also agrees with the microstructure of the tissue and mechanical properties of its components. If the two matrices had been modeled as one phase with compressive modulus E_m and tensile modulus E_t, we would also have been able to simulate certain mechanical behavior (e.g., the load versus time). However, it would not have been predicted, for example, that the matrix stress in radial direction remains mostly compressive in unconfined geometry; only the average over the two matrices could have been determined.

Further work on fibril reinforced models is required. The maximum strain involved in Fig. 1 is 4% and the maximum nominal strain involved in Figs 2–6 is 8%. For larger strains, the representation of the fibril modulus (1) might need to be refined based on further measurements. On the other hand, collagen fibrils are often considered as viscoelastic [13]. Efforts must be made to identify the role of fibril viscoelasticity in the mechanism of fibril reinforcement, while the proteoglycan matrix is considered elastic.

Acknowledgement

The financial support from the Natural Sciences and Engineering Research Council of Canada is appreciated.

References

[1] C.G. Armstrong, W.M. Lai and V.C. Mow, An analysis of the unconfined compression of articular cartilage, *J. Biomech. Eng.* **106** (1984), 165–173.

[2] P.A. Huijing, Muscle as a collagen fiber reinforced composite: a review of force transmission in muscle and whole limb, *J. Biomech.* **32** (1999), 329–345.

[3] G.E. Kempson, H. Muir, C. Pollard and M. Tuke, The tensile properties of the cartilage of human femoral condyles related to the content of collagen and glycosaminoglycans, *Biochim. Biophys. Acta* **297** (1973), 456–472.

[4] S.M. Klisch and J.C. Lotz, Application of a fiber-reinforced continuum theory to multiple deformations of the annulus fibrosus, *J. Biomech.* **32** (1999), 1027–1036.

[5] W.M. Lai and V.C. Mow, Drag induced compression of articular cartilage during a permeation experiment, *Biorheology* **17** (1980), 111–123.

[6] L.P. Li, M.D. Buschmann and A. Shirazi-Adl, A fibril reinforced nonhomogeneous poroelastic model for articular cartilage: inhomogeneous response in unconfined compression, *J. Biomech.* **33** (2000), 1533–1541.

[7] L.P. Li, M.D. Buschmann and A. Shirazi-Adl, Fibril stiffening accounts for strain-dependent stiffness of articular cartilage in unconfined compression, *Trans. Orthop. Res. Soc.* **26** (2001), 425.

[8] L.P. Li, A. Shirazi-Adl and M.D. Buschmann, Effects of material heterogeneity on mechanical behavior of articular cartilage, *Trans. Orthop. Res. Soc.* **25** (2000), 130.

[9] L.P. Li, J. Soulhat, M.D. Buschmann and A. Shirazi-Adl, Nonlinear analysis of cartilage in unconfined ramp compression using a fibril reinforced poroelastic model, *Clin. Biomech.* **14** (1999), 673–682.

[10] G.D. Pins, E.K. Huang, D.L. Christiansen and F.H. Silver, Effects of static axial strain on the tensile properties and failure mechanisms of self-assembled collagen fibers, *J. Appl. Polymer Sci.* **63** (1997), 1429–1440.

[11] V. Roth and V.C. Mow, The intrinsic tensile behavior of the matrix of bovine articular cartilage and its variation with age, *J. Bone Joint Surg.* **62** (1980), 1102–1117.

[12] M.H. Schwartz, P.H. Leo and J.L. Lewis, A microstructural model for the elastic response of articular cartilage, *J. Biomech.* **27** (1994), 865–873.

[13] J.L. Wang, M. Parnianpour, A. Shirazi-Adl and A.E. Engin, Failure criterion of collagen fiber: viscoelastic behavior simulated by using load control data, *Theoret. Appl. Fracture Mechan.* **27** (1997), 1–12.

[14] S.L.Y. Woo, P. Lubock, M.A. Gomez, G.F. Jemmott, S.C. Kuei and W.H. Akeson, Large deformation nonhomogeneous and directional properties of articular cartilage in uniaxial tension, *J. Biomech.* **12** (1979), 437–446.

Biorheology 39 (2002) 97–108
IOS Press

Differential responses of chondrocytes from normal and osteoarthritic human articular cartilage to mechanical stimulation

D.M. Salter [a,*], S.J. Millward-Sadler [a], G. Nuki [b] and M.O. Wright [c]

[a] *Department of Pathology, Edinburgh University Medical School, Teviot Place, Edinburgh EH8 9AG, UK*

[b] *Department of Medical Sciences, Rheumatology Section, Western General Hospital, Edinburgh EH4 2XU, UK*

[c] *Department of Biomedical Sciences, Edinburgh University Medical School, Teviot Place, Edinburgh EH8 9AG, UK*

Abstract. Mechanical stimulation is critically important for the maintenance of normal articular cartilage integrity. Molecular events regulating responses of chondrocytes to mechanical forces are beginning to be defined. Chondrocytes from normal human knee joint articular cartilage show increased levels of aggrecan mRNA following 0.33 Hz mechanical stimulation whilst at the same time relative levels of MMP3 mRNA are decreased. This anabolic response, associated with membrane hyperpolarisation, is activated via an integrin-dependent interleukin (IL)-4 autocrine/paracrine loop. Work in our laboratory suggests that this chondroprotective response may be aberrant in osteoarthritis (OA). Chondrocytes from OA cartilage show no changes in aggrecan or MMP3 mRNA following 0.33 Hz mechanical stimulation. $\alpha 5\beta 1$ integrin is the mechanoreceptor in both normal and OA chondrocytes but downstream signalling pathways differ. OA chondrocytes show membrane depolarisation following 0.33 Hz mechanical stimulation consequent to activation of an IL1β autocrine/paracrine loop. IL4 signalling in OA chondrocytes is preferentially through the type I (IL4α/cγ) receptor rather than via the type II (IL4α/IL13R) receptor. Altered mechanotransduction and signalling in OA may contribute to changes in chondrocyte behaviour leading to increased cartilage breakdown and disease progression.

1. Introduction

Osteoarthritis (OA) is a degenerative disease of diarthrodial joints in which degradative and repair processes in articular cartilage, subchondral bone and synovium occur concurrently. The pathological features of cartilage including changes in matrix composition, chondrocyte clustering, cartilage fissuring and flaking are well recognised. These changes are generally accepted to be secondary to the effects of mechanical forces on the joint. Although mechanical forces are important in maintaining cartilage, in a situation where normal mechanical forces act on structurally abnormal cartilage or abnormal mechanical forces act on normal cartilage, chondrocyte responses appear to be deleterious and lead to progressive cartilage loss [1,3,26]. Chondrocytes respond to what they perceive as injurious mechanical stimuli by activating 'reparative responses', altering synthesis of matrix macromolecules and degradative enzymes.

*Address for correspondence: D.M. Salter, MD, Department of Pathology, Edinburgh University Medical School, Teviot Place, Edinburgh EH8 9AG, UK. Tel.: +44 1316502946; Fax: +44 01316506528; E-mail: Donald.Salter@ed.ac.uk.

0006-355X/02/$8.00

The result is a cartilage which has different structural and biochemical characteristics to normal articular cartilage and is unable to support normal joint function leading to steady progression of the osteoarthritic process.

Mechanical stimulation of chondrocytes in culture and of articular cartilage explants stimulates a number of physiological and biochemical responses resulting in changes in expression of matrix re-modelling genes. Such changes include altered proteoglycan synthesis [5,24,25,28] induction of aggrecan gene expression, and production of fibronectin which would have significant effects on cartilage structure if reflected *in vivo*. The nature of the response in part depends on the nature of the mechanical stimulus. In general dynamic stimuli result in an anabolic-response whereas static compression more frequently inhibits chondrocytes activity. Thus chondrocytes appear able to differentiate between mechanical stimuli and respond appropriately. The means by which different mechanical stimuli are recognized by chondrocytes and how these signals are transduced into biochemical responses are an area of intense study at present and mechanotransduction pathways are beginning to be defined.

When applied to cartilage, mechanical forces induce numerous physiochemical changes such as deformation of chondrocytes and the surrounding extracellular matrix, production of hydrostatic pressure gradients, fluid flow, streaming potentials, and alterations in matrix water content, fixed charge density, mobile ion concentrations, and changes in pH and osmotic pressure [20,27]. These changes may separately and collectively influence chondrocyte metabolism. It is likely that the response of a chondrocyte in cartilage to mechanical load will be a result of integration of a number of stimuli being perceived by the cell. Thus a mechanical stimulus to cartilage may be recognised by a chondrocyte via a number of 'mechanoreceptors' including mechanosensitive ion channels. Recent studies have however suggested major roles for members of the integrin family of cell adhesion molecules as chondrocyte mechanoreceptors [8,23,30].

Integrins are a family of α/β heterodimeric cell surface adhesion receptors that can bind a wide variety of extracellular matrix and cell surface ligands [7]. Most integrins are expressed on a wide range of cells, and most cells express several integrins. Thus far, 17α and 8β subunits have been identified, which can associate in a restricted manner to form at least 23 different combinations. There is mild variation in the results of studies from different laboratories on integrin expression by chondrocytes in normal and osteoarthritic adult human articular cartilage which may be explained by differences in technique and sampling [10,13,17,18,21]. In general $\alpha1\beta1$ (receptor for collagen), $\alpha5\beta1$ (receptor for fibronectin) and $\alpha V\beta5$ (receptor for vitronectin) heterodimers are consistently identified and are likely to have significant functional effects. Integrin expression appears to be altered in osteoarthritis, with increased expression of other heterodimers such as $\alpha2\beta1$.

Activation of integrins is known to initiate a cascade of intracellular signalling events which have the potential to modulate expression of genes necessary for maintenance of cartilage content and integrity. Recent work from the authors' laboratory has supported the idea that integrins act as chondrocyte mechanoreceptors in both normal and osteoarthritic cartilage. However the response of chondrocytes from osteoarthritic cartilage is significantly different from that of normal chondrocytes suggesting that altered sensing of the mechanical environment and inappropriate responses of the resident cell population may be important in disease progression.

2. Materials and methods

2.1. Mechanical stimulation of chondrocytes in monolayer culture

The technique used for mechanical stimulation of primary monolayer cultures of human articular chondrocytes has been described in detail previously [31–33]. In brief, flexible, plastic tissue culture dishes are placed in a sealed pressure chamber with inlet and outlet ports. The chamber is pressurised with an inert gas from a cylinder, the frequency of pressurisation being dictated by an electronic timer controlling the inlet and outlet valves. The standard regime of cyclical pressurisation used consists of pressure pulses of 50 kPa above atmospheric pressure at a frequency of 0.33 Hz (2 seconds on, 1 second off) for 20 minutes. Cyclical pressurisation in this system has been shown to induce deformation and strain on the base of the plastic tissue culture dish and its adherent cells [31]. A pressure pulse of 50 kPa results in 3700 microstrain on the base of the dish with a maximum strain rate of 10,000 microstrain per second.

2.2. Isolation and culture of chondrocytes

Chondrocytes were isolated by sequential enzymatic digestion from cartilage obtained from human adult knee joints following arthroplasty, resection for peripheral vascular disease or at autopsy with patients or relatives consent. In all cases articular cartilage was assessed and graded macroscopically for the presence or absence of osteoarthritis using the Collins/McElligott system and samples processed for histological examination. Articular cartilage from different anatomical regions of the knee joint including the patella, tibial plateau, and femoral condyle were pooled. Normal and osteoarthritic cartilage were kept separately. Isolated cells were seeded in Ham's F12 medium supplemented with 10% fetal calf serum, 50 μg/ml ascorbic acid, 2 mM L-glutamine, 100 IU/ml penicillin and 100 μg/ml streptomycin, at a concentration of 5×10^4 cells/ml into 55 mm diameter tissue culture dishes (Nunc). Primary, non-confluent cultures of chondrocytes were used in all experiments. Morphologically the cells from both normal and osteoarthritic cartilage were typically flattened with a polygonal cell shape. RT-PCR and immunohistochemistry showed expression of cartilage specific molecules including type II collagen and aggrecan.

2.3. Chondrocyte response to mechanical stimulation

Three major cellular responses have been used in the authors' studies to identify chondrocyte mechanoreceptors and mechanotransduction pathways. These have been (i) a change in membrane potential [31], (ii) protein tyrosine phosphorylation [11], and (iii) changes in the relative levels of matrix and matrix metalloprotease gene expression [16]. For electrophysiology, membrane potentials of cells were recorded using a single electrode bridge circuit and calibrator. Microelectrodes having tip resistances of between 40 to 60 meg ohms and tip potentials of approximately 3 mV were used to impale the cells. Results from a cell were accepted if on impalement there was a rapid change in membrane voltage potential level in the cell and if this membrane potential remained constant for at least 60 seconds.

Protein tyrosine phosphorylation was assessed by standard western blotting and immunoprecipitation techniques. For assessment of relative levels of gene expression, PCR products following RTPCR using specific primers for aggrecan, matrix metalloproteinase (MMP)-1, MMP-3, TIMP-1, tenascin and GAPDH were analysed by electrophoresis using a 1% (w/v) agarose gel stained with ethidium bromide. The intensity of the bands was measured under UV fluorescence using EASY Image analysis software. The ratio of intensities of the bands for the product of interest to the housekeeping gene GAPDH was then calculated.

3. Results

3.1. *Differential responses of normal and osteoarthritic chondrocytes to 0.33 Hz mechanical stimulation*

3.1.1. *Electrophysiology*

Stimulation of human chondrocytes from normal and osteoarthritic articular cartilage at a frequency of 0.33 Hz (2 second on/1 second off) for 20 minutes at 37°C results in changes in membrane potential as a result of activation of membrane ion channels (Fig. 1). Chondrocytes from normal articular cartilage show a membrane hyperpolarization which secondary to activation of small conductance Ca^{2+}-activated K^+ channels [32]. In contrast stimulation of chondrocytes from osteoarthritic cartilage under the same conditions induces a membrane depolarisation response [15] which is blocked by tetrodotoxin a specific inhibitor of sodium channel function. The membrane response of both normal and osteoarthritic chondrocytes is blocked by gadolinium (Fig. 1) suggesting roles for stretch activated ion channels [19].

3.1.2. *Gene expression*

Twenty minutes of mechanical stimulation at 0.33 Hz leads to an increase in relative levels of aggrecan mRNA and a decrease in relative levels of MMP3 mRNA in normal articular chondrocytes ([16], Table 1). These changes peak within 1 and 3 hours post stimulation and return to baseline levels by 24 hours. No changes in relative levels of TIMP, MMP1 or tenascin are seen. Osteoarthritic chondrocytes however show no changes in relative levels of mRNA of any of these molecules under the same conditions.

The differential changes in membrane potential and gene expression following mechanical stimulation have been used to identify integrin associated signalling pathways and cytokine autocrine/paracrine loops in mechanotransduction pathways and comparisons between cells from normal and osteoarthritic cartilage.

3.2. *Integrin signalling*

RGD-containing oligopeptides and function-blocking antibodies against integrin subunits expressed by normal and osteoarthritic chondrocytes *in vivo* were used to investigate possible roles for integrins in the mechanotransduction pathway [15,16,33]. Exposure of normal and osteoarthritic chondrocytes to GRGDSP peptide, which competes for integrin–matrix ligand binding sites, resulted in inhibition of the hyperpolarization and depolarisation membrane responses. Similarly the membrane potential response of normal and osteoarthritic chondrocytes to 0.33 Hz mechanical stimulation was considerably reduced when cells were preincubated for 30 minutes with anti-$\beta 1$ integrin or anti-$\alpha 5$ integrin antibodies (Fig. 2). In contrast antibodies against other integrin subunits had no effect on the membrane response. These observations are consistent with a role for $\alpha 5\beta 1$ integrin, the classical fibronectin receptor, in the mechanotransduction process, potentially as a mechanoreceptor. This idea is supported by the observations that RGD containing peptides prevent the changes in aggrecan and MMP3 mRNA levels that occur following mechanical stimulation of normal chondrocytes [16].

Studies using pharmacological inhibitors of integrin-associated signalling pathways indicate that downstream pathways activated by mechanical stimulation differ in normal and osteoarthritic chondrocytes [15,32,33]. Tyrosine kinase activity, phospholipase C and inositol triphosphate appear to be necessary for the membrane potential response of both normal and osteoarthritic chondrocytes (Fig. 3). Analysis of protein tyrosine phosphorylation shows a similar pattern in normal and osteoarthritic chondrocytes early following mechanical stimulation. In both types of cells there is rapid tyrosine phosphorylation of

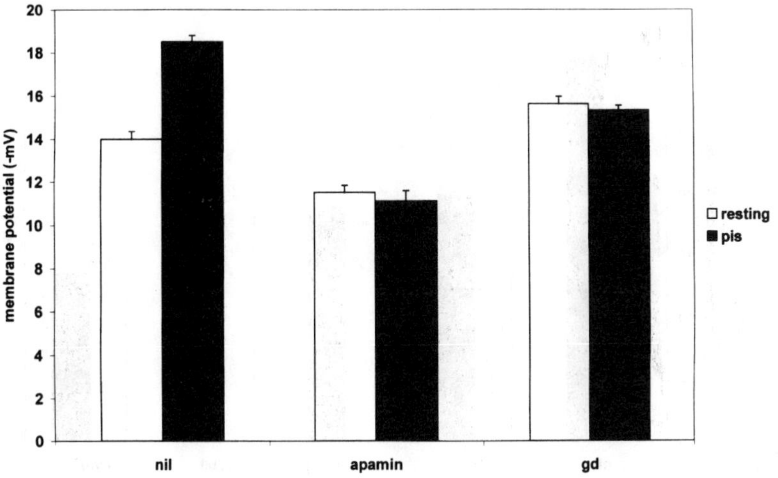

(a)

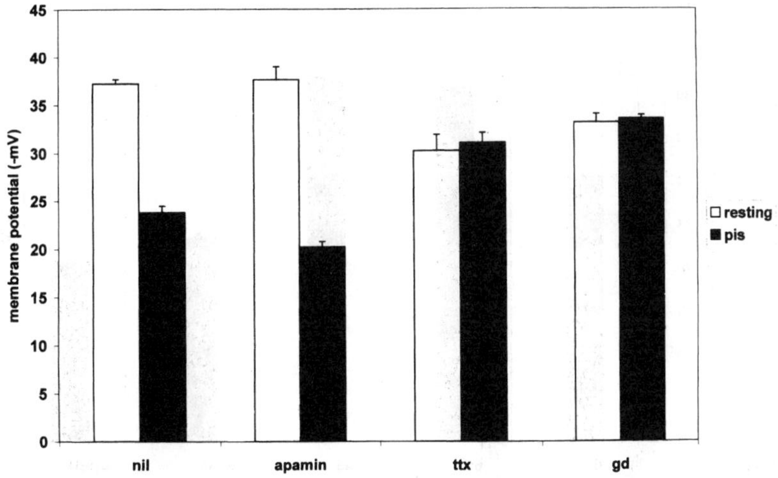

(b)

Fig. 1. Effect of ion channel blockade on the membrane potential response of chondrocytes from (a) normal and (b) osteoarthritic human articular cartilage to 20 minutes mechanical stimulation at 0.33 Hz. pis = post mechanical stimulation. gd = gadolinium, ttx = tetrodotoxin.

Table 1

Changes in relative levels of mRNA in chondrocytes from normal and osteoarthritic human articular cartilage following 20 min mechanical stimulation at 0.33 Hz

		Time points (hours post-stimulation)				
		0	1	3	6	24
Normal	Aggrecan	0.54 ± 0.08	$1.21 \pm 0.17^*$	$1.11 \pm 0.16^*$	$0.71 \pm 0.27^*$	0.40 ± 0.51
	MMP3	1.73 ± 0.17	$0.84 \pm 0.14^*$	$0.84 \pm 0.15^*$	$1.04 \pm 0.37^*$	2.12 ± 0.46
OA	Aggrecan	0.65 ± 0.16	0.57 ± 0.09	0.68 ± 0.12	0.70 ± 0.07	0.68 ± 0.10
	MMP3	1.07 ± 0.35	1.18 ± 0.59	1.05 ± 0.32	1.01 ± 0.34	1.07 ± 0.34

* Significant difference ($p < 0.05$) when compared to resting cells.

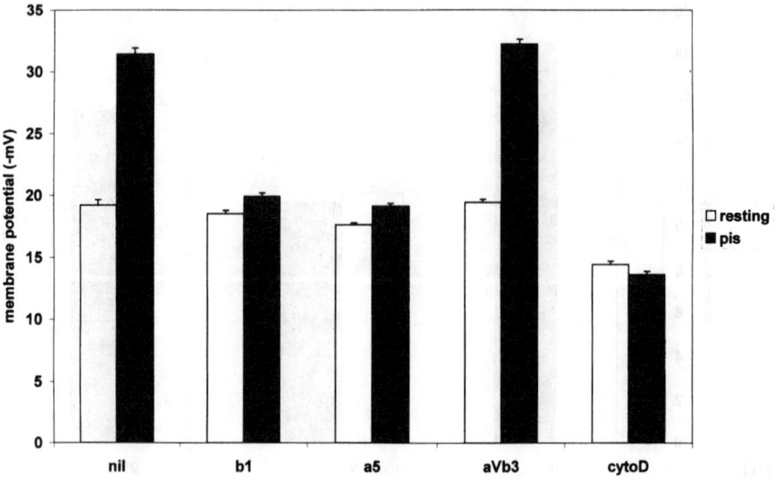

(a)

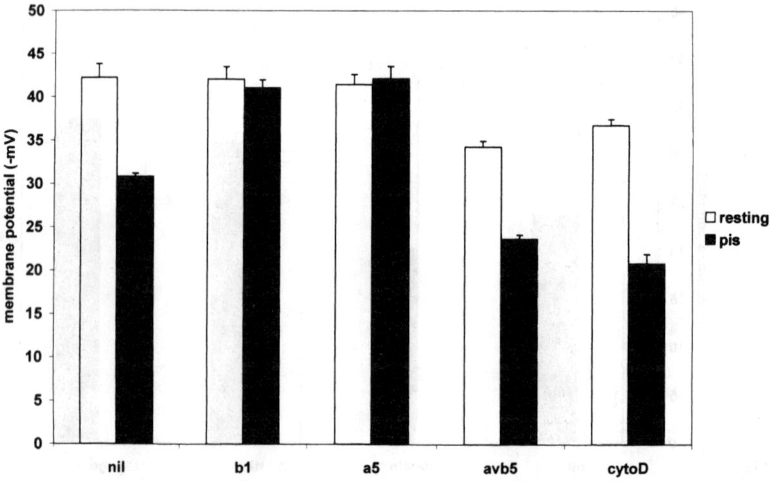

(b)

Fig. 2. Effect of anti-integrin antibodies and cytochalasin D on the membrane potential response of chondrocytes from (a) normal and (b) osteoarthritic human articular cartilage to 20 minutes mechanical stimulation at 0.33 Hz. pis = post mechanical stimulation. b1 = anti-β1-integrin, a5 = anti-α5-integrin, aVb3 = anti-αVβ3-integrin, aVb5 = anti-αVβ5-integrin, cyto D = cytochalasin D.

pp125FAK, paxillin and β-catenin via an RGD-dependent pathway [11]. In contrast, experiments using cytochalasin D [15,33] demonstrate that, whereas disruption of the actin cytoskeleton prevents the membrane hyperpolarisation response of normal chondrocytes, it has no effect on the depolarisation response of osteoarthritic chondrocytes (Fig. 2). Furthermore inhibitors of protein kinase C block the electrophysiological response of normal but not osteoarthritic chondrocytes (Fig. 3).

3.3. *Cytokine-signalling loops in the mechanotransduction pathway*

Conditioned media transferred from dishes of chondrocytes mechanically stimulated for 20 minutes at 0.33 Hz induces a membrane potential response of resting chondrocytes consistent with the release of a preformed soluble factor from the mechanically-stimulated cells [14]. In the case of mechanically stimulated normal chondrocytes the conditioned media has membrane hyperpolarising activity whereas

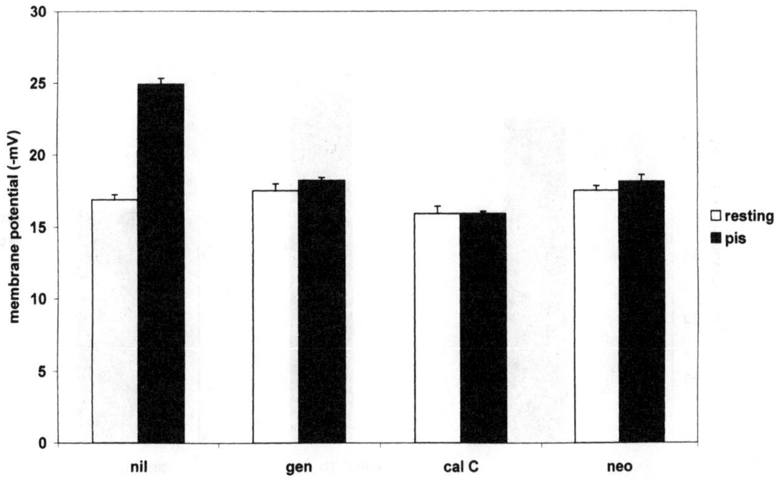

(a)

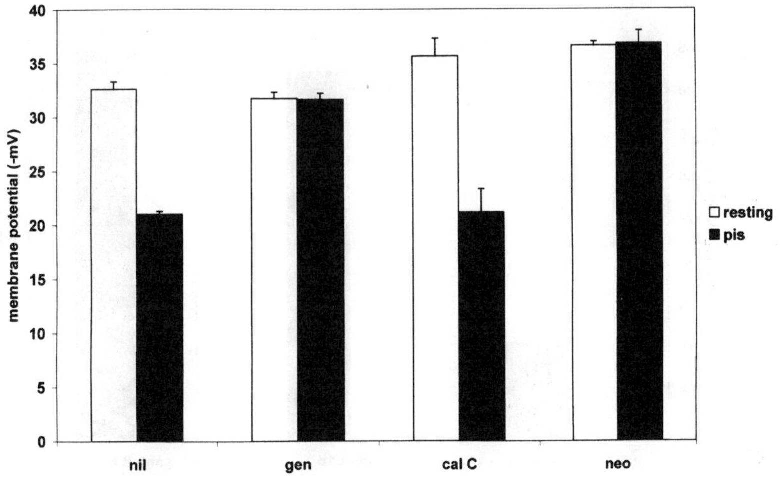

(b)

Fig. 3. Effect of blockade of intracellular signalling on the membrane potential response of chondrocytes from (a) normal and (b) osteoarthritic human articular cartilage to 20 minutes mechanical stimulation at 0.33 Hz. pis = post mechanical stimulation. gen = genestein an inhibitor of tyrosine kinase activity; cal C = calphostin C; an inhibitor of protein kinase C activity; neo = neomycin, an inhibitor of phospholipase C.

the conditioned media from osteoarthritic chondrocytes results in membrane depolarisation. These differences appear to be the result of activation of autocrine/paracrine cytokine signalling loops involving different cytokine–receptor pairs.

3.3.1. Interleukin-4 signalling in normal chondrocytes

Addition of neutralizing IL-4 antibodies to the culture medium before mechanical stimulation abolished the hyperpolarization response (Fig. 4). Neutralizing antibodies to IL-1β had no effect. These results are consistent with IL-4 being the active autocrine or paracrine signalling molecule in the mechanotransduction pathway [16]. This idea is supported by the observations that anti IL-4 antibodies when added to conditioned medium after mechanical stimulation but before transfer of that medium to unstimulated cells, prevented the subsequent hyperpolarization of resting cells. Anti IL-4Rα antibodies prevent the hyperpolarization response of chondrocytes to mechanical stimulation, whereas inhibitory antibodies

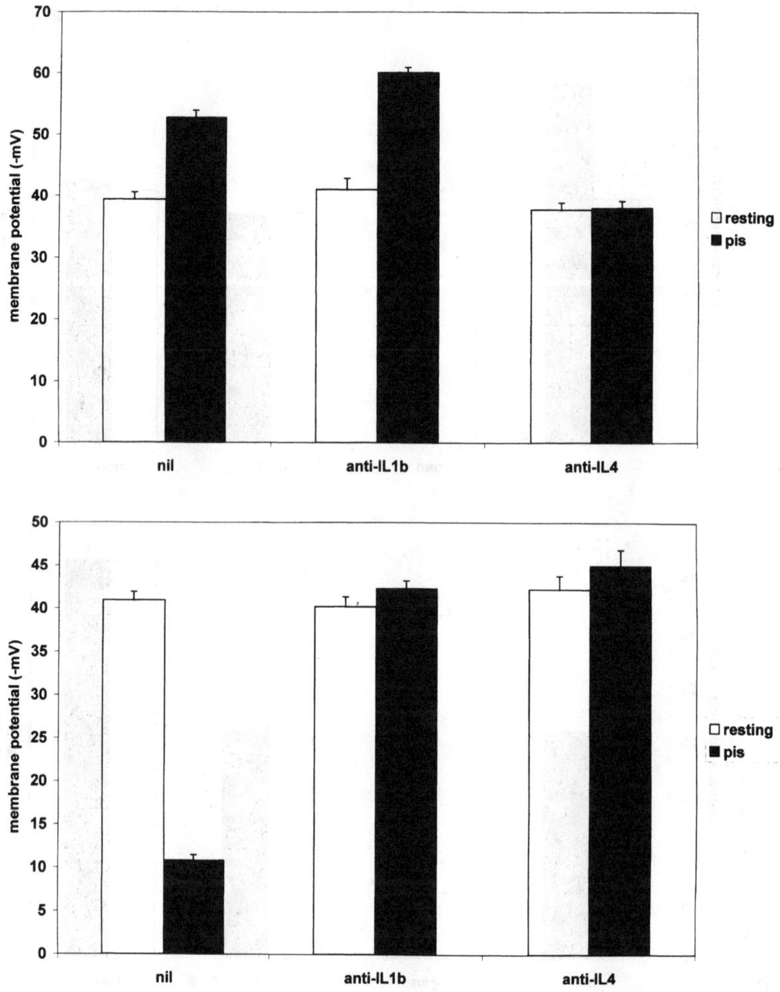

(a)

(b)

Fig. 4. Effect of anti-IL-4 and IL-1β antibodies on the membrane potential response of chondrocytes from (a) normal and (b) osteoarthritic human articular cartilage to 20 minutes mechanical stimulation at 0.33 Hz. pis = post mechanical stimulation.

to the cγ receptor subunit had no effect on the response consistent with signalling through the type II IL-4 receptor. Release of IL-4 and subsequent signalling events are downstream of initial integrin-associated mechanotransduction as the hyperpolarization response of normal human articular chondrocytes to recombinant IL-4 is unaffected by antiβ1 integrin antibodies, genistein (a tyrosine kinase inhibitor) and gadolinium [14]. As well as having an effect on the membrane potential response IL-4 antibodies also prevent changes in the mRNA ratios for aggrecan and MMP3 at 1 or 3 hours following mechanical stimulation [16].

3.3.2. Alternative interleukin-4/interleukin-1β signalling in osteoarthritic chondrocytes

The cytokine loop activated by mechanical stimulation is only beginning to be defined. IL1-β antibodies and IL-4 antibodies at a concentration of 1 μg/ml both inhibit the depolarisation response of osteoarthritic chondrocytes to mechanical stimulation (Fig. 4) suggesting additional involvement of IL-1β as an autocrine/paracrine signalling molecule. Chondrocytes from osteoarthritic cartilage also show significant differences in their responses to recombinant IL-4. Addition of IL-4 to normal human articular chondrocytes results in a membrane hyperpolarization response that is blocked by function inhibitory antibodies against the IL-4Rα subunit of the IL-4 receptor but not by antibodies against the cγ subunit. In contrast, chondrocytes from osteoarthritic cartilage show a membrane depolarisation response to IL-4 that is inhibited by antibodies to cγ suggesting that IL-4 may preferentially signal through the type I IL-4 receptor in chondrocytes from osteoarthritic cartilage.

4. Discussion

We have shown that chondrocytes from normal and osteoarthritic human articular cartilage, in primary monolayer culture show differential responses to 0.33 Hz mechanical stimulation. This difference is seen in both membrane potential response and in changes in relative levels of aggrecan and MMP3 mRNA. These responses are the result of integrin-dependent mechanotransduction pathways which lead to activation of autocrine/paracrine cytokine signalling cascades. An anabolic response, involving IL-4 signalling via type II IL-4 receptors and increased aggrecan mRNA and decreased MMP3 mRNA, is present in normal chondrocytes (Fig. 5(a)) but absent in chondrocytes from osteoarthritic cartilage. Indeed this chondroprotective response is replaced in osteoarthritic chondrocytes by a potentially catabolic pathway involving interleukin-1β (Fig. 5(b)).

Chondrocytes in osteoarthritis are known to show significant phenotypic differences from those of normal cartilage [2]. These differences are reflected in the differences in matrix production and expression of surface receptors for extracellular matrix molecules and cytokines [17]. Changes in response to mechanical stimulation are therefore yet another manifestation of the increasingly recognised difference between normal and osteoarthritic chondrocytes. Such differences may have significant effects on cartilage structure and functions if anabolic mechanical stimuli fail to be recognised or elicit inappropriate responses.

The mechanotransduction pathway we have identified in chondrocytes involves recognition of the mechanical stimulus by integrins and activation of integrin signalling pathways with the generation of a cytokine loop resulting in a second cascade of intracellular signalling events. Normal and osteoarthritic chondrocytes show differences at multiple stages of the mechanotransduction cascade. Early events are similar, involve $\alpha5\beta1$ integrin and stretch activated ion channels and are associated with similar rapid tyrosine phosphorylation events. The actin cytoskeleton is required for the integrin-dependent mechanotransduction in normal but not osteoarthritic chondrocytes. The reasons for this are not clear but may be related to the known ability of the proarthritic molecule nitric oxide to inhibit actin polymerisation and prevent focal adhesion kinase and rho A translocation to the site of $\alpha5\beta1$ integrin ligation [6].

Importantly, autocrine/paracrine signalling cascades generated following mechanical stimulation also differ in respect of the cytokines produced and the effect of these cytokines on cell responses. These changes are difficult to explain at present but are clearly related to cytokine receptor expression and function, which in turn may be influenced by the environment of the cell.

These studies have been carried out using an *in vitro* model culture system and it will be important to ascertain whether similar responses are present *in vivo*. Modified responses of chondrocytes in osteoarthritic cartilage to mechanical stimulation may be an appropriate, adaptive response to the altered mechanical environment to which the cell is exposed. Alternatively, dysregulated mechanotransduction secondary to proarthritic agents such as NO and IL-1β which influence integrin and IL-4 signalling [4] may be important in continued cartilage breakdown and progression of osteoarthritis.

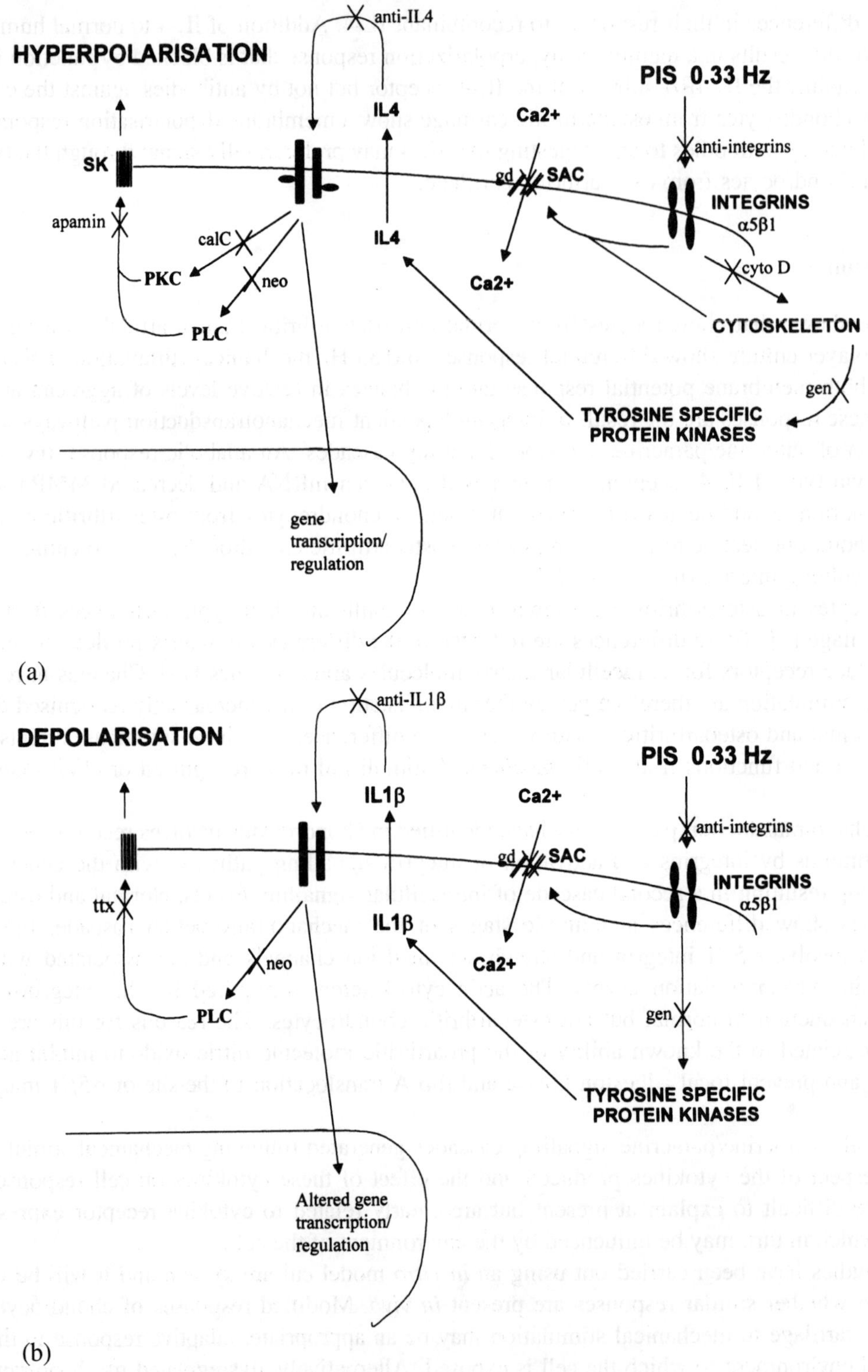

Fig. 5. Mechanotransduction pathways in chondrocytes from (a) normal and (b) osteoarthritic human articular cartilage activated by 0.33 Hz mechanical stimulation.

References

[1] M.E. Adams and K. Brandt, Hypertrophic repair of canine articular cartilage in osteoarthritis after anterior cruciate ligament transection, *J. Rheumatol.* **18** (1991), 428–435.

[2] T. Aigner and J. Dudhia, Phenotypic modulation of chondrocytes as a potential therapeutic target in osteoarthritis: a hypothesis, *Ann. Rheum. Dis.* **56** (1997), 287–291.

[3] F. Behrens, E.L. Kraft and T.R. Oegema, Jr., Biochemical changes in articular cartilage after joint immobilization by casting or external fixation, *J. Orthop. Res.* **7** (1989), 335–343.

[4] Y.R. Boisclair, J. Wang, J. Shi, K.R. Hurst and G.T. Ooi, Role of the suppressor of cytokine signaling-3 in mediating the inhibitory effects of interleukin-1beta on the growth hormone-dependent transcription of the acid-labile subunit gene in liver cells, *J. Biol. Chem.* **275** (2000), 3841–3847.

[5] M.D. Buschmann, Y.A. Gluzband, A.J. Grodzinsky and E.B. Hunziker, Mechanical compression modulates matrix biosynthesis in chondrocyte/agarose culture, *J. Cell Sci.* **108** (1995), 1497–1508.

[6] R.M. Clancy, J. Rediske, X. Tang, N. Nijher, S. Frenkel, M. Philips and S.B. Abramson, Outside-in signaling in the chondrocyte. Nitric oxide disrupts fibronectin-induced assembly of a subplasmalemmal actin/rho A/focal adhesion kinase signaling complex, *J. Clin. Invest.* **100** (1997), 1789–1796.

[7] R.O. Hynes, Integrins: Versatility, modulation and signaling in cell adhesion, *Cell* **69** (1992), 11–25.

[8] D. Ingber, Integrins as mechanochemical transducers, *Curr. Opin. Cell. Biol.* **3** (1991), 841–848.

[9] M.O. Jortikka, R.I. Inkinen, M.I. Tammi, J.J. Parkkinen, J. Haapala, I. Kiviranta, H.J. Helminen and M.J. Lammi, Immobilisation causes long lasting matrix changes both in the immobilised and contralateral joint cartilage, *Ann. Rheum. Dis.* **56** (1997), 255–261.

[10] G. Lapadula, F. Iannone, C. Zuccaro, V. Grattagliano, M. Covelli, V. Patella, G. Lo Bianco and V. Pipitone, Chondrocyte phenotyping in human osteoarthritis, *Clin. Rheumatol.* **17** (1998), 99–104.

[11] H.S. Lee, S.J. Millward-Sadler, M.O. Wright, G. Nuki and D.M. Salter, Integrin and mechanosensitive ion channel-dependent tyrosine phosphorylation of focal adhesion proteins and beta-catenin in human articular chondrocytes after mechanical stimulation, *J. Bone Miner. Res.* **15** (2000), 1501–1509.

[12] A. Lischke, R. Moriggl, S. Brandlein, S. Berchtold, W. Kammer, W. Sebald, B. Groner, X. Liu, L. Hennighausen and K. Friedrich, The interleukin-4 receptor activates STAT5 by a mechanism that relies upon common gamma-chain, *J. Biol. Chem.* **273** (1998), 31 222–31 229.

[13] R.F. Loeser, C.S. Carlson and M.P. McGee, Expression of beta 1 integrins by cultured articular chondrocytes and in osteoarthritic cartilage, *Exp. Cell. Res.* **217** (1995), 248–257.

[14] S.J. Millward-Sadler, M.O. Wright, H.S. Lee, K. Nishida, H. Caldwell, G. Nuki and D.M. Salter, Integrin-regulated secretion of interleukin 4: A novel pathway of mechanotransduction in human articular chondrocytes, *J. Cell. Biol.* **145** (1999), 183–189.

[15] S.J. Millward-Sadler, M.O. Wright, H. Lee, H. Caldwell, G. Nuki and D.M. Salter, Altered electrophysiological responses to mechanical stimulation and abnormal signalling through alpha5beta1 integrin in chondrocytes from osteoarthritic cartilage, *Osteoarthritis Cartilage* **8** (2000), 272–278.

[16] S.J. Millward-Sadler, M.O. Wright, L.W. Davies, G. Nuki and D.M. Salter, Mechanotransduction via integrins and interleukin-4 results in altered aggrecan and matrix metalloproteinase 3 gene expression in normal but not osteoarthritic human articular chondrocytes, *Arthritis Rheum.* **43** (2000), 2091–2099.

[17] K. Ostergaard and D.M. Salter, Immunohistochemistry in the study of normal and osteoarthritic human articular cartilage, *Progr. Histochem. Cytochem.* **33** (1998), 93–168.

[18] K. Ostergaard, D.M. Salter, J. Petersen, K. Bendtzen, J. Hvolris and C.B. Andersen, Expression of alpha and beta subunits of the integrin superfamily in articular cartilage from macroscopically normal and osteoarthritic human femoral heads, *Ann. Rheum. Dis.* **57** (1998), 303–308.

[19] J. Sadoshima, T. Takahashi, L. Jahn and S. Izumo, Roles of mechano-sensitive ion channels, cytoskeleton, and contractile activity in stretch-induced immediate-early gene expression and hypertrophy of cardiac myocytes, *Proc. Natl. Acad. Sci. USA* **89** (1992), 9905–9909.

[20] R.L. Sah, Y.J. Kim, J.Y. Doong, A.J. Grodzinsky, A.H. Plaas and J.D. Sandy, Biosynthetic response of cartilage explants to dynamic compression, *J. Orthop. Res.* **7** (1989), 619–636.

[21] D.M. Salter, D.E. Hughes, R. Simpson and D.L. Gardner, Integrin expression by human articular chondrocytes, *Br. J. Rheumatol.* **31** (1992), 231–234.

[22] D.M. Salter, J.E. Robb and M.O. Wright, Electrophysiological responses of human bone cells to mechanical stimulation: Evidence for specific integrin function in mechanotransduction, *J. Bone Miner. Res.* **12** (1997), 1133–1141.

[23] J.Y.-J. Shyy and S. Chien, Role of integrins in cellular responses to mechanical stress and adhesion, *Curr. Opin. Cell. Biol.* **9** (1997), 707–713.

[24] J. Steinmeyer, B. Ackermann and R.X. Raiss, Intermittent cyclic loading of cartilage explants modulates fibronectin metabolism, *Osteoarthritis Cartilage* **5** (1997), 331–341.

[25] J. Steinmeyer and S. Knue, The proteoglycan metabolism of mature bovine articular cartilage explants superimposed to continuously applied cyclic mechanical loading, *Biochem. Biophys. Res. Commun.* **240** (1997), 216–221.

[26] M. Tammi, A.M. Saamanen, A. Jauhiainen, O. Malminen, I. Kiviranta and H. Helminen, Proteoglycan alterations in rabbit knee articular cartilage following physical exercise and immobilization, *Connect. Tiss. Res.* **11** (1983), 45–55.

[27] J.P.G. Urban, The chondrocyte: A cell under pressure, *Br. J. Rheumatol.* **33** (1994), 901–908.

[28] W.B. Valhmu, E.J. Stazzone, N.M. Bachrach, F. Saed-Nejad, S.G. Fischer, V.C. Mow and A. Ratcliff, Load-controlled compression of articular cartilage induces a transient stimulation of aggrecan gene expression, *Arch. Biochem. Biophys.* **353** (1998), 29–36.

[29] J.P. Veldhuijzen, L.A. Bourret and G.A. Rodan, In vitro studies of the effect of intermittent compressive forces on cartilage cell proliferation, *J. Cell. Physiol.* **98** (1979), 299–306.

[30] N. Wang, J.P. Butler and D.E. Ingber, Mechanotransduction across the cell surface and through the cytoskeleton, *Science* **260** (1993), 1124–1127.

[31] M.O. Wright, R.A. Stockwell and G. Nuki, Response of the plasma membrane to applied hydrostatic pressure in chondrocytes and fibroblasts, *Connect. Tiss. Res.* **28** (1992), 49–70.

[32] M.O. Wright, P. Jobanputra, C. Bavington, D.M. Salter and G. Nuki, The effects of intermittent pressurisation on the electrophysiology of cultured human articular chondrocytes: Evidence for the presence of stretch-activated membrane ion channels, *Clin. Sci.* **90** (1996), 61–71.

[33] M.O. Wright, K. Nishida, C. Bavington, J.L. Godolphin, E. Dunne, S. Walmsley, P. Jobanputra, G. Nuki and D.M. Salter, Hyperpolarisation of cultured human chondrocytes following cyclical pressure-induced strain: Evidence of a role for $\alpha 5\beta 1$ integrin as a chondrocyte mechanoreceptor, *J. Orthop. Res.* **15** (1997), 742–747.

PART II

Mechanobiology of Chondrocytes

Biorheology 39 (2002) 111–117
IOS Press

High pressure effects on cellular expression profile and mRNA stability. A cDNA array analysis

Reijo K. Sironen [a,*], Hannu M. Karjalainen [a,*], Kari Törrönen [a], Mika A. Elo [a], Kai Kaarniranta [a], Masaharu Takigawa [b], Heikki J. Helminen [a] and Mikko J. Lammi [a,**]

[a] *Department of Anatomy, University of Kuopio, P.O. Box 1627, 70211 Kuopio, Finland*

[b] *Department of Biochemistry and Molecular Dentistry, Okayama University Graduate School of Medicine and Dentistry, Okayama 700-8525, Japan*

Abstract. Hydrostatic pressure has a profound effect on cartilage tissue and chondrocyte metabolism. Depending on the type and magnitude of pressure various responses can occur in the cells. The mechanisms of mechanotransduction at cellular level and the events leading to specific changes in gene expression are still poorly understood. We have previously shown that induction of stress response in immortalized chondrocytes exposed to high static hydrostatic pressure increases the stability of heat shock protein 70 mRNA. In this study, our aim was to examine the effect of high pressure on gene expression profile and to study whether stabilization of mRNA molecules is a general phenomenon under this condition. For this purpose a cDNA array analysis was used to compare mRNA expression profile in pressurized vs. non-pressurized human chondrosarcoma cells (HCS 2/8). mRNA stability was analyzed using actinomycin-treated and nontreated samples collected after pressure treatment. A number of immediate-early genes, and genes regulating cell cycle and growth were up-regulated due to high pressure. Decrease in osteonectin, fibronectin, and collagen types VI and XVI mRNAs was observed. Also bikunin, cdc37 homologue and Tiam1, genes linked with hyaluronan metabolism, were down-regulated. In general, stability of down-regulated mRNA species appeared to increase. However, no increase in mRNA above control level due to stabilization was noticed in the genes available in the array. On the other hand, mRNAs of certain immediate-early genes, like c-jun, jun-B and c-myc, became destabilized under pressure treatment. Increased accumulation of mRNA on account of stabilization under high pressure conditions seems to be a tightly regulated, specific phenomenon.

1. Introduction

Articular cartilage functions as load-bearing tissue and has to withstand high compressive loads. The extracellular matrix produced by chondrocytes is resilient and mediates forces between moving bones. Rising from a chair produces a nearly 20 MPa pressure peak and during walking pressures cycle between atmospheric and 3–4 MPa at a frequency of 1 Hz [7]. During other activities the pressure can be even higher. Chondrocytes respond to mechanical forces by altering their production of extracellular components and by remodeling the cartilage tissue. Thus immobilization leads to loss of cartilage proteoglycans within a few weeks, and remobilization can for the most part reverse this event [12]. On the other hand, excessive load can permanently damage cartilage and alter the cellular activity which has been shown in various models of joint instability *in vivo* [4].

*R.K.S. and H.M.K. contributed equally to this work.
**Address for correspondence: Mikko Lammi, Department of Anatomy, University of Kuopio, P.O. Box 1627, 70211 Kuopio, Finland. Tel.: +358 17 163 027; Fax: +358 17 163 032; E-mail: mikko.lammi@uku.fi.

Moderate level of hydrostatic pressure increases the synthesis of cartilage matrix molecules [6,11,14], while high continuous pressure results in inhibition of synthesis [6,10]. Under high hydrostatic pressure, heat shock protein 70 accumulates in the pressurized cells [8,16]. This accumulation involves stabilization of the corresponding mRNA [8], an event that appears to require protein neosynthesis [9]. Expression profiling of chondrocytic cells under pressure revealed a number of other genes whose expression was affected by high hydrostatic pressure [13].

For the moment, studies of mechanical forces on chondrocyte metabolism have mainly focused on measuring the changes in the synthetic activity of chondrocytes in tissue or cell cultures in response to various loading protocols. Many hypotheses concerning the possible mechanotransduction and mechanoreceptors have been suggested, however, analyses of how mechanical force affects gene expression have received less attention. In this study we have analyzed the alterations of gene expression in a human chondrosarcoma cell line on account of high continuous hydrostatic pressure and its possible effects on mRNA stability using a cDNA array.

2. Materials and methods

2.1. Cell culture and pressure treatment

HCS2/8 human chondrosarcoma cell line [17] was cultured in a humidified 5% CO_2/95% air atmosphere at 37°C in DMEM (Gibco, Paisley, UK) supplemented with 10% fetal calf serum (PAA, Linz, Austria), penicillin (50 units/ml, PAA), streptomycin (50 units/ml, PAA), and 3 mM glutamine (PAA). Cells were grown to subconfluent state before the experiments. Before exposure to hydrostatic pressure, medium was changed and 15 mM Hepes (pH 7.3) was added. To expose the cells to pressure, the culture dishes were filled with the medium described above and sealed with a plastic membrane. The apparatus for hydrostatic pressurization of the cells has been previously described in detail [11]. The pressure level of the test chamber was selected to be 30 MPa and static mode of pressure loading was used. Samples from control and pressurized cultures were collected after pressurization for 6 hours. To study the mRNA stability, actinomycin D (2.5 mg/ml, Sigma, St. Louis, MO, USA) was dissolved in methanol and added to control and pressurized cultures at final concentration of 10 μM for 5 hours, and the samples were collected for mRNA isolation.

2.2. Expression profiling with a cDNA array

The Atlas Human Cancer 1.2 cDNA array kit was purchased from Clontech Laboratories (Palo Alto, CA, USA). All procedures for labeling and purification of the probes were accomplished according to manufacturer's recommendations. α-^{32}P-dATP-labeled cDNA probes were generated by reverse transcription of mRNA from untreated and pressurized monolayer cultures using gene-specific primers. Unincorporated label was removed by column chromatography. The membranes were hybridized in ExpressHyb^{TM} solution overnight at 68°C, washed twice in 0.2 × SSC and 0.1% sodium dodecyl sulfate (SDS), and twice in 0.1 × SSC, 0.5% SDS, for 20 min each. Autoradiography signals were quantified by using a PhosphorImager^{TM} (Molecular Dynamics, Sunnyvale, CA, USA), and the values obtained were normalized against several housekeeping genes.

3. Results

Continuous 30 MPa hydrostatic pressure was applied on chondrosarcoma cells to compare the mRNA expression profile in the pressurized cells with control cultures using the cDNA array technique. Since our goal was to investigate the mRNA stability as well, we pressurized the cells for 6 hours, a time interval needed to reach maximal heat shock protein 70 mRNA level and stabilization [8]. A number of changes in mRNA expression levels was observed due to high pressure. The up-regulated genes are presented in Table 1, and down-regulated ones in Table 2. The house-keeping genes (GAPDH, α-tubulin, β-actin, 40S ribosomal protein, 23-kDa highly basic protein and HLAC) were expressed consistently, and their average was used for the normalization.

To study changes in mRNA stability, possibly affected by high hydrostatic pressure, mRNA synthesis was stopped by addition of actinomycin D in control and pressurized cultures, and expression profiles were compared with corresponding RNA samples which were collected immediately after pressurization. A number of up-regulated (Table 1) and down-regulated (Table 2) genes were observed. Changes in signal levels between actinomycin D-treated and untreated array samples were considered to indicate that mRNA half-life differed from the average half-life of total mRNA pool (Tables 1 and 2). Ratio 1.0 represents the case where the mRNA half-life did not change.

4. Discussion

The immediate early genes, such as c-jun, c-myc, jun-B, some G-proteins and transcription factors were observed to be induced effectively during pressure treatment. Ets transcription factors are involved in cartilage and bone development, and type X collagen, e.g., has a conserved Ets-binding sequence in its promoter region [2]. Therefore, increase in Elf mRNA may have relevance in chondrocyte's response to pressure. mRNAs for c-jun, c-myc and jun-B were present at rather low level in control cell cultures, however, they had a long half-life, as indicated by increase of their relative abundance in actinomycin D treated cultures. Interestingly, even though pressurization resulted in a notable increase of these mRNA species, their half-life appeared to be shortened in pressurized cells. This seems to indicate faster turnover rate of the early phase messages. The signal of these genes seemed to be intense even 6 hours after the beginning of pressurization indicating ongoing activation of other target genes.

Under high hydrostatic pressure, a number of genes involved in cell cycle regulation were up-regulated. Effects of pressure on the cell cycle have been previously studied with *Saccharomyces cerevisiae*, where hydrostatic pressure in the range of 15 to 25 MPa caused arrest of the cell cycle in G_1 phase, while pressure of 50 MPa did not [1]. Although the expression profiling suggests that 30 MPa hydrostatic pressure would result in cell cycle arrest in the chondrosarcoma cells, as indicated by increased signal of p21 and Gadd45, no direct evidence is available for an involvement of apoptotic process. In fact, many of the mRNA species involved in apoptosis were decreased in the pressurized cells.

Histone H4 signal was slightly increased in pressurized chondrosarcoma cells under study, while in HeLa cells all histone mRNA levels dropped at 41 MPa hydrostatic pressure applied for 10 min [15]. However, the differences in cell type and duration of the loading may explain these contradicting results. Destabilization of histone mRNA was noticed in both experiments. Continuous 30 MPa pressure decreases aggrecan production in primary chondrocytes [10], and similar effect on mRNAs of matrix molecules osteonectin, fibronectin and collagen type VI was seen here. Hyaluronan has an important role

Table 1

Genes detected on cDNA array which were up-regulated and/or whose mRNA stability was changed due to high hydrostatic pressure

Gene	Fold of change			Function	GB access
	a	b	c		number
c-jun	41.9	20.4	0.2	proto-oncogene	J04111
E4BP4 gene	13.7	5.2	0.7	transcription factor	X64318
growth arrest and DNA damage protein 45 (Gadd 45)	9.7	1.9	0.9	DNA repair, stress response	S40706
heparin-binding EGF-like growth factor (HBEGF)	8.9	3.6	1.4	growth factor	M60278
c-myc oncogene	8.1	8.8	0.2	proto-oncogene	V00568
nerve growth factor-inducible PC4 homologue	7.2	2.9	0.3	growth factor, cytokine	Y10313
jun-B	5.9	4.3	0.3	proto-oncogene	M29039
early growth response gene alpha (EGR-α)	5.9	4.6	0.7	transcription regulator	S81439
CDC-like kinase (CLK-1)	5.0	3.0	0.6	cell cycle-regulating kinase	L29222
NGF-inducible anti-proliferative protein PC3	4.4	6.4	0.6	cell cycle-regulating kinase	U72649
Rho8 protein	4.0	4.1	1.1	G-protein, signal transduction	X95282
EGF-response factor 1 (ERF-1)	3.6	3.3	0.8	transcription regulator	X79067
early growth response protein 1 (hEGR1, KROX 24)	3.3	2.8	0.6	transcription regulator	X52541
Gem (Ras-like protein KIR)	3.1	3.9	0.8	G-protein, signal transduction	U10550
DNA-binding protein CPBP	2.8	1.4	0.4	DNA binding protein	U44975
RBQ1 retinoblastoma binding protein	2.7	1.1	0.5	cell cycle, transcription	X85133
tyrosine-protein kinase ABL2 (tyrosine kinase ARG)	2.7	1.8	2.2	proto-oncogene	M35296
B4-2 protein	2.7	1.9	0.7	morphogenesis	U03105
MCL-1	2.5	3.1	1.7	bcl family protein	L08246
vascular endothelial growth factor (VEGF)	2.5	2.4	0.9	growth factor	M32977
CDK-interacting protein (cip1, waf1, p21)	2.4	1.4	1.1	cell cycle-regulating protein	U09579
platelet-derived growth factor A (PDGF-1)	2.2	1.4	0.9	growth factor	X06374
fos-related antigen (fra-1)	2.1	2.7	1.0	proto-oncogene	X16707
cyclin-dependent kinases regulatory subunit 2 (CKS-2)	2.1	1.3	1.0	cell cycle-regulating kinase	X54942
fos-related antigen 2 (fra-2)	2.1	3.2	1.0	proto-oncogene	X16706
ets-related transcription factor Elf-1	2.0	1.4	1.5	transcription activator	M82882
activator 1 37-kDa subunit	1.6	2.4	0.5	replication factor	M87339
STAT induced STAT inhibitor-3	1.5	4.7	0.6	kinase inhibitor	AB004904
wee-1 like protein kinase	1.5	2.6	0.4	cell cycle-regulating protein	U10564
histone H4	1.3	1.7	0.5	histone	X67081
interleukin-17 receptor	1.2	2.1	0.9	interferon receptor	U58917
bone morphogenetic protein 4 (BMP-4)	1.2	6.2	1.6	growth factor	D30751
polyhomeotic 2 homolog (HPH2)	1.2	2.2	1.1	transcriptional regulator	U89278

a) Fold of increase in expression under 30 MPa hydrostatic pressure for 6 h.
b) Relative mRNA stability in control sample (actinomycin-treated control vs. control).
c) Relative mRNA stability in pressurized sample (actinomycin-treated pressurized sample vs. pressurized sample).

in cartilage matrix linking tens to hundreds of aggrecan molecules into proteoglycan aggregates. mRNAs coding for three hyaluronan-associated proteins were down-regulated by high pressure, too. Placental bikunin is essential for biosynthesis of inter-α-trypsin inhibitor heavy chain-hyaluronan complex [18], and cdc37 has been shown to bind to hyaluronan [5]. Tiam1 and hyaluronan receptor CD44 interaction stimulates Rac1 signaling and hyaluronan-mediated breast tumor cell migration [3]. Decrease in mRNAs coding both H and M chains of lactate dehydrogenase suggest changes in carbohydrate metabolism in pressurized cells.

Table 2

Genes detected on cDNA array which were down-regulated and/or whose mRNA stability was changed due to high hydrostatic pressure

Gene	Fold of change			Function	GB access
	a	b	c		number
protein kinase C inhibitor	0.6	0.6	1.4	kinase inhibitor	U51004
tissue inhibitor of matrix metalloproteinase 2 (TIMP-2)	0.6	0.8	1.7	protease inhibitor	J05593
c-myc binding protein	0.6	0.7	1.4	inhibitor of myc-activity	D89667
osteonectin (SPARC, BM-40)	0.6	0.9	2.0	extracellular matrix protein	J03040
cyclin-dependent kinases regulatory subunit 1 (CKS-1)	0.5	0.8	1.7	cell cycle-regulating kinase	X54941
guanylate kinase (GMP kinase)	0.5	0.8	1.6	cGMP cycling	L76200
active breakpoint cluster region-related protein	0.5	0.6	1.3	GTPase-activating protein	U01147
hepatoma-derived growth factor (HDGF)	0.5	0.9	2.0	growth factor	D16431
signal transducer and activator of transcription 3 (STAT3)	0.5	1.1	2.0	transcription factor	L29277
L-lactate dehydrogenase H chain	0.5	0.6	1.5	carbohydrate metabolism	Y00711
fibronectin	0.5	0.8	2.0	extracellular matrix protein	X02761
TAX1-binding protein TXB151	0.5	0.7	2.1	transcriptional activator	U33821
procollagen α_1(VI)	0.5	0.8	2.8	extracellular matrix protein	X15879
von Hippel–Lindau tumor suppressor protein	0.5	1.2	1.8	tumor suppressor	L15409
extracellular signal-regulated kinase 6 (ERK-6)	0.5	0.8	0.9	signal transduction	X79483
GTP-binding protein ras-associated with diabetes (RAD1)	0.5	0.7	1.5	G-protein, signal transduction	L24564
procollagen α_1(XVI)	0.5	0.7	1.8	extracellular matrix protein	M92642
fatty acid synthase	0.5	1.0	1.8	lipid metabolism	S80437
acid finger protein	0.5	1.9	2.3	DNA binding	U09825
CHD3	0.5	0.8	1.9	transcription factor	AF006515
γ-tubulin	0.5	0.9	1.5	cytoskeletal protein	M61764
L-lactate dehydrogenase M chain	0.5	1.0	1.7	carbohydrate metabolism	X02152
placental bikunin, inter-α-trypsin inhibitor light chain	0.5	0.8	1.9	protease inhibitor	U78095
replication protein A 70 kDa subunit	0.5	0.7	1.7	replication factor	M63488
nucleoside 5'-diphosphate phosphotransferase	0.5	0.8	1.5	nucleotide metabolism	Y07604
PRSM1 metalloproteinase	0.4	1.0	1.5	metalloproteinase	U58048
growth-arrest specific protein (GAS)	0.4	0.7	1.9	growth suppression	L13720
procollagen α_3(VI)	0.4	1.0	2.0	extracellular matrix protein	X52022
insulin-like growth factor binding protein 6 (IGFBP6)	0.4	0.7	1.5	modulation of IGF activity	M62402
CDC37 homolog	0.4	0.8	1.5	cell cycle regulation	U63131
BCL7B protein	0.4	1.3	2.0	cytoskeletal protein	X89985
smoothened	0.4	0.8	1.6	morphogenesis	U84401
dishevelled 1 (segment polarity protein)	0.4	0.8	1.6	morphogenesis	U46461
IMP synthetase (inosinicase)	0.4	1.2	1.4	purine metabolism	U37436
inosine phosphorylase (PNP)	0.4	0.9	2.3	nucleotide metabolism	X00737
extracellular signal-regulated kinase 2 (ERK-2)	0.4	0.8	1.4	signal transduction	L11285
TRAP1	0.4	0.8	1.6	death-receptor	U12595
transforming growth factor-α (TGF-α)	0.4	1.0	1.7	growth factor	K03222
chromatin assembly factor 1 p48 subunit	0.4	0.8	1.6	chromatin protein	X74262
branched-chain amino acid aminotransferase	0.4	0.8	1.8	amino acid metabolism	U68418
T-lymphoma invasion and metastasis inducing (Tiam1)	0.4	0.7	2.0	oncogene	U16296
death-associated protein-1 (DAP-1)	0.3	0.7	2.4	death-receptor	X76105
Wnt-5a	0.3	1.0	2.1	morphogenesis	L20861

Table 2
(Continued)

Gene	Fold of change			Function	GB access number
	a	b	c		
PCNA	0.3	1.1	1.6	cell cycle-regulation	M15796
Fas-activated serine/threonine (FAST) kinase	0.3	0.8	2.1	death kinase	X86779
vaccinia-related kinase 2	0.3	1.1	1.8	intracellular kinase network	AB000450
integrin-linked kinase (ILK)	0.2	0.9	2.1	signal transduction	U40282
apoptosis regulator bax	0.2	0.5	4.5	apoptosis	L22474

a) Fold of decrease in expression under 30 MPa hydrostatic pressure for 6 h.
b) Relative mRNA stability in control sample (actinomycin-treated control vs. control).
c) Relative mRNA stability in pressurized sample (actinomycin-treated pressurized sample vs. pressurized sample).

Expression of heat shock protein 70 is an example of accumulation of mRNA solely due to mRNA stabilization without increased transcription in cells grown under high pressure [8]. In this experiment that kind of stabilization pattern was not observed with other genes, which indicates the uniqueness of the heat shock response under pressure. The overall picture given by this array analysis thus indicates that high static hydrostatic pressure influence on the amount of expression products involved in cell cycle regulation, transcription factors regulating various other genes and also various matrix molecules. The stability of up-regulated genes as a whole is somewhat decreased and involves no clear stabilization. In contrast, the stability of down-regulated genes seems to be somewhat increased.

References

[1] F. Abe and K. Horikoshi, Tryptophan permease gene TAT2 confers high-pressure growth in Saccharomyces cerevisiae, *Mol. Cell. Biol.* **20** (2000), 8093–8102.

[2] F. Beier, I. Eerola, E. Vuorio, P. Luvalle, E. Reichenberger, W. Bertling, K. von der Mark and M.J. Lammi, Variability in the upstream promoter and intron sequences of the human, mouse and chick type X collagen genes, *Matrix Biol.* **15** (1996), 415–422.

[3] L.Y. Bourguignon, H. Zhu, L. Shao and Y.W. Chen, CD44 interaction with Tiam1 promotes Rac1 signaling and hyaluronic acid-mediated breast tumor cell migration, *J. Biol. Chem.* **275** (2000), 1829–1938.

[4] K.D. Brandt, Animal models: insights into osteoarthritis (OA) provided by the cruciate-deficient dog, *Br. J. Rheumatol.* **30** (1991), 5–9.

[5] N. Grammatikakis, A. Grammatikakis, M. Yoneda, Q. Yu, S.D. Banerjee and B.P. Toole, A novel glycosaminoglycan-binding protein is the vertebrate homologue of the cell cycle control protein, Cdc37, *J. Biol. Chem.* **270** (1995), 16 198–16 205.

[6] A.C. Hall, J.P. Urban and K.A. Gehl, The effects of hydrostatic pressure on matrix synthesis in articular cartilage, *J. Orthop. Res.* **9** (1991), 1–10.

[7] W.A. Hodge, R.S. Fijan, K.L. Carlson, R.G. Burgess, W.H. Harris and R.W. Mann, Contact pressures in the human hip joint measured in vivo, *Proc. Natl. Acad. Sci. USA* **83** (1986), 2879–2883.

[8] K. Kaarniranta, M. Elo, R. Sironen, M.J. Lammi, M.B. Goldring, J.E. Eriksson, L. Sistonen and H.J. Helminen, Hsp70 accumulation in chondrocytic cells exposed to high continuous hydrostatic pressure coincides with mRNA stabilization rather than transcriptional activation, *Proc. Natl. Acad. Sci. USA* **95** (1998), 2319–2324.

[9] K. Kaarniranta, C.I. Holmberg, H.J. Helminen, J.E. Eriksson, L. Sistonen and M.J. Lammi, Protein synthesis is required for stabilization of hsp70 mRNA upon exposure to both hydrostatic pressurization and elevated temperature, *FEBS Lett.* **475** (2000), 283–286.

[10] M.J. Lammi, R. Inkinen, J.J. Parkkinen, T. Häkkinen, M. Jortikka, L.O. Nelimarkka, H.T. Järveläinen and M.I. Tammi, Expression of reduced amounts of structurally altered aggrecan in articular cartilage chondrocytes exposed to high hydrostatic pressure, *Biochem. J.* **304** (1994), 723–730.

[11] J.J. Parkkinen, J. Ikonen, M.J. Lammi, J. Laakkonen, M. Tammi and H.J. Helminen, Effects of cyclic hydrostatic pressure on proteoglycan synthesis in cultured chondrocytes and articular cartilage explants, *Arch. Biochem. Biophys.* **300** (1993), 458–465.

[12] A.M. Säämänen, M. Tammi, J. Jurvelin, I. Kiviranta and H.J. Helminen, Proteoglycan alterations following immobilization and remobilization in the articular cartilage of young canine knee (stifle) joint, *J. Orthop. Res.* **8** (1990), 863–873.
[13] R. Sironen, M. Elo, K. Kaarniranta, H.J. Helminen and M.J. Lammi, Transcriptional activation in chondrocytes submitted to hydrostatic pressure, *Biorheology* **37** (2000), 85–93.
[14] R.L. Smith, J. Lin, M.C. Trindade, J. Shida, G. Kajiyama, T. Vu, A.R. Hoffman, M.C. van der Meulen, S.B. Goodman, D.J. Schurman and D.R. Carter, Time-dependent effects of intermittent hydrostatic pressure on articular chondrocyte type II collagen and aggrecan mRNA expression, *J. Rehabil. Res. Dev.* **37** (2000), 153–161.
[15] A.L. Symington, S. Zimmerman, J. Stein, G. Stein and A.M. Zimmerman, Hydrostatic pressure influences histone mRNA, *J. Cell Sci.* **98** (1991), 123–129.
[16] K. Takahashi, T. Kubo, K. Kobayashi, J. Imanishi, M. Takigawa, Y. Arai and Y. Hirasawa, Hydrostatic pressure influences mRNA expression of transforming growth factor-beta 1 and heat shock protein 70 in chondrocyte-like cell line, *J. Orthop. Res.* **15** (1997), 150–158.
[17] M. Takigawa, K. Tajima, H.O. Pan, M. Enomoto, A. Kinoshita, F. Suzuki, Y. Takano and Y. Mori, Establishment of a clonal human chondrosarcoma cell line with cartilage phenotypes, *Cancer Res.* **49** (1989), 3996–4002.
[18] L. Zhuo, M. Yoneda, M. Zhao, W. Yingsung, N. Yoshida, Y. Kitagawa, K. Kawamura, T. Suzuki and K. Kimata, Defect in SHAP-hyaluronan complex causes severe female infertility. A study by inactivation of the bikunin gene in mice, *J. Biol. Chem.* **276** (2001), 7693–7696.

Biorheology 39 (2002) 119–124
IOS Press

Integrins and cell signaling in chondrocytes

Richard F. Loeser *
Departments of Internal Medicine and Biochemistry, Section of Rheumatology, Rush Medical College of Rush-Presbyterian-St. Luke's Medical Center, Chicago, IL 60612, USA

Abstract. Integrins are adhesion receptor heterodimers that transmit information from the extracellular matrix (ECM) to the cell through activation of cell signaling pathways. Chondrocytes express several members of the integrin family including $\alpha_5\beta_1$ which is the primary chondrocyte receptor for fibronectin. Cell signaling mediated through integrins regulates several chondrocyte functions including differentiation, matrix remodeling, responses to mechanical stimulation and cell survival. Integrin-mediated activation of members of the mitogen-activated protein kinase family likely plays a key role in transmitting signals regulating chondrocyte gene expression. Upstream mediators of mitogen-activated protein kinase (MAP kinase) activation include focal adhesion kinase (FAK) and proline-rich tyrosine kinase 2 (pyk2) which are both expressed by chondrocytes. A better understanding of chondrocyte integrin signaling is needed to define the mechanisms by which the ECM regulates chondrocyte function.

1. Introduction

The extracellular matrix (ECM) of adult articular cartilage is not just an inert framework for holding chondrocytes and for covering the ends of the bones. Rather the ECM provides signals to the chondrocyte which regulate a number of important cellular functions. The integrins are the major family of ECM receptors which can transmit information from the matrix to the cell. Integrin binding of ECM ligands results in the formation of signaling complexes which play a key role in the regulation of cell survival, proliferation, differentiation, and matrix remodeling [1,14,15]. Integrin signaling is coordinated with both the organization of the cytoskeleton and with signaling by growth factor and cytokine receptors [3,14]. The long-term goal of our studies is to determine how signals generated through integrins regulate chondrocyte function with a particular focus on changes in function which are relevant to the development of cartilage destruction.

At least 20 different integrin heterodimers have been described resulting from the combination of 9 types of β subunits (designated $\beta1 \ldots \beta9$) with 14 types of α subunits [18]. Initial studies were performed to characterize which integrins were expressed by adult articular chondrocytes and which cartilage ECM proteins served as chondrocyte integrin ligands. Work from several laboratories has revealed that chondrocytes express at least the $\alpha1\beta1$, $\alpha2\beta1$, $\alpha3\beta1$, $\alpha5\beta1$, $\alpha6\beta1$, $\alpha10\beta1$, $\alpha v\beta3$, and $\alpha v\beta5$ integrins [8,10,11,20,27,30]. Using immunohistochemical staining of adult human articular cartilage, Salter et al. [27] noted that the $\alpha5\beta1$ integrin was a prominent chondrocyte integrin with variable and weaker expression of $\alpha1\beta1$ and $\alpha3\beta1$. Using a combination of immunofluorescence, immunoprecipitation, and FACS analysis, Woods et al. [30] demonstrated that adult human chondrocytes express $\alpha1\beta1$, $\alpha5\beta1$, and $\alpha v\beta5$ integrins accompanied by weak expression of $\alpha3\beta1$ and $\alpha v\beta3$.

We have seen a similar pattern of integrin expression in cultured human and bovine chondrocytes and have noted an increase in integrin expression *in situ* in OA cartilage compared to normal cartilage [20].

*Address for correspondence: Dr. Richard F. Loeser, MD, Rheumatology, Rush-Presbyterian-St. Luke's Medical Center, 1725 W. Harrison, Suite 1017, Chicago, IL 60612, USA. Tel.: +1 312 942 8994; Fax: +1 312 942 3053; E-mail: rloeser@rush.edu.

Relative to adult chondrocytes, fetal chondrocytes and chondrosarcoma cells have higher levels of $\alpha2\beta1$ and $\alpha6\beta1$ integrins [16,26]. In comparing primary adult human articular chondrocytes to human articular chondrocytes which had been immortalized using the SV-40 T-antigen we noted little to no $\alpha2$ by flow cytometry analysis on primary cells while immortalized cells were clearly positive [21]. When studies were recently repeated using a more sensitive secondary antibody, primary chondrocytes were found to express $\alpha2$ but at low levels compared to the immortalized cells and also compared to $\alpha1$ on primary cells (unpublished observations). This suggests that chondrocytes do express $\alpha2$ at low levels but expression is stimulated in immortalized and fetal cells.

The $\alpha1\beta1$, $\alpha2\beta1$ and $\alpha10\beta1$ integrins can all serve as receptors for type II collagen [8,11,12,20,21]. The $\alpha1\beta1$ integrin also mediates adhesion of chondrocytes to type VI collagen [21] and to cartilage matrix protein (matrilin-1) [22]. In addition to collagen, $\alpha2\beta1$ can mediate binding to chondroadherin [7]. The $\alpha5\beta1$ integrin serves as the primary chondrocyte fibronectin (FN) receptor [20] while αV-containing integrins bind to vitronectin and osteopontin [19] and may serve as alternative FN receptors. It is likely that not all of the integrins expressed by chondrocytes have been described and the potential exists for additional integrin–matrix protein interactions to be discovered.

2. Integrin signaling

Cell binding of ECM proteins which serve as integrin ligands results in receptor clustering and stimulation of signaling pathways ("outside-in" signaling) [18]. Studies of the integrin-activated signaling pathways have been extensively reported in many cell types other than chondrocytes. Integrins do not have intrinsic kinase activity but instead they initiate cell signaling through the recruitment of adapter proteins including Shc, Crk, p130^{Cas}, paxillin, talin, viniculin, and caveolin [3,6,14,18,24]. These proteins and others such as focal adhesion kinase (FAK) and pyk2 interact with the integrin cytoplasmic tails when integrins are bound to ECM ligands which induce a conformational change in the receptor subunits. The formation of these integrin-associated signaling complexes is closely linked to the cytoskeleton. Many of the signaling proteins recruited into the complex are activated through tyrosine phosphorylation. Additional proteins found in or closely related to the integrin signaling complexes include Src, RhoA, Rac1, Ras, Raf1, Grb2, Sos, MEK kinase, MEK1, and members of the MAP kinase family including ERK1,2, JNK, and p38. New findings related to integrin signaling are appearing on almost a daily basis and a complete description is beyond the scope of this review. Importantly, the MAP kinases appear to play a central role in mediating downstream signals from integrins which can regulate gene expression through activation of transcription factors such as AP-1 and NFκB. A model of a portion of the major integrin signaling pathways is shown in Fig. 1.

3. Adhesion-mediated signaling studies in chondrocytes

Preliminary experiments were performed to evaluate tyrosine phosphorylation of chondrocyte proteins in adherent cells compared to cells placed in suspension. Chondrocytes isolated from bovine articular cartilage as described [19] were grown to confluence in monolayer cultures in media containing 10% FBS. Cells were either maintained in monolayer or isolated with trypsin and placed in suspension overnight in alginate bead cultures [20]. After overnight incubation some monolayer cultures were harvested and the cells were plated on fibronectin (20 μg/ml) coated dishes or on dishes coated with 3% BSA and cultured for an additional hour. After one hour the cells plated on fibronectin had adhered and were beginning

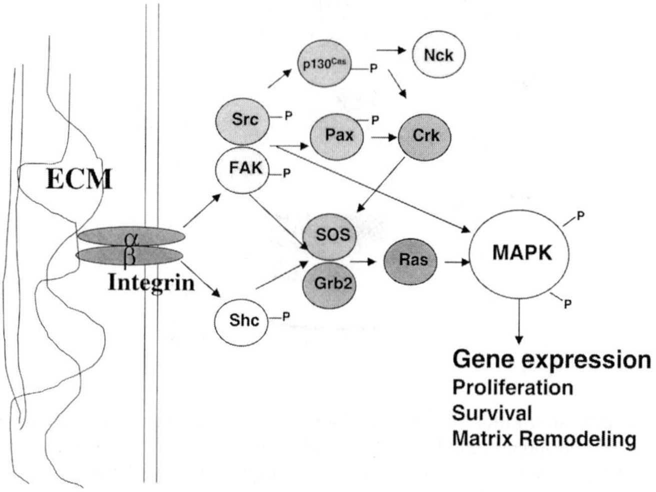

Fig. 1. A model for extracellular matrix (ECM) signaling via integrins which regulate activity of the mitogen activated protein kinases (MAP kinase). The MAP kinases in turn modulate the activity of transcription factors affecting gene expression. The model is not meant to be a complete listing of all of the proteins involved.

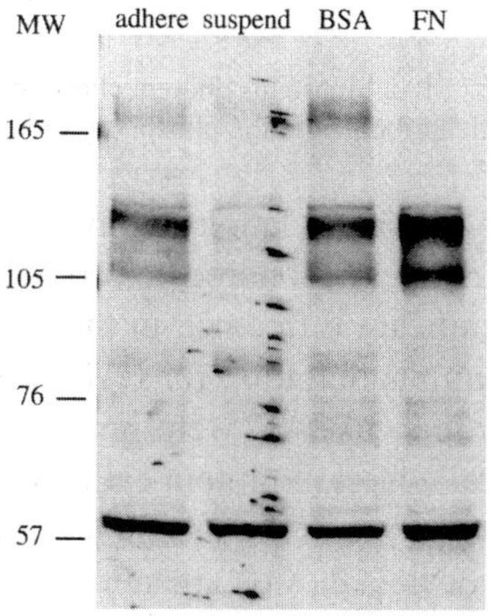

Fig. 2. Tyrosine phosphorylation of chondrocyte proteins in response to adhesion. Primary articular chondrocytes were cultured to confluency in monolayer (adhere) or harvested from monolayer and placed in suspension culture overnight in alginate (suspend) or plated for 1 hour on culture dishes coated with 3% BSA or 20 μg/ml fibronectin (FN). Cell lysates were immunoblotted with antibodies to phosphotyrosine.

to spread while those in dishes coated with BSA stayed in suspension. Cell lysates were prepared from all cultures and then samples containing equal amounts of total protein were separated by SDS-PAGE on 7.5% gels and transferred to nitrocellulose for immunoblotting with an antibody (PY-20) to phosphotyrosine. As shown in Fig. 2, several proteins were phosphorylated to a greater extent in monolayer cultures (adherent) compared to overnight suspension culture. The monolayer cells that were harvested and re-plated on FN also increased phosphorylation of proteins particularly around 120 kDa compared to cells placed in suspension in BSA coated plates (Fig. 2). Interestingly, a higher MW protein just above 165 kDa showed decreased tyrosine phosphorylation after adhesion to FN.

Since FAK has been shown to be a prominent protein in integrin signaling and has a MW of about 120 kDa [28], FAK antibodies were used to immunoprecipitate proteins from adherent and suspended

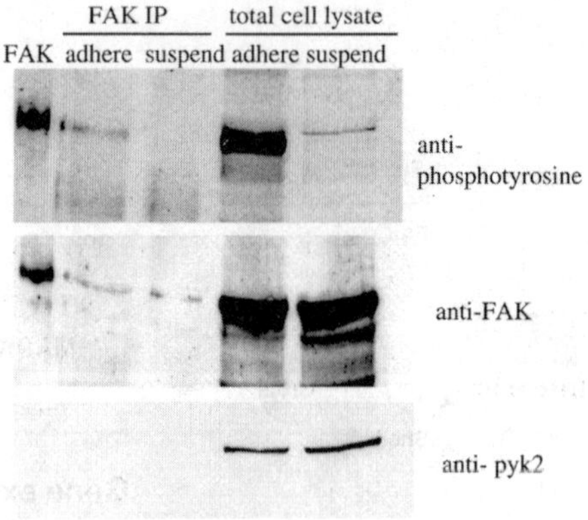

Fig. 3. Phosphorylation of focal adhesion kinase (FAK) in adherent chondrocytes. Cell lysates from chondrocytes cultured as described in Fig. 1 were either immunoprecipitated with anti-FAK antibodies (FAK IP) or used without immunoprecipitation (total cell lysate). The samples were immunoblotted with anti-phosphotyrosine, anti-FAK, or anti-pyk2 antibodies as indicated. A positive control for FAK was run in the first lane (FAK).

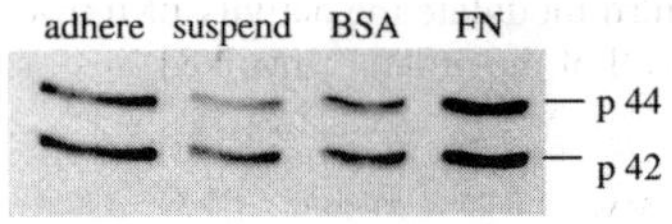

Fig. 4. Phosphorylation of the ERK 1 and ERK 2 MAP kinase in adherent chondrocytes. Cell lysates from chondrocytes cultured as described in Fig. 1 were immunoblotted with antibodies to the phosphorylated forms of ERK 1 and 2 (p44 and p42).

cells followed by analysis for tyrosine phosphorylation by immunoblotting. This showed that FAK was phosphorylated to a greater extent in adherent chondrocytes compared to chondrocytes in suspension in alginate (Fig. 3). There was not a significant difference in the total amount of FAK between adherent and suspension cultures as shown by immunoblotting for FAK (Fig. 3). Analysis of the total cell lysate suggested that phosphoproteins in addition to FAK were present at 120 kDa. Therefore, we evaluated lysates for the presence of another focal adhesion protein called CADTK or pyk2 [32] which has a MW similar to FAK. Pyk2 was also found to be present in the chondrocyte cell lysate (Fig. 3) suggesting that this protein also participates in chondrocyte adhesion signaling.

As detailed above, signals downstream of integrins include activation of the MAP kinases. We evaluated for activation of the ERK1, ERK2 member of the MAP kinase family using phosphospecific antibodies and found that phosphorylation was greater in adherent cells in monolayer culture relative to cells in alginate (Fig. 4). When monolayer cells were replated over BSA or allowed to adhere for 1 hour to fibronectin ERK 1 and 2 phosphorylation was greater in the cells on FN. We have noted similar activation of ERK1,2 phosphorylation when chondrocytes were treated with an activating antibody to the $\alpha5\beta1$ integrin as well as phosphorylation of the other MAP kinases, JNK and p38 (manuscript submitted for publication). These results demonstrate that like other cell types chondrocyte integrin signaling utilizes the MAP kinase pathway.

4. Integrin signaling works in concert with growth factors and cytokines

The cooperation between growth factors/cytokines and integrins in cell signaling is becoming increasingly apparent. Integrin aggregation and occupancy has been shown to synergize with growth factor stimulation in the phosphorylation of growth factor receptors and in the activation of MAP kinase [2,25]. The requirement for cell anchorage in the mitogenic response to growth factors is thought to be due to a synergy between integrin and growth factor signaling [17]. The proliferative response of rabbit sternal chondrocytes to FGF required fibronectin binding to the $\alpha5\beta1$ integrin [13]. Adhesion to fibronectin was also shown to increase FGF and IGF-I stimulated sulfate incorporation which was inhibited by treatment with nitric oxide, an agent which may disrupt the formation of focal adhesion signaling complexes [9]. Further evidence for cooperativity between IGF-I signaling and integrins comes from a study showing co-immunoprecipitation of the $\beta1$ integrin subunit and the IGF-I receptor in IGF-I treated chondrocytes [29].

There is also evidence for synergy between integrin and cytokine signaling in chondrocytes. Stimulation of metalloproteinase synthesis by RGD peptides or fibronectin fragments was enhanced by IL-1 and inhibited by the IL-1 receptor antagonist [4] and chondrocyte adhesion to fibronectin was shown to increase production of IL-6 and GM-CSF [31]. Stimulation of the chondrocyte $\alpha5\beta1$ integrin with an activating monoclonal antibody was found to increase production of IL-1β [5]. Finally, mechanical signals which induce chondrocyte membrane hyperpolarization appear to involve cooperative signaling between the $\alpha5\beta1$ integrin and IL-4 [23].

5. Conclusion

We are just beginning to understand how the ECM functions in regulating various activities of chondrocytes which are important in cartilage homeostasis and remodeling. Because integrins "integrate" the ECM with cytoskeletal signaling complexes and integrate matrix signals with signals from growth factor and cytokine receptors, this family of cell adhesion receptors will likely play a central role in transmitting signals from the ECM. It will be a great challenge to unravel the complex signaling networks which integrins use to regulate the activity of transcription factors which in turn regulate chondrocyte gene expression and function.

Acknowledgements

The author would like to thank past and present members of his laboratory who have contributed to this work including Greg Sigmon, Sagypash Sadiev, and Christopher Forsyth. This work was funded by grants from the Arthritis Foundation and the National Institutes of Health.

References

[1] J.C. Adams and F.M. Watt, Regulation of development and differentiation by the extracellular matrix, *Development* **117** (1993), 1183–1198.
[2] A.E. Aplin and R.L. Juliano, Integrin and cytoskeletal regulation of growth factor signaling to the MAP kinase pathway, *J. Cell Sci.* **112** (1999), 695–706.
[3] A.E. Aplin et al., Anchorage-dependent regulation of the mitogen-activated protein kinase cascade by growth factors is supported by a variety of integrin alpha chains, *J. Biol. Chem.* **274** (1999), 31 223–31 228.
[4] E.C. Arner and M.D. Tortorella, Signal transduction through chondrocyte integrin receptors induces matrix metalloproteinase synthesis and synergizes with interleukin-1, *Arthritis Rheum.* **38** (1995), 1304–1314.
[5] M.G. Attur et al., Functional genomic analysis in arthritis-affected cartilage: Yin-Yang regulation of inflammatory mediators by alpha5beta1 and alphaVbeta3 Integrins, *J. Immunol.* **164** (2000), 2684–2691.

[6] K. Burridge and M. Chrzanowska-Wodnicka, Focal adhesions, contractility, and signaling, *Annu. Rev. Cell. Dev. Biol.* **12** (1996), 463–518.

[7] L. Camper et al., Integrin alpha2beta1 is a receptor for the cartilage matrix protein chondroadherin, *J. Cell. Biol.* **138** (1997), 1159–1167.

[8] L. Camper et al., Isolation, cloning, and sequence analysis of the integrin subunit alpha10, a beta1-associated collagen binding integrin expressed on chondrocytes, *J. Biol. Chem.* **273** (1998), 20 383–20 389.

[9] R.M. Clancy et al., Outside-in signaling in the chondrocyte. Nitric oxide disrupts fibronectin-induced assembly of a subplasmalemmal actin/rho A/focal adhesion kinase signaling complex, *J. Clin. Invest.* **100** (1997), 1789–1796.

[10] J. Durr et al., Identification and immunolocalization of laminin in cartilage, *Exp. Cell. Res.* **222** (1996), 225–233.

[11] J. Durr et al., Localization of beta 1-integrins in human cartilage and their role in chondrocyte adhesion to collagen and fibronectin, *Exp. Cell. Res.* **207** (1993), 235–244.

[12] M. Enomoto et al., Beta 1 integrins mediate chondrocyte interaction with type I collagen, type II collagen, and fibronectin, *Exp. Cell. Res.* **205** (1993), 276–285.

[13] M. Enomoto-Iwamoto et al., Involvement of alpha5beta1 integrin in matrix interactions and proliferation of chondrocytes, *J. Bone Miner. Res.* **12** (1997), 1124–1132.

[14] F.G. Giancotti, Complexity and specificity of integrin signalling, *Nat. Cell. Biol.* **2** (2000), E13–E14.

[15] B.M. Gumbiner, Cell adhesion: the molecular basis of tissue architecture and morphogenesis, *Cell* **84** (1996), 345–357.

[16] K. Holmvall et al., Chondrocyte and chondrosarcoma cell integrins with affinity for collagen type II and their response to mechanical stress, *Exp. Cell. Res.* **221** (1995), 496–503.

[17] A. Howe et al., Integrin signaling and cell growth control, *Curr. Opin. Cell. Biol.* **10** (1998), 220–231.

[18] R.O. Hynes, Integrins: versatility, modulation, and signaling in cell adhesion, *Cell* **69** (1992), 11–25.

[19] R.F. Loeser, Integrin-mediated attachment of articular chondrocytes to extracellular matrix proteins, *Arthritis Rheum.* **36** (1993), 1103–1110.

[20] R.F. Loeser et al., Expression of beta 1 integrins by cultured articular chondrocytes and in osteoarthritic cartilage, *Exp. Cell. Res.* **217** (1995), 248–257.

[21] R.F. Loeser et al., Integrin expression by primary and immortalized human chondrocytes: evidence of a differential role for alpha1beta1 and alpha2beta1 integrins in mediating chondrocyte adhesion to types II and VI collagen, *Osteoarthritis Cartilage* **8** (2000), 96–105.

[22] S. Makihira et al., Enhancement of cell adhesion and spreading by a cartilage-specific noncollagenous protein, cartilage matrix protein (CMP/Matrilin-1), via integrin alpha1beta1, *J. Biol. Chem.* **274** (1999), 11 417–11 423.

[23] S.J. Millward-Sadler et al., Integrin-regulated secretion of interleukin 4: A novel pathway of mechanotransduction in human articular chondrocytes, *J. Cell. Biol.* **145** (1999), 183–189.

[24] S. Miyamoto et al., Integrin function: molecular hierarchies of cytoskeletal and signaling molecules, *J. Cell. Biol.* **131** (1995), 791–805.

[25] S. Miyamoto et al., Integrins can collaborate with growth factors for phosphorylation of receptor tyrosine kinases and MAP kinase activation: roles of integrin aggregation and occupancy of receptors, *J. Cell. Biol.* **135** (1996), 1633–1642.

[26] D.M. Salter et al., Chondrocyte heterogeneity: immunohistologically defined variation of integrin expression at different sites in human fetal knees, *J. Histochem. Cytochem.* **43** (1995), 447–457.

[27] D.M. Salter et al., Integrin expression by human articular chondrocytes, *Br. J. Rheumatol.* **31** (1992), 231–234.

[28] D.D. Schlaepfer and T. Hunter, Integrin signalling and tyrosine phosphorylation: just the FAKs?, *Trends Cell. Biol.* **8** (1998), 151–157.

[29] M. Shakibaei et al., Signal transduction by beta1 integrin receptors in human chondrocytes in vitro: collaboration with the insulin-like growth factor-I receptor, *Biochem. J.* **342** (1999), 615–623.

[30] V.L. Woods, Jr. et al., Integrin expression by human articular chondrocytes, *Arthritis Rheum.* **37** (1994), 537–544.

[31] I. Yonezawa et al., VLA-5-mediated interaction with fibronectin induces cytokine production by human chondrocytes, *Biochem. Biophys. Res. Commun.* **219** (1996), 261–265.

[32] H. Yu et al., Activation of a novel calcium-dependent protein – tyrosine kinase. Correlation with c-Jun N-terminal kinase but not mitogen-activated protein kinase activation, *J. Biol. Chem.* **271** (1996), 29 993–29 998.

Biorheology 39 (2002) 125–132
IOS Press

The effects of pressure on chondrocyte tumour necrosis factor receptor expression

C.I. Westacott [a,*], J.P.G. Urban [b], M.B. Goldring [c] and C.J. Elson [a]
[a] *Department of Pathology, University of Bristol Medical School, Bristol BS8 1TD, UK*
[b] *Department of Physiology, University of Oxford, Oxford OX1 3PT, UK*
[c] *Harvard Medical School Division of Rheumatology, Beth Israel Deaconess Medical Centre, Boston, MA, USA*

Abstract. This work was performed to determine whether one aspect of load, pressure, could alter tumour necrosis factor (TNF) receptor type I (RI) expression on chondrocytes. Encapsulated tsT/AC62, osteoarthritic (OA) or non-arthritic (NA) chondrocytes were centrifuged at speeds representing 5 or 20 MPa, incubated for specific periods, released from alginate and TNFRI and II (TNFRII) expression determined by flow cytometry. Significant ($p < 0.05$, $n = 4$) changes in tsT/AC62 chondrocyte TNFRI expression were apparent 24 hours after application of 20 MPa. Five or 20 MPa increased OA chondrocyte TNFRI expression; chondrocytes from some OA patients were markedly sensitive to 20 MPa. NA chondrocyte TNFRI expression usually decreased in response to 5 and 20 MPa. Significant pressure-induced differences in TNFRI expression between NA and OA groups were apparent at 5, but not 20 MPa. Pressure did not significantly alter TNRFII expression on tsT/AC62, NA or OA chondrocytes. These results suggest a mechanism whereby sensitivity of chondrocytes to the effects of TNFα may be increased, in susceptible individuals, in regions of the joint that experience peak loading.

1. Introduction

The processes leading to focal loss of cartilage from OA knee joints are not clear. Turnover of cartilage is regulated by the catabolic cytokines tumour necrosis factor alpha (TNFα) and interleukin-1beta (IL-1β) and the anabolic cytokines insulin-like growth factor and transforming growth factor-β, all of which act via specific membrane-bound receptors on chondrocytes. Catabolic cytokines stimulate chondrocytes to produce proteinases, which cleave aggrecan [2] and type II collagen [12], the major components of articular cartilage. Some time ago we showed that susceptibility of cartilage to the effects of tumour necrosis factor alpha (TNFα) varies with location on the knee joint, as does chondrocyte expression of the type I signalling receptor (TNFRI). Significantly, the response of biopsies to the effects of TNFα is directly related to TNFRI expression [24] suggesting a role for TNFα in focal loss of cartilage in OA. Thus, in the presence of the amounts of TNFα (0.25–5 ng) produced within the OA joints [27], only those regions of the knee where chondrocyte TNFR expression is high will be susceptible to its effects. However, how regional differences in TNFR expression occur is not known.

In OA joints, cartilage is lost from regions of peak loading [5] and mechanical loading, *per se*, can damage cartilage directly [1]. However, loading is also known to have a profound effect on chondrocyte metabolism. For example, removal of load by short-term immobilisation of an animal joint prevented IL-1 induced suppression of proteoglycan synthesis [22], whereas long-term immobilisation lead to loss

*Address for correspondence: Carole I. Westacott, PhD, Department of Pathology and Microbiology, School of Medical Sciences, University of Bristol, University Walk, Bristol BS8 1TD, UK. Tel.: +44 117 9288605; Fax: +44 117 9287896; E-mail: Carole.Westacott@bristol.ac.uk.

0006-355X/02/$8.00

of cartilage proteoglycan content which was gradually restored once the joint was remobilised [14]. As cytokines act via specific receptors to control cartilage matrix turnover [24], these observations raise the possibility that loading induces reversible changes in chondrocyte cytokine receptor expression, thereby temporarily altering cartilage matrix turnover.

Chondrocytes perceive mechanical stress through changes to the extracellular environment and events such as load-induced cell or matrix deformation, fluid flow and rise in pressure are suggested to act as the signals [17] that result in changes to chondrocyte metabolism. A sudden rise in pressure is one of the major changes experienced by chondrocytes in articular cartilage during loading [3]. Moreover, application of pressure is known to have a significant effect on cartilage metabolism, influencing proteoglycan synthesis [8,15,21]. In addition, hydrostatic pressure applied to chondrocytes alters mRNA expression for TGFβ [19] IL-6 [18,20] and TNFα [18] the cytokines that control matrix turnover. However, to our knowledge, the effect of pressure on expression of the chondrocyte receptors through which such cytokines act has not been investigated. The aims of this work were therefore to determine whether one aspect of load, pressure, could alter expression of chondrocyte TNFR.

2. Methods and materials

2.1. Chondrocyte preparation

Macroscopically normal cartilage was obtained from the femoral condyles of OA patients ($n = 11$, age 65.5 ± 10.7 years) or non-arthritic (NA) subjects ($n = 8$, age 81 ± 8.9 years) undergoing surgery for total knee or hip replacement, respectively. Cartilage was washed and chondrocytes released by enzymatic digestion as described previously [26]. Isolated chondrocytes, or temperature sensitive immortalised chondrocytes, tsT/AC62, derived from the macroscopically normal cartilage on an OA knee [16], were encapsulated in alginate to maintain phenotypic stability [7]. Beads were incubated (24 hours, 37°C, 5% CO_2) in 6 well plates containing DMEM/F12 supplemented with 10% FBS (Gibco), 2 mM glutamine, 100 U/ml penicillin, 100 μg/ml streptomycin, 25 mM HEPES.

2.2. Application of pressure

Cartilage, *in vivo*, is exposed to peak pressures of up to 20 MPa during activities such as rising from a chair [10], whilst during walking the average peak pressure is around 5 MPa. Other studies have shown exposure to high pressure for up to 2 hours inhibits metabolic response, whereas short applications of low physiological pressure stimulates metabolic activity [8,9,15]. Since chondrocyte responses appear to depend on duration as well as magnitude of applied pressure, two loading regimes were adopted. Cells were exposed to either high pressure of long duration (20 MPa, 1 hour) or low pressure of short duration (5 MPa, 5 min). Eight beads, each containing approximately 4×10^4 chondrocytes, were placed into sterile 2 ml tubes (Eppendorf) containing 1 ml fresh tissue culture medium. Pressure was applied by centrifugation [11] in a Sigma 1-15 Microfuge at speeds calculated, according to the manufacturer's instructions, to represent the desired pressures. For the high pressure protocol, beads were subjected to centrifugation for 60 minutes at $200 \times g$ (20 MPa) whereas for the low pressure protocol, beads were subjected to centrifugation for 5 minutes at $50 \times g$ (5 MPa). Unloaded control beads were incubated in medium on the bench alongside the centrifuge. At the end of the test period, beads were resuspended in medium by gently tapping and inverting tubes several times, and incubation (37°C, 5% CO_2) continued for up to 48 hours.

2.3. Flow cytometry

Chondrocytes were released from alginate [7] at approximately 4, 24, or 46 hours after loading, stained with specific non-crossreacting antibodies to the two receptor types and TNFR expression determined by flow cytometry (FACscan) [24].

2.4. Statistical analyses

The significance of the difference in TNFR expression on tsT/AC62 chondrocytes subjected to pressure (test) as compared with TNFR expression on untreated (control) chondrocytes was determined by Student's paired t test. Pressure-induced changes in TNFR expression on chondrocytes isolated from NA and OA cartilage were compared by expressing results as a ratio of receptor expression on test chondrocytes with that on control chondrocytes. The significance of the difference in ratios of receptor expression between NA and OA chondrocytes subjected to low or high loading regimes was analysed by Student's t test with Welch's correction for unequal variance.

3. Results

3.1. Response of immortalised chondrocytes to high pressure protocol (20 MPa, 1 hour)

To determine whether pressure could alter TNFRI expression, we first investigated the effects of centrifugation on cells of the tsT/AC62 cell line. Chondrocytes were subjected to high pressure and TNFR expression determined at 4, 24 and 46 hours. Figure 1 illustrates the results of two experiments performed to determine TNFR expression on tsT/AC62 cells with time after pressurisation. Changes in receptor expression on test (pressured) as compared with unpressured control cells were apparent at each time point for both TNFRI and TNFRII. Figure 1 shows the greatest difference between test and control populations in the proportion of chondrocytes expressing TNFRI as well the number of TNFRI/cell occurred approximately 24 hours after application of pressure. In all four experiments were performed at 20 MPa and the effects on TNFR expression at 24 hours after application of pressure are summarised in Table 1. It can be seen that a significant pressure-induced difference was observed in both the proportion of chondrocytes expressing TNFRI ($p = 0.045$) and the number of TNFRI/cell ($p = 0.036$). By contrast, there was no significant difference in either the proportion of chondrocytes expressing TNFRII ($p = 0.55$) or the number of TNFRII/cell ($p = 0.64$).

3.2. Response of isolated OA chondrocytes to high pressure protocol (20 MPa, 1 hour)

The effects of pressure were next investigated on chondrocytes isolated from OA cartilage. Figure 2 (A and B) shows the response of chondrocytes from OA patient 1 to the effects of 20 MPa pressure. Differences were apparent as early as 4 hours after application of pressure in both (i) the proportion of chondrocytes expressing TNFRI as well as (ii) TNFRI/cell and were still apparent at after 24 hours. In all, chondrocytes isolated from 7 different OA knee joints were subjected to 20 MPa. TNFRI expression on chondrocytes from some knees (4/7) changed dramatically in response to the effects of pressure. However, TNFR expression on chondrocytes from other populations (3/7) were relatively unchanged. By contrast, TNFRII expression was similar whether chondrocytes were loaded or not (data not shown).

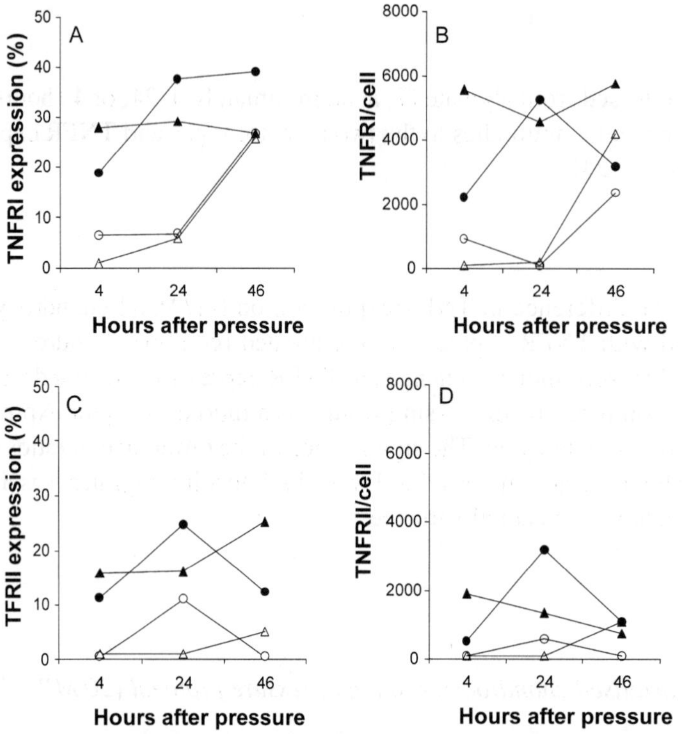

Fig. 1. The results of two experiments showing the response of tsT/AC62 chondrocytes to pressure generated by centrifugation. Encapsulated tsT/AC62 chondrocytes were centrifuged for 1 hour at 200 × g, rested (37°C, 5% CO_2), then released from alginate 4, 24 or 46 hours after exposure to pressure. Chondrocytes were stained using specific antibodies to the two receptor types and TNFR expression determined by flow cytometry. (A) Proportion of chondrocytes expressing TNFRI. (B) Number of TNFRI/cell. (C) Proportion of chondrocytes expressing TNFRII. (D) Number of TNRFII/cell. Closed symbols represent TNFR expression on test chondrocytes subjected to pressure, open symbols TNFR expression on corresponding untreated control chondrocytes measured at the same time point.

3.3. *Response of isolated NA chondrocytes to high pressure protocol (20 MPa, 1 hour)*

Chondrocytes derived from 6 NA joints were subjected to 20 MPa. Figure 2 (C and D) shows the response of chondrocytes from NA patient 1 to the effects of 20 MPa pressure. Slight variation was observed in response to load. Overall, net TNFRI expression was decreased on NA chondrocytes 24 hours after application of pressure. By comparison, TNFRII measurements were at or near the lowest detection limit of the assay whether chondrocytes were loaded or not.

3.4. *Comparison of the effects of high and low pressure regimes on isolated OA and NA chondrocyte TNFR expression*

Chondrocytes isolated from both NA and OA cartilage were also subjected to the low pressure protocol (5 MPa, 5 min). Small variation in TNFRI expression was observed on chondrocytes from both groups, 24 hours after application of pressure. In order to compare pressure-induced changes in TNFR expression on chondrocytes isolated from NA as compared with OA joints, the results were expressed as a ratio of receptor expression on test versus control chondrocytes. Table 2 shows TNFRI expression 24 hours after application of pressure. Pressure-induced differences in TNFR expression were significantly greater on

Table 1

TNF-R expression on tsTAC/62 cells 24 hours after exposure to pressure

Experiment	Test				Control			
	TNFRI (%)	TNFRI per cell	TNFRII (%)	TNFRII per cell	TNFRI (%)	TNFRI per cell	TNFRII (%)	TNFRII per cell
1	22.2	13698	16.3	2893	12.0	7171	27.0	2567
2	23.5	3200	5.0	100	15.6	2224	6.8	2228
3	37.6	5250	24.8	3177	6.8	100	10.9	547
4	29.0	4564	16.2	1350	5.8	181	1.0	100
Mean ± SEM	*28.1 ± 3.5	*6678 ± 2378	15.6 ± 4.1	1880 ± 716	10.1 ± 2.3	2419 ± 2872	11.4 ± 5.6	1361 ± 610

Results are expressed as mean ± SEM. A significant difference was observed in chondrocytes exposed to pressure in both the proportion expressing TNFRI ($p = 0.045$) and the number of TNFRI/cell ($p = 0.036$) when compared with untreated (control) cells. No significant pressure-induced difference was observed in either the proportion of chondrocytes expressing TNFRII ($p = 0.55$) or the number of TNFRII/cell ($p = 0.64$) (Student's t test).

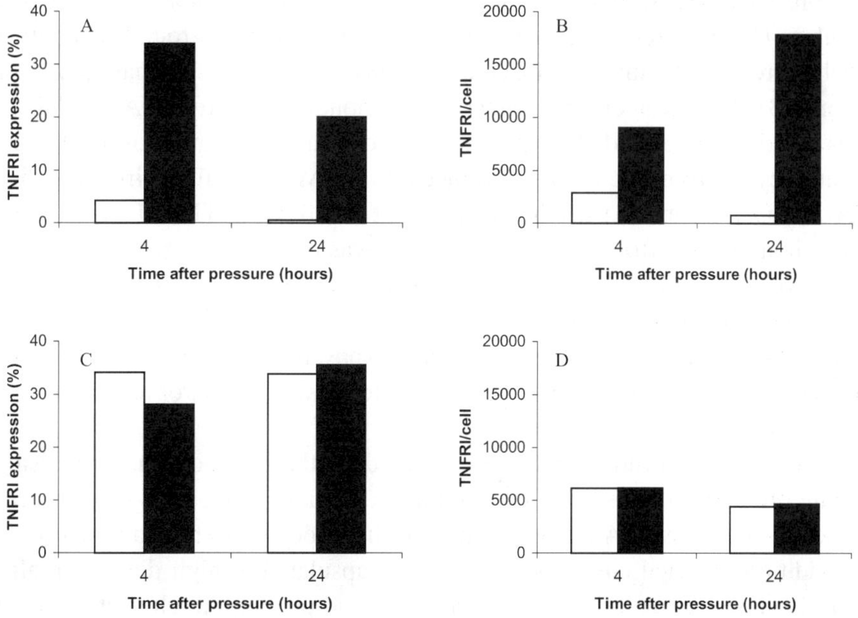

Fig. 2. The response of chondrocytes isolated from osteoarthritic (OA) or non-arthritic (NA) cartilage to pressure generated by centrifugation. Chondrocytes isolated from OA and NA cartilage by enzymatic digestion and encapsulated in alginate were subjected to pressure by centrifugation for 1 hour at 20 MPa (200 × g). After resting (37°C, 5% CO_2), chondrocytes were released from alginate 4 or 24 hours after exposure to pressure. Chondrocytes were stained using specific antibodies to the two receptor types and TNFR expression determined by flow cytometry. The proportion of chondrocytes from OA patient 1 expressing TNFRI after exposure to 20 MPa is shown in (A), and the number of TNFRI/cell in (B). The proportion of chondrocytes from NA subject 1 expressing TNFRI after exposure to 20 MPa are shown in (C), and the number of TNFRI/cell (D). Black bars represent TNFR expression on test chondrocytes subjected to pressure, white bars TNFR expression on corresponding untreated control chondrocytes measured at the same time point.

chondrocytes from OA populations exposed to 5 MPa pressure as compared with similar chondrocytes from NA subjects. However, for 20 MPa, the pressure-induced difference in TNFRI expression between groups was not statistically significant.

Table 2

TNFRI expression on isolated chondrocytes 24 hours after exposure to pressure

Pressure	NA		OA	
	TNFRI (%)	TNFRI/cell	TNFRI (%)	TNFRI/cell
5 MPa	0.68 ± 0.09 ($n = 4$)	0.49 ± 0.14 ($n = 4$)	$1.21 \pm 0.15^{*}$ ($n = 8$)	$1.73 \pm 0.52^{*}$ ($n = 8$)
20 MPa	0.86 ± 0.14 ($n = 6$)	0.74 ± 0.15 ($n = 6$)	6.72 ± 5.61 ($n = 7$)	12.88 ± 8.79 ($n = 7$)

Results are expressed as the mean±SEM of the ratio of TNF-R expression on test (pressured): control (untreated) chondrocytes, $^{*}p = 0.01$ (% TNFRI) and $p = 0.03$ (TNFRI/cell) respectively (unpaired test with Welch's correction for unequal variance).

4. Discussion

The current results show that application of pressure in the physiological range can upregulate TNFR expression on both a cell line derived from an OA joint and on articular chondrocytes taken directly from OA joints. Moreover, only TNFRI, not TNFRII, expression was significantly altered suggesting that the effects of pressure are selective, and that only the signalling receptor, TNFRI, is sensitive to such mechanical stimulus. The peak response in TNFRI expression occurred around 24 hours after exposure to pressure, which corresponds with the time course of sheer-stress induced change in chondrocyte IL-6 protein production [13]. The mean effect of pressure on chondrocytes from NA cartilage at both 5 MPa and 20 MPa was a reduction in TNFRI expression. By contrast, up-regulation of TNFRI was usually observed on chondrocytes from OA cartilage subjected to 5 MPa, resulting in a significant difference between the NA and OA groups. At the higher pressure of 20 MPa, TNFRI expression on tsT/AC62 chondrocytes and chondrocytes from 4 of 7 OA patients was markedly increased. Thus tsT/AC62 cells and chondrocytes from some OA patients respond differently to pressure than chondrocytes from NA cartilage. However, not all chondrocytes derived from the cartilage of OA patients responded to higher pressure. In chondrocytes from 3 of 7 OA preparations, exposure to 20 MPa reduced TNFRI expression suggesting differences between OA patients in chondrocyte sensitivity and/or response to pressure higher in the physiological range.

It may be argued that these results could be explained by the isolation and processing of chondrocytes or the loading protocol. Against the first possibility, previous work shows differences exist in TNFRI expression between NA and OA subjects whether chondrocytes were examined *in vitro* [24,26] or *ex vivo* [27]. In addition, isolated chondrocytes were encapsulated at high density in alginate beads, a technique previously shown to maintain the morphologic and phenotypic characteristics of primary [7] and immortalised chondrocytes [16] for long periods of time. Furthermore, chondrocytes supported in alginate were pre-incubated, centrifuged and rested in the same tubes thereby eliminating undue manipulation, so that cells were subjected only to the effects of pressure. Moreover, differences have been reported in the effects of compressive load on OA, as compared with NA, chondrocytes embedded in agarose gel [6]. Against the second possibility, pressures applied to chondrocytes were calculated to fall within the physiological range, although the time course of the experiments did not. Under laboratory conditions, it is not possible to reproduce accurately all the changes to the microenvironment that occur *in vivo* when joints are subjected to mechanical load. These experiments were therefore designed not to mimic cartilage physiology, but to determine whether one aspect of load, pressure, can alter chondrocyte TNFR expression. Nevertheless, the results show the effect of pressure on TNFR expression was more marked on tsT/AC62 chondrocytes, derived from an OA joint and chondrocytes from OA cartilage, than those from NA cartilage.

Overall, these findings suggest that conditions exist within some OA joints under which chondrocytes become more sensitive to pressure, resulting in increased TNFRI expression. Cytokines produced within the joint, such as IL-1β and IL-6, can also enhance TNFRI expression on chondrocytes [25]. Moreover, chondrocyte production of IL-6 protein is up-regulated by sheer stress [13], and expression of TNFα mRNA is increased by hydrostatic pressure [18]. Thus, it seems probable that chondrocyte production of the cytokines that influence catabolic events may also be enhanced by load. Loading may therefore induce concomittant changes in chondrocyte cytokine production as well as receptor expression and thereby potentiate focal loss of cartilage in the joints of susceptible individuals [4].

Acknowledgement

This work was supported by the Wellcome Trust.

References

[1] M.A. Adams, A.J. Kerin and M.R. Wisnom, Sustained loading increases the compressive strength of articular cartilage, *Conn. Tiss. Res.* **39** (2000), 245–256.

[2] E.C. Arner, C.E. Hughes, C.P. Deciccio, B. Caterson and M.D. Tortorella, Cytokine-induced cartilage proteoglycan degradation is mediated by aggrecanase, *Osteoarthritis and Cartilage* **6** (1998), 214–228.

[3] D.R. Carter, T.E. Orr, D.P. Fyhrie and D.J. Schurman, Influences of mechanical stress on prenatal and postnatal skeletal development, *Clin. Orthop.* **219** (1987), 237–250.

[4] D. Coggan, P. Croft, S. Kellingray, D. Barret, M. McLaren and C. Cooper, Occupational physical activities and osteoarthritis of the knee, *Arthritis Rheum.* **43** (2000), 1443–1449.

[5] P.A. Dieppe and J.R. Kirwan, The localisation of osteoarthritis, *Brit. J. Rheumatol.* **33** (1994), 201–204.

[6] S.H. Elder, J.H. Kimura, J. Soslowsky, M. Lavagino and S.A. Goldstein, Effect of compressive loading on chondrocyte differentiation in agarose cultures of chick limb-bud cells, *J. Orthop. Res.* **18** (2000), 78–86.

[7] J. Guo, G.W. Jourdian and D.K. MacCallum, Culture and growth characteristics of chondrocytes encapsulated in alginate beads, *Conn. Tiss. Res.* **19** (1989), 277–297.

[8] A.C. Hall, J.P.G. Urban and K.A. Gehl, The effects of hydrostatic pressure on matix synthesis in articular cartilage, *J. Orthop. Res.* **9** (1991), 1–10.

[9] T. Handa, H. Ishihara, R. Ohshima, H. Osada, H. Tsuji and K. Obata, Effects of hydrostatic pressure on matrix synthesis and matrix metalloproteinase production in the human lumbar intervertebral disc, *Spine* **97** (1997), 1085–1091.

[10] W.A. Hodge, R.S. Fuan, K.L. Carlson, R.G. Burgess, W.H. Harris and R.W. Mann, Contact pressures in the human hip joint measured in vivo, *Proc. Natl. Acad. Sci. USA* **83** (1986), 2879–2883.

[11] H. Inoue, K. Hiasa, Y. Samma et al., Stimulation of proteoglycan and DNA synthesis in chondrocytes by centrifugation, *J. Dent. Res.* **69** (1990), 1560–1563.

[12] L.D. Kozaci, D.J. Buttle and A.P. Hollander, Degradation of type II collagen, but not proteoglycan, correlates with matrix metalloproteinase activity in cartilage explant cultures, *Arthritis Rheum.* **40** (1997), 164–174.

[13] M. Mohtai, M.K. Gupta, B. Donlan et al., Expression of interleukin-6 in osteoarthritic chondrocytes and effects of fluid-induced shear on this expression in normal human chondrocytes in vivo, *J. Bone Joint Surg.* **14** (1993), 67–73.

[14] M.J. Palmoski, R.A. Colyer and K.D. Brandt, Joint motion in the absence of normal loading does not maintain normal articular cartilage, *Arthritis Rheum.* **23** (1979), 325–334.

[15] J.J. Parkkinen, J. Ikonen, M.J. Lammi, J. Laakonen and H.J. Helminen, Effects of cyclic hydrostatic pressure on proteoglycan synthesis in cultured chondrocytes and articular cartilage explants, *Arch. Biochem. Biophys.* **300** (1993), 458–465.

[16] J.R. Robbins, B. Thomas, L. Tan, B. Choy, J.L. Arbiser, F. Berenbaum and M.B. Goldring, Immortalised human adult articular chondrocytes maintain cartilage-specific phenotype and resonses to interleukin-1β, *Arthritis Rheum.* (in press).

[17] R. Sah, A. Grodzinsky, A. Plaas and J. Sandy, Effects of static and dynamic compression on matrix metabolism in cartilage explants, in: *Articular Cartilage and Osteoarthritis*, K. Kuettner, R. Shleyerbach, J. Peyton and V. Hascall, eds, Raven Press, New York, 1992, pp. 373–3392.

[18] K. Takahashi, T. Kubo, Y. Arai, I. Kitijama, M. Takigawa, J. Imanishi and Y. Hirasawa, Hydrostatic pressure induces expression of interleukin-6 and tumour necrosis factor alpha mRNA's in a chondrocyte-like cell line, *Ann. Rheum. Dis.* **57** (1998), 231–236.

[19] K. Takahashi, T. Kubo, K. Kobayashi, J. Imanishi, M. Takigawa, Y. Arai and Y. Hirasawa, Hydrostatic pressure influences mRNA expression of transforming growth factorβ1 and heat shock protein 70 in chondrocyte-like cell line, *J. Orthop. Res.* **15** (1997), 150–158.
[20] M.C.D. Trindade, J. Shida, T. Ikenoue, S. Yerby, S.B. Goodman, D.J. Schurman and R.L. Smith, Intermittent hydrostatic pressure inhibits interleukin-6 and monocyte chemoattractant protein-1 expression by human osteoarthritic chondrocytes in vitro, *Trans. Orthop. Res. Soc.* **25** (2000), 176.
[21] Y. Uchio, M. Ochi, T. Toyoda, J.Q. Yao and B.B. Seedhom, Different effects of static and cyclic hydrostatic pressure on proteoglycan biosynthesis of articular chondrocytes between sub-populations, *Trans. Orthop. Res. Soc.* **25** (2000), 648.
[22] J.P.G. Urban, The chondrocyte: a cell under pressure, *Brit. J. Rheumatol.* **33** (1994), 901–908.
[23] P.L.E.M. Van Lent, F.A.J. Van der Loo, L. Van den Bersselaar and W.B. Van den Berg, Chondrocytes nonresponsiveness of arthritic articular cartilage caused by short-term immobilisation, *J. Rheumatol.* **18** (1991), 193–200.
[24] G.R. Webb, C.I. Westacott and C.J. Elson, Chondrocyte tumor necrosis factor receptors and focal loss of cartilage in osteoarthritis, *Osteoarthritis and Cartilage* **5** (1997), 427–437.
[25] G.R. Webb, C.I. Westacott and C.J. Elson, Osteoarthritic synovial fluid and synovium supernatants up-regulate tumor necrosis factor receptors on human articular chondrocytes, *Osteoarthritis and Cartilage* **6** (1998), 167–176.
[26] C.I. Westacott, R.M. Atkins, P.A. Dieppe and C.J. Elson, Tumor necrosis factor-alpha receptor expression on chondrocytes isolated from human articular cartilage, *J. Rheumatol.* **21** (1994), 1710–1715.
[27] C.I. Westacott, A.F. Barakat, L. Wood, M.J. Perry, P. Neison, I. Bisbinas, L. Armstrong, A.B. Millar and C.J. Elson, Tumor necrosis factor alpha can contribute to focal loss of cartilage in osteoarthritis, *Osteoarthritis and Cartilage* **8** (2000), 213–221.

Biorheology 39 (2002) 133–143
IOS Press

The effect of mechanical stress on cartilage energy metabolism

R.B. Lee, R.J. Wilkins, S. Razaq and J.P.G. Urban *
Physiology Laboratory, Oxford University, Oxford, UK

Abstract. Cartilage is routinely subjected to varying mechanical stresses which are known to affect matrix turnover by a variety of pathways. Here we show that mechanical loads which suppress sulphate incorporation or protein synthesis by articular chondrocytes, also inhibit rates of oxygen uptake and of lactate production. Although the mechanisms have not been definitively identified, it has been shown that high hydrostatic pressures reduce the activity of the glucose transporter GLUT. Furthermore, fluid expression consequent on static loading changes intracellular pH and ionic strength; intracellular changes which would reduce the activity of glycolytic enzymes. Both pathways would thus lead to a fall in rates of glycolysis and a reduction in intracellular ATP, and – since ATP concentrations directly affect sulphation of proteoglycans – a rapid fall in sulphate incorporation. Our results suggest that load-induced changes in matrix synthesis in cartilage can occur by means other than changes in gene expression.

Keywords: Proteoglycan, ATP, lactate, oxygen, GLUT, MCT

1. Introduction

Articular chondrocytes are exposed to high mechanical loads during normal movement. It has long been known that cartilage chondrocytes respond to these mechanical stresses; cartilage in unloaded areas is softer and has different mechanical properties than cartilage from areas which are routinely exposed to high loads suggesting that mechanical stress has an important role in regulating matrix composition [51]. There has thus been a growing interest in understanding how mechanical loads influence matrix biosynthesis and turnover. Numerous *in vitro* tests have shown that matrix production by chondrocytes, either isolated or in the matrix, is very sensitive to a variety of mechanical signals [50]. Sulphate incorporation as a marker of proteoglycan biosynthesis has been extensively investigated and is upregulated under some mechanical signals (pressure, flow [9,21,31]) and inhibited under others (fluid expression or high sustained hydrostatic pressure [24,27]) but the pathways for mechanotransduction remain poorly understood. Little is at present known about the effects of load on other aspects of cellular behaviour.

Here we summarise the effect of mechanical stress on cartilage energy metabolism. Matrix production is an energy demanding process and the effect of mechanical stress on matrix turnover could be modulated in part by its effect on chondrocyte ATP turnover. In this brief review we will summarise present understanding of matrix energy metabolism, the role of glucose and lactate transport in the process, show that mechanical stresses can influence energy production and hence matrix synthesis and discuss possible mechanisms.

*Address for correspondence: Dr J. Urban, Physiology Laboratory, Oxford University, Parks Road, Oxford OX1 3PT, UK. Tel.: +44 1865 272509; Fax: +44 1865 272469; E-mail: jpgu@physiol.ox.ac.uk.

2. Chondrocyte energy metabolism

In the typical mammalian cell, carbohydrates are broken down through the Embden–Meyerhof–Parnas (E–M–P) pathway of glycolysis, in which glucose is converted to pyruvate, followed by entry of the latter as acetyl-coenzyme A into the Krebs cycle, in which the remaining carbon is liberated as CO_2. In such cells under normal aerobic conditions, the O_2-dependent mitochondrial process of oxidative phosphorylation predominates as the source of the cell's energy currency, ATP, which is formed in high overall yield per unit of carbohydrate consumed (Fig. 1).

Articular cartilage is characterised, on the other hand, by meagre O_2 consumption and minimal release of CO_2. Carbohydrate breakdown in this tissue is dominated by a near-quantitative conversion of glucose to lactate as an end-product of glycolysis [29,32,38], a reaction sequence which of course consumes no O_2. Moreover, experiments with ^{14}C-labelled glucose indicate that little pyruvate enters the Krebs cycle [33,47], and CO_2 release from articular cartilage is small [17]. Current opinion is therefore that substrate-level phosphorylations in the glycolytic pathway form the principal source of ATP in articular cartilage [23], even though the efficiency is low: a net gain of only 2 mol ATP per mol of glucose metabolised (Fig. 1). On the evidence of low specific O_2 consumption alone, oxidative phosphorylation appears to make only a minor contribution to ATP formation in this tissue [23]. Certainly when compared on a "per cell" basis to correct for its low cellularity, articular cartilage uses remarkably little O_2 compared with most other tissues: only 2–5% as much O_2 per cell as liver or kidney, for example (Table 1 in [49]). It has also been reported that the mitochondria of articular chondrocytes *in situ* lack particular cytochromes [34].

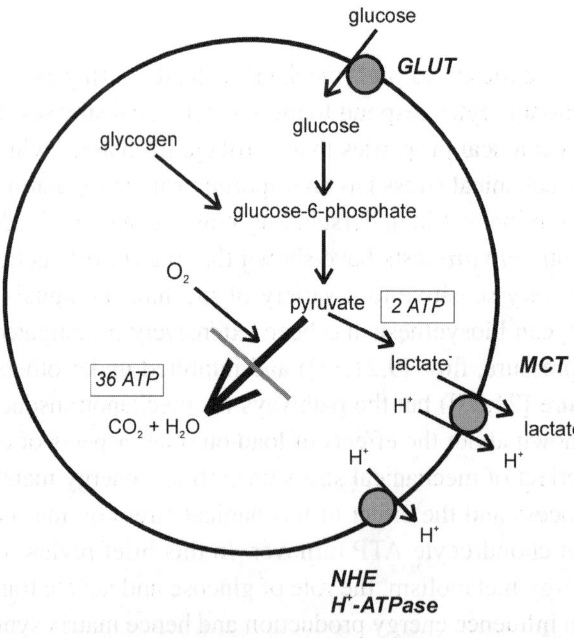

Fig. 1. A schematic view of the paths for glucose metabolism in chondrocytes. Glucose enters the chondrocyte via the transporter GLUT. It does not undergo oxidative phosphorylation even in the presence of oxygen. Each glucose molecule is broken down through the glycolytic pathway to form 2 lactic acid molecules with the production of 2 molecules of ATP. The lactic acid is almost completely ionised to the lactate ion and a proton and both are removed from the cell through the coupled MCT-proton transporter.

2.1. Oxygen concentration influences cartilage metabolism

Although articular chondrocytes contain no major O_2-consuming metabolic process which can be readily identifed, many studies have shown that the cells are sensitive to ambient O_2 concentration. Near-atmospheric concentrations of O_2 were optimum for matrix synthesis by explants of rabbit or bovine articular cartilage [5,60], and by bovine growth plate chondrocytes in culture [10]. Under hypoxia, intracellular ATP concentrations of articular cartilage slices fell as did rates of lactate production [29] (Fig. 2(a)). Bovine articular chondrocytes in culture showed evidence of lower acid (probably lactic) production and formed less matrix during 7 d under severe hypoxia than during a similar period in air, though cell survival was unaffacted [16]. Hyperoxia was as deleterious for matrix synthesis as severe hypoxia [22,60]. But responses in culture are not always clear cut and there are reports of increased glucose

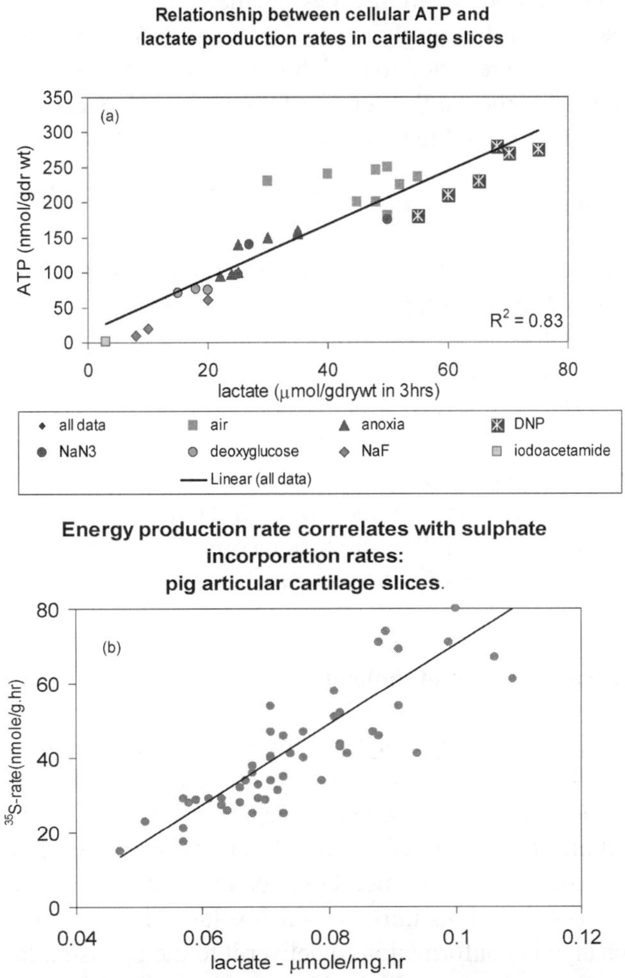

Fig. 2. The relationship between ATP, glycolysis rate and sulphate incorporation in articular chondrocytes. (a) ATP varies with rate of lactate production. Cartilage was incubated in air or under hypoxia or in the presence of a variety of glycolytic inhibitors for 3 hours and intracellular ATP and the rate of lactate production measured (adapted from [29]). (b) Sulphate incorporation varies with the rate of glycolysis. Cartilage slices were incubated for 4 hours in medium whose osmolality ranged from 280–550 mOsm. The rate of sulphate incorporation and lactate production were measured.

breakdown and lactate production by rabbit articular chondrocytes or cartilage explants after 6–7 d under low O_2 concentrations [28,32].

2.2. *Gradients of metabolites across cartilage*

As articular cartilage is avascular, nutrients such as oxygen and glucose diffuse from the synovial fluid and possibly from the blood vessels in the subchondral bone through the cartilage matrix to the cells. Lactic acid which is produced by the cells is removed from the matrix by the reverse route. Little is known about the resulting metabolite gradients across the tissue. The oxygen tension in the synovial fluid is reported to be 4–10% pO_2 [14]. Calculations based on rates of oxygen consumption by chondrocytes indicate that gradients through the depth of cartilage are steep; pO_2 has been estimated to be as low as 1% pO_2 in the deeper zones. Proton concentrations are higher in the cartilage matrix than in the synovial fluid because along with other cations, they balance the negative charge of the matrix [15]. In addition, production of lactic acid during glycolysis has a strong influence on tissue pH [11]. Lactate concentrations in synovial fluid are reported to be 5–8 mM [26] compared to 1 mM in plasma and would be expected to be considerably higher in the depths of the tissue. Thus cells in the deep zone are under low pO_2 and glucose concentrations and low pH compared to the surface cells.

2.3. *Energy metabolism and matrix production*

It has long been known that sulphate incorporation has a requirement for ATP. In cartilage, complete suppression of glycolysis was found to suppress matrix synthesis [3]. Suppression of glycolysis by low oxygen or by glycolytic inhibitors or by acid pH, leads to a fall in intracellular ATP and also a dose-dependent fall in sulphate incorporation [29], even though sulphate incorporation requires only a small fraction of intracellular ATP. Oxygen levels, which appear to affect ATP concentrations (Fig. 2(a)), have been shown to influence aggrecan aggregation and rates of synthesis and collagen production [10,37,60]; it has been suggested that they influence the pattern of glycosaminoglycan synthesis [44]. These studies indicate that factors causing a reduction in intracellular ATP concentration through suppression of glycolysis appear to cause a corresponding fall in sulphate incorporation and to a lesser extent, to the protein synthesis as measured by incorporation of radioactive amino-acids such as proline or leucine.

3. Membrane transport and energy metabolism

One of the key regulating steps in the glycolytic pathway is the transport of glucose from the extracellular space into the cell. The lipid nature of the plasma membrane which bounds the chondrocyte renders its impermeable to all but the smallest, uncharged solutes. In most cases, therefore, the exchange of nutrients such as glucose and metabolic products such as lactate between the chondrocyte and surrounding extracellular matrix requires the presence of membrane-bound protein permeation pathways. These proteins can be channels – water-filled pores through which solutes diffuse – or carriers, which bind their substrate and undergo a change in conformation to deliver it to the opposing face of the membrane [48]. Transport proteins, especially carriers, can be highly specific conferring membrane permeability on a very limited range of substrates. Moreover, for each type of carrier, there can be families of related proteins ("isoforms") which although performing the same broad function show subtle differences in their properties to suit their tissue location. Such specific protein carrier families have been identified for glucose [6], and for the product of its anaerobic metabolism, lactic acid [18] (Fig. 1).

3.1. The glucose transporter GLUT

Given the essential role of glucose in the provision of ATP for chondrocyte metabolism, there has been surprisingly little characterisation of the membrane transport pathways responsible for glucose uptake by these cells. In other vertebrate cells, two glucose transporters have been identified, which have been named GLUT and SGLT [6]. Both systems are carrier proteins, which bind glucose at one membrane face, undergo a conformational change and release the substrate at the opposing face. GLUT performs facilitated diffusion, which – as the name suggests – is a passive process enabling lipophobic glucose to move along its concentration gradient across the plasma membrane. Since its first identification in erythrocytes [36], a family of GLUT proteins have been cloned, exhibiting tissue-specific distribution and functional properties tailored to their location. Several isoforms of SGLT have also been identified although, in contrast, they are almost exclusively expressed in the brush border membranes of epithelial cells, such as those of the intestine, where they mediate the active accumulation of glucose from the lumen which the cells define [6]. This process is classed as secondary active transport, since the uptake of glucose is energised by the coupled movement of Na^+ ions along an inwardly directed electrochemical gradient, maintained by the Na^+-K^+ ATPase. The accumulated glucose subsequently effluxes the cell on GLUT at the contraluminal membrane.

Although active Na^+-driven accumulation of glucose might seem sensible given the extracellular surroundings of the chondrocyte, there have been no reports of SGLT in chondrocytes, reiterating the very limited sites of expression of this transporter. Instead, it appears that high affinity, high capacity passive uptake on GLUT is sufficient to match metabolic needs. Western immunoblotting studies using human articular chondrocytes show that in addition to GLUT1, the erythrocyte isoform found at low levels in a host of other tissues, GLUT3, an isoform previously described in neurons, placenta and testes, is also present [35]. While GLUT1 is a relatively low affinity system (exhibiting a K_m typically in the range 20 mM), GLUT3 operates at lower concentrations of glucose ($K_m < 10$ mM), ensuring that the supply of glucose is maintained to the essential tissues in which it is expressed [6]. GLUT3 mediated glucose transport would therefore be in keeping with the circumstances of the chondrocyte, where large amounts of glucose must be scavenged, yet, on account of the avascular nature of cartilage, glucose is a relatively scarce resource in the extracellular matrix. Low levels of GLUT2 (the hepatocyte isoform) and of GLUT4 (the insulin-sensitive muscle form) have also been detected in bovine articular chondrocytes (A.C. Hall, personal communication). Using radiolabelled glucose analogues, we have characterised the uptake of glucose into bovine articular chondrocytes, and shown that – in agreement with the immunoblotting studies in human cells – a saturable uptake with a high affinity (low K_m) for glucose can be measured which does not require the presence of extracellular Na^+ ions and can be inhibited by the specific GLUT inhibitor phloretin (R.J. Wilkins, unpublished observations).

3.2. Lactic acid transport

The lactic acid produced by chondrocytes must diffuse from the cells, through the extracellular matrix to the synovial fluid and thereafter to capillaries and the systemic circulation. For a predominantly anaerobic cell type, there must be an effective efflux from the cell of the large quantities of lactic acid produced if intracellular pH (pH_i) is to be maintained. Given that the pK_a of lactic acid means that it exists almost entirely as H^+ and lactate$^-$ ions at physiological pH, transport exchange between the cytoplasm and the extracellular environment is commonly mediated by a family of membrane-bound H^+-lactate$^-$ cotransport carrier proteins, called the monocarboxylate transporters, or MCT. To date, nine MCT isoforms have

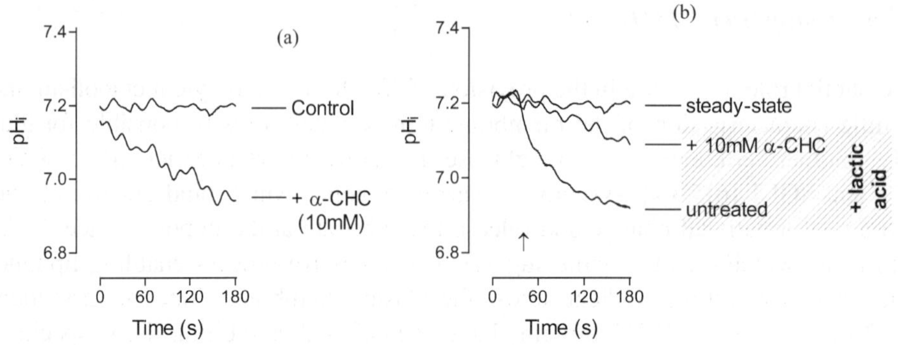

Fig. 3. Lactate is removed from chondrocytes via the H^+-lactate$^-$ cotransport carrier protein, the monocarboxylate transporter (MCT). Isolated chondrocytes were incubated with the intracellular pH-sensitive fluorescent dye BCECF-AM and their pH estimated from changes in fluorescent emission [58]. (a) Removal of lactic acid is necessary to prevent acidification of intracellular pH. Under control conditions, intracellular pH remained steady at near neutral pH values. In the presence of the specific MCT inhibitor α-cyanohydroxycinnamic acid (α-CHC), lactic acid accumulated in the cytosol and the chondrocytes steadily acidified. (b) Entry of extracellular lactate is mediated by MCT. When extracellular lactic acid was added to the external medium, chondrocytes gradually acidified as lactic acid entered the cells. This acidification was inhibited by blocking MCT with α-CHC in which case the chondrocytes maintained near neutral intracellular pH levels even in the presence of external lactic acid.

been reported, with tissue-specific distribution [18]. For other highly glycolytic tissues, such as "white" skeletal muscle fibres and leukocytes, the MCT-4 isoform is the predominant isoform. MCT-4 demonstrates a low affinity for lactate$^-$ which – coupled with a high transport capacity – makes it well-suited to perform H^+-lactate$^-$ efflux for cells producing large quantities of lactic acid.

Despite the long-recognised reliance upon anaerobic glycolysis in chondrocytes, until recently, nothing was known about the way in which the lactic acid produced in this pathway leaves the cell. We have employed a H^+-sensitive dye to characterise H^+-lactate cotransport; although ordinarily operating as an efflux pathway *in vivo*, the transport can be studied by measuring pH_i changes which result upon the extracellular addition of lactic acid to chondrocyte suspensions. Addition of increasing concentrations of lactic acid acidifies chondrocytes in a saturable fashion, which is indicative of a carrier-mediated process. Furthermore, the acidification can be inhibited by α-cyanohydroxycinnamic acid (α-CHC), a well-established inhibitor of MCT, confirming that H^+-lactate$^-$ cotransport mediated by this transporter is responsible for the acidification observed. The kinetic parameters of H^+-lactate$^-$ influx indicate that the carrier is a high-capacity system with a relatively low affinity for lactate$^-$ ($K_m = 11$ mM), properties consistent with a contribution to lactic acid transport from MCT4. Additional experiments show that chondrocytes rapidly acidify when treated with α-CHC, indicating that there is a high basal level of lactic acid extrusion from these cells under resting conditions (Fig. 3). Western immunoblotting studies of human and bovine articular chondrocytes confirm the presence of MCT4, together with small amounts of MCT-1, a finding which is consistent with the mixed kinetic properties revealed by the transport studies described above.

3.3. Removal of intracellular H^+ ions

The maintenance of near-neutral intracellular pH faces challenges on two fronts. In addition to the cellular acidosis which will accompany the production of lactic acid, there is a constant leak of H^+ ions into the cell. This background acid loading is common to all vertebrate cells where – despite a mildly alkaline extracellular environment – the inside negative membrane potential establishes an inwardly-directed electrochemical gradient for H^+ ions [52]. For cartilage cells, surrounded by an acidic extracellular matrix, this problem will, however, be especially acute. The high basal level of H^+-lactate$^-$ on MCT which is revealed by treatment with α-CHC is only one of a number of systems which could operate to hold intracellular pH steady. In chondrocytes, we have previously shown that other H^+ extrusion systems also operate to counter background acid loading [8,58]. Both systems are active transporters, moving H^+ ions against the electrochemical gradient, one energised directly by the hydrolysis of ATP (primary active transport, H^+-ATPase) the other using energy released by the movement of Na^+ ions into the cell along their electrochemical gradient ($Na^+ \times H^+$ exchange, a secondary active transporter). The allosteric activation of $Na^+ \times H^+$ by intracellular H^+ ions makes this system an especially powerful regulator of chondrocyte pH [30].

An influx of $HCO_3{}^-$ ions across the plasma membrane affords another way by which intracellular H^+ ions could be removed from the cytosol. Given that $HCO_3{}^-$ ions react with H^+ ions to generate carbonic acid, which in turn dissociates to CO_2 and H_2O, the influx of the anion is effectively the equivalent of H^+ efflux. In other cells, $HCO_3{}^-$ ions are transported across the plasma membrane by a variety of carrier proteins, including Na^+-$HCO_3{}^-$ cotransport, $Na^+[HCO_3{}^-]_2 \times Cl^-$ exchange, and $Cl^- \times HCO_3{}^-$ [8,40,56]. In articular chondrocytes, however, the role of these systems is, to varying degrees, attenuated. Such an observation is in accord with reports that carbonic anhydrase activity in these cells is low [12] (R.J. Wilkins, unpublished observations) such that H^+ buffering by $HCO_3{}^-$ will be restricted, the low rates of CO_2 production by articular chondrocytes, and the exclusion of free anions from the extracellular matrix.

4. Mechanical stress can affect cartilage energy metabolism

Several studies have shown that static loading of cartilage inhibits matrix synthesis in a dose-dependent manner [24] and that this fall in synthesis rates appears to be due in part at least, from increases in extracellular osmolality or proton concentrations consequent on fluid expression from the matrix [15,39,42,43,54]. Long-term application of hydrostatic pressures greater than 20 MPa also suppress matrix synthesis [21,27]. We find that these conditions also lead to a fall in the rate of glycolysis (Fig. 4). Chondrocytes or cartilage slices subjected to high hydrostatic pressures (20–50 MPa) or an increase in osmolality or exposure to acid pH for 2–4 hours, produce less lactate and consume less oxygen than chondrocytes held at physiological osmolalities (350–450 mOsm) or pH (pH 6.9–7.1) (Fig. 4). Since ATP levels appear dependent on glycolysis rate and sulphate incorporation appears linked to glycolysis and hence ATP concentrations (Fig. 2), the rapid fall in sulphate incorporation seen under some loading regimes may arise from the conseqeuent loss of ATP.

4.1. How does mechanical stress lead to a fall in glycolytic rate?

The mechanisms relating mechanical stress to a fall in glycolysis and ATP are unknown at present but possible mechanisms do suggest themselves.

Under static loading, fluid is expressed from cartilage, increasing extracellular H^+, Na^+ and osmolality [15,42,53] and hence altering intracellular ionic composition and pH [55,57]. The activity of enzymes involved in glycolysis particularly phosphofructokinase are very sensitive to cytolosic pH and ionic composition [1,4,46]. Changes in intracellular ions consequent on cell shrinkage as a result of static loading could thus lead to a fall in activity of glycolytic enzymes and the rate of glycolysis and hence lead to a reduction in the concentration of intracellular ATP.

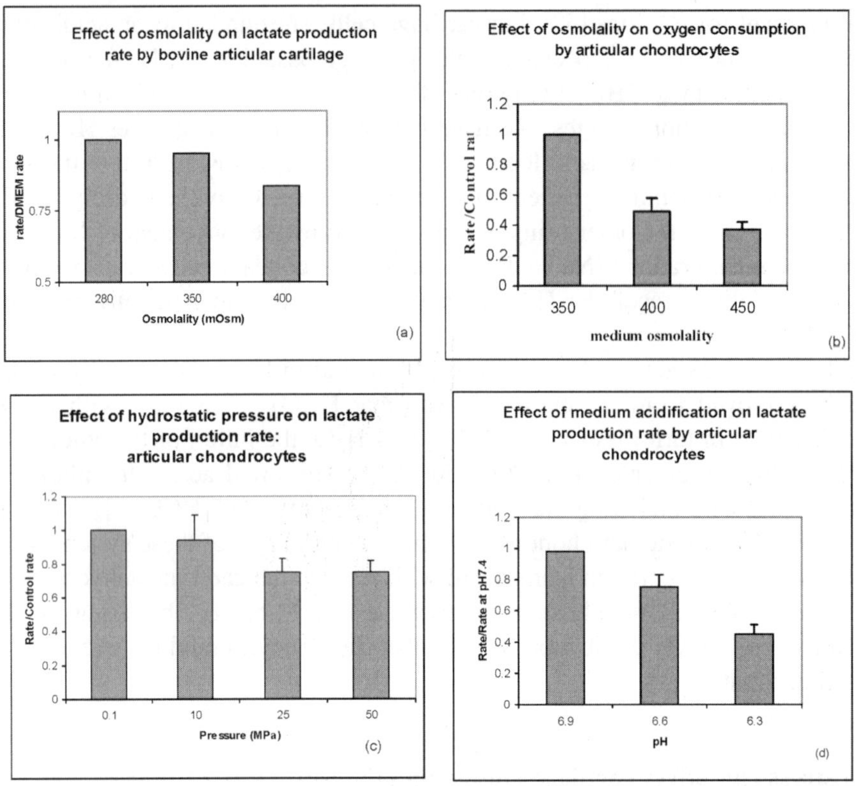

Fig. 4. The effect of physical stresses on lactate production and oxygen consumption. Isolated articular chondrocytes were incubated at 37°C under a range of medium osmolalities or levels of pH for 4 hours [54] and rates of lactate production or oxygen consumption measured. Alternatively isolated chondrocytes were exposed to high hydrostatic pressures for 20 sec at 37°C, then incubated at 0.1 MPa for 2 hours [21], and rates of lactate production measured. (a) Effect of osmolality on lactate production. (b) Effect of osmolality on oxygen consumption. (c) Effect of hydrostatic pressure on lactate production. (d) Effect of medium pH on lactate production.

Hydrostatic pressure is now known to modify a variety of membrane transport processes, including the Na^+-K^+ ATPase, K^+-Cl^- cotransport, $Na^+ \times H^+$ exchange, $Cl^- \times HCO_3^-$ exchange (the erythrocyte "chloride shift"), and nucleoside and amino acid carriers [2,20]. However, of particular interest in the present context is the observation that hydrostatic pressure can inhibit glucose transport in erythrocytes, an effect which is attributed to it being harder under pressure for GLUT to attain the increase in volume necessary for the rate-limiting conformational change to occur. While these studies in erythrocytes primarily have served to validate a kinetic model of glucose translocation, for chondrocytes the effects of hydrostatic pressure on glucose transport may have significant ramifications. We have therefore begun to characterise the effects of hydrostatic pressure on glucose transport in chondrocytes using an optical technique adapted from the light scattering method used for erythrocytes by [45]. Cells are loaded with glucose by suspension in a solution containing 100 mM of the sugar, and the rapid efflux of glucose on GLUT upon resuspension in glucose-free solution is visualised as a volume change. Our studies indicate that high hydrostatic pressures (100–400 ATA) retard the rate at which chondrocytes shrink under these conditions, confirming that pressure similarly inhibits glucose transport in these cells. Since glucose transport into the cell is the first regulatory step for glycolysis (Fig. 1), its inhibition could be one factor causing the fall in glycolysis rate seen under such pressures.

Thus although the mechanisms have not been investigated directly, it appears that both static loading and high hydrostatic pressures could inhibit glycolysis, albeit through different mechanisms.

5. Conclusions

It is now clear that there is no single mechanotransduction process linking mechanical load to chondrocyte biosynthesis. Even simple signals such as hydrostatic pressure can affect many different processes; pressure applied to chondrocytes has been reported to activate the sodium–hydrogen exchanger, NHE, suppress the sodium pump and depolymerise cytoskeletal components [7,19,25] depending on pressure level and duration of signal. These changes can have down-stream effects on cell behaviour, activating second messengers and altering transcriptional expression profiles. Here we suggest that some loading conditions cause a fall in glycolysis rate and hence intracellular ATP concentrations and that the fall in ATP is causally linked to the rapid inhibition of sulphate incorporation seen under these loading conditions. Thus although it is clear that mechanical stress can affect gene expression [13,41,59], other pathways may also be involved in the changes in biosynthesis arising from mechanical loading.

Acknowledgement

This work was supported by the Arthritis Research Campaign (U0507, U0508, W0604).

References

[1] R.P. Aaronson and C. Frieden, Rabbit muscle phosphofructokinase: studies on the polymerization. The behavior of the enzyme at pH 8, pH 6, and intermediate pH values, *J. Biol. Chem.* **247** (1972), 7502–7509.

[2] I. Afzal, J.J. Browning, J.C. Ellory, R.J. Naftalin and R.J. Wilkins, Effects of high hydrostatic presssure on net glucose flux in human red blood cells, *J. Physiol.* **531** (2001), 130p.

[3] M.S. Baker, J. Feigan and D.A. Lowther, The mechanism of chondrocyte hydrogen peroxide damage. Depletion of intracellular ATP due to suppression of glycolysis caused by oxidation of glyceraldehyde-3-phosphate dehydrogenase, *J. Rheumatol.* **16** (1989), 7–14.

[4] P.E. Bock and C. Frieden, Phosphofructokinase. I. Mechanism of the pH-dependent inactivation and reactivation of the rabbit muscle enzyme, *J. Biol. Chem.* **251** (1976), 5630–5636.

[5] C.T. Brighton, J.M. Lane and J.K. Koh, In vitro rabbit articular cartilage organ model. II. 35S incorporation in various oxygen tensions, *Arthritis Rheum.* **17** (1974), 245–252.

[6] G.K. Brown, Glucose transporters: structure, function and consequences of deficiency, *J. Inherit. Metab. Dis.* **23** (2000), 237–246.

[7] J.A. Browning, R.E. Walker, A.C. Hall and R.J. Wilkins, Modulation of $Na^+ \times H^+$ exchange by hydrostatic pressure in isolated bovine articular chondrocytes, *Acta Physiol. Scand.* **166** (1999), 39–45.

[8] J.A. Browning and R.J. Wilkins, The characterisation of mechanisms regulating intracellular pH in a transformed human articular chondrocyte cell line, *J. Physiol.* **513** (1998), 54p.

[9] M.D. Buschmann, Y.J. Kim, M. Wong, E. Frank, E.B. Hunziker and A.J. Grodzinsky, Stimulation of aggrecan synthesis in cartilage explants by cyclic loading is localized to regions of high interstitial fluid flow, *Arch. Biochem. Biophys.* **366** (1999), 1–7.

[10] C.C. Clark, B.S. Tolin and C.T. Brighton, The effect of oxygen tension on proteoglycan synthesis and aggregation in mammalian growth plate chondrocytes, *J. Orthop. Res.* **9** (1991), 477–484.

[11] B. Diamant, J. Karlsson and A. Nachemson, Correlation between lactate levels and pH in discs of patients with lumbar rhizopathies, *Experientia* **24** (1968), 1195–1196.

[12] S. Ekman and Y. Ridderstrale, Carbonic anhydrase localization in normal and osteochondrotic joint cartilage of growing pigs, *Vet. Pathol.* **29** (1992), 308–315.

[13] M.A. Elo, R.K. Sironen, K. Kaarniranta, S. Auriola, H.J. Helminen and M.J. Lammi, Differential regulation of stress proteins by high hydrostatic pressure, heat shock, and unbalanced calcium homeostasis in chondrocytic cells, *J. Cell. Biochem.* **79** (2000), 610–619.

[14] W.R. Ferrell and H. Najafipour, Changes in synovial pO_2 and blood flow in the rabbit knee joint due to stimulation of the posterior articular nerve, *J. Physiol.* **449** (1992), 607–617.

[15] M. Gray, A. Pizzanelli, A. Grodzinsky and R. Lee, Mechanical and physiochemical determinants of the chondrocyte biosynthetic response, *J. Orthop. Res.* **6** (1988), 777–792.

[16] M.J. Grimshaw and R.M. Mason, Bovine articular chondrocyte function in vitro depends upon oxygen tension, *Osteoarthritis Cartilage* **8** (2000), 386–392.

[17] C. Guri and D. Bernstein, Rat epiphyseal cartilage; V. Glucose-C14 metabolism as related to growth and to various anatomical areas, in vitro, *Proc. Soc. Exp. Biol. Med.* **124** (1967), 386–391.

[18] A.P. Halestrap and N.T. Price, The proton-linked monocarboxylate transporter (MCT) family: structure, function and regulation, *Biochem. J.* **343** (1999), 281–299.

[19] A.C. Hall, Differential effects of hydrostatic pressure on cation transport pathways of isolated articular chondrocytes, *J. Cell. Physiol.* **178** (1999), 197–204.

[20] A.C. Hall, D.M. Pickles and A.G. Macdonald, Aspects of eurkaryotic cells, *Adv. Comp. Environ. Physiol.* **17** (1993), 30–85.

[21] A.C. Hall, J.P.G. Urban and K.A. Gehl, The effects of hydrostatic pressure on matrix synthesis in articular cartilage, *J. Orthop. Res.* **9** (1991), 1–10.

[22] R.K. Jacoby and M.I. Jayson, Organ culture of adult human articular cartilage. I. The effect of hyperoxia on synthesis of glycosaminoglycan, *J. Rheumatol.* **2** (1975), 270–279.

[23] K. Johnson, A. Jung, A. Murphy, A. Andreyev, J. Dykens and R. Terkeltaub, Mitochondrial oxidative phosphorylation is a downstream regulator of nitric oxide effects on chondrocyte matrix synthesis and mineralization, *Arthritis Rheum.* **43** (2000), 1560–1570.

[24] I.L. Jones, A. Klamfeldt and T. Sandstrom, The effect of continuous mechanical pressure upon the turnover of articular cartilage proteoglycans in vitro, *Clin. Orthop.* **165** (1982), 283–289.

[25] M.O. Jortikka, J.J. Parkkinen, R.I. Inkinen, J. Karner, H.T. Jarvelainen, L.O. Nelimarkka et al., The role of microtubules in the regulation of proteoglycan synthesis in chondrocytes under hydrostatic pressure, *Arch. Biochem. Biophys.* **374** (2000), 172–180.

[26] P. Kortekangas, O. Peltola, A. Toivanen and H.T. Aro, Synovial fluid L-lactic acid in acute arthritis of the adult knee joint, *Scand. J. Rheumatol.* **24** (1995), 98–101.

[27] M.J. Lammi, R. Inkinen, J.J. Parkkinen, T. Hakkinen, M. Jortikka, L.O. Nelimarkka et al., Expression of reduced amounts of structurally altered aggrecan in articular-cartilage chondrocytes exposed to high hydrostatic-pressure, *Biochem. J.* **304** (1994), 723–730.

[28] J. Lane, C. Brighton and B. Menkowitz, Anaerobic and aerobic metabolism in articular cartilage, *J. Rheumatol.* **4**(4) (1977), 334–342.

[29] R.B. Lee and J.P. Urban, Evidence for a negative Pasteur effect in articular cartilage, *Biochem. J.* **321** (1997), 95–102.

[30] S.A. Levine, M.H. Montrose, C.M. Tse and M. Donowitz, Kinetics and regulation of three cloned mammalian Na^+/H^- exchangers stably expressed in a fibroblast cell line, *J. Biol. Chem.* **268** (1993), 25 527–25 535.

[31] L. Lipiello, C. Kaye, T. Neumata and H.J. Mankin, In vitro metabolic response of articular cartilage segments to low levels of hydrostatic pressure, *Conn. Tiss. Res.* **13** (1985), 99–107.

[32] R. Marcus, The effect of low oxygen concentration on growth, glycolysis, and sulfate incorporation by articular chondrocytes in monolayer culture, *Arthritis and Rheumatism* **16** (1973), 646–656.

[33] R.M. Mason, C.A. Spencer and N.T. Palmer, Chondrocyte energy metabolism and its modulation by iodoacetaete and diclofenac. in: *Diclofenac (Volataren) and Cartilage in Osteoarthritis*, R. Moskowitz and K. Hirohata, eds, 1989, pp. 33–41.

[34] F. Mignotte, A.M. Champagne, B. Froger-Gaillard, L. Benel, M. Gueride, M. Adolphe et al., Mitochondrial biogenesis in rabbit articular chondrocytes transferred to culture, *Biol. Cell.* **71** (1991), 67–72.

[35] A. Mobasheri, G. Neama, S. Bell and D. Carter, Molecular evidence for two glucose transporters (GLUT1 and GLUT3) in adult human articular cartilage, *Int. J. Exp. Path.* (2001) (in press).

[36] M. Mueckler, C. Caruso, S.A. Baldwin, M. Panico, I. Blench, H.R. Morris et al., Sequence and structure of a human glucose transporter, *Science* **229** (1985), 941–945.

[37] S.W. O'Driscoll, J.S. Fitzsimmons and C.N. Commisso, Role of oxygen tension during cartilage formation by periosteum, *J. Orthop. Res.* **15** (1997), 682–687.

[38] P. Otte, Basic cell metabolism of articular cartilage. Manometric studies, *Z. Rheumatol.* **50** (1991), 304–312.

[39] M. Palmoski and K. Brandt, Effects of static and cyclic compressive loading on articular cartilage plugs in vitro, *Arth. Rheum.* **27** (1984), 675–681.

[40] M. Puceat, pHi regulatory ion transporters: an update on structure, regulation and cell function, *Cell. Mol. Life Sci.* **55** (1999), 1216–1229.

[41] P.M. Ragan, A.M. Badger, M. Cook, V.I. Chin, M. Gowen, A.J. Grodzinsky et al., Down-regulation of chondrocyte aggrecan and type-II collagen gene expression correlates with increases in static compression magnitude and duration, *J. Orthop. Res.* **17** (1999), 836–842.

[42] R. Sah, A. Grodzinsky, A. Plaas and J. Sandy, Effects of static and dynamic compression on matrix metabolism in cartilage explants. in: *Articular Cartilage and Osteoarthritis*, K. Kuettner, R. Shleyerbach, J. Peyron and V. Hascall, eds, Raven Press, New York, 1992, pp. 373–392.

[43] R. Schneiderman, D. Keret and A. Maroudas, Effects of mechanical and osmotic pressure on the rate of glycosaminoglycan synthesis in the human adult femoral head cartilage: an in vitro study, *J. Orthop. Res.* **4** (1986), 393–408.

[44] J.E. Scott, R.A. Stockwell, C. Balduini and G. De Luca, Keratan sulphate: a functional substitute for chondroitin sulphate in O_2 deficient tissues?, *Pathol. Biol. (Paris)* **37** (1989), 742–745.

[45] H.K. Sen and W.F. Widdas, Determination of the temperature and pH dependence of glucose transfer across the human erythrocyte membrane measured by glucose exit, *J. Physiol.* **160** (1962), 392–403.

[46] K. Shearwin, C. Nanhua and C. Masters, Interactions between glycolytic enzymes and cytoskeletal structure – the influence of ionic strength and molecular crowding, *Biochem. Int.* **21** (1990), 53–60.

[47] C.A. Spencer, T.N. Palmer and R.M. Mason, Intermediary metabolism in the swarm rat chondrosarcoma chondrocyte, *Biochem. J.* **265** (1990), 911–914.

[48] W.D. Stein, Channels, *Carriers and Pumps*, Academic Press, 1990.

[49] R.A. Stockwell, in: *Cartilage. Molecular Aspects*, B.K. Hall, ed., Academic Press, New York, 1983, pp. 253–280.

[50] J.-F. Stoltz, ed., *Mechanobiology: Cartilage and Chondrocyte*, IOS Press, Amsterdam, 2001.

[51] M. Tammi, K. Paukkonen, I. Kiviranta, J. Jurvelin, A.-M. Saamenen and H.J. Helminen, Joint loading induced alterations in articular cartilage, in: *Joint Loading. Biology and Health of Articular Cartilage*, H.J. Helminen, I. Kiviranta, A.-M. Saamanen, M. Tammi, K. Paukkonen and J. Jurvelin, eds, Wright, Bristol, 1987, pp. 64–88.

[52] R.C. Thomas, Intracellular pH, in: *Acid-Base Balance*, R. Hainsworth, ed., Manchester University Press, Manchester, 1986, pp. 50–74.

[53] J.P.G. Urban and M.T. Bayliss, Regulation of proteoglycan synthesis rate in cartilage in vitro: influence of extracellular ionic composition, *Biochem. Biophys. Acta* **992** (1989), 59–65.

[54] J.P.G. Urban, A.C. Hall and K.A. Gehl, Regulation of matrix synthesis rates by the ionic and osmotic environment of articular chondrocytes, *J. Cell. Physiol.* **154** (1993), 262–270.

[55] J.P.G. Urban and R.J. Wilkins, Extracellular ions and hydrostatic pressure: their influence on chondrocyte intracellular ionic composition, in: *Advances in Osteoarthritis*, S. Tanaka and C. Hamanishi, eds, Springer-Verlag, Tokyo, 1998, pp. 3–20.

[56] R. Wilkins and A.C. Hall, Bovine articular chondrocytes demonstrate only minimal bicarbonate-dependent recovery from changes to intracellular pH, *J. Physiol.* **459** (1993), 289p.

[57] R. Wilkins, A.C. Hall and J.P.G. Urban, The correlation between changes in intracellular pH and changes in matrix synthesis rates in chondrocytes, *Trans. Am. Orthop. Res. Soc.* **17** (1992), 182–182.

[58] R.J. Wilkins and A.C. Hall, Measurement of intracellular pH in isolated bovine articular chondrocytes, *Exp. Physiol.* **77** (1992), 521–524.

[59] M. Wong, M. Siegrist and X. Cao, Cyclic compression of articular cartilage explants is associated with progressive consolidation and altered expression pattern of extracellular matrix proteins, *Matrix Biol.* **18** (1999), 391–399.

[60] G.E. Ysart and R.M. Mason, Responses of articular cartilage explant cultures to different oxygen tensions, *Biochim. Biophys. Acta* **1221** (1994), 15–20.

Biorheology 39 (2002) 145–152
IOS Press

Human chondrocyte senescence and osteoarthritis

James A. Martin and Joseph A. Buckwalter *
University of Iowa Department of Orthopaedic Surgery, Iowa City, IA, USA

Abstract. Although osteoarthritis (OA) is not an inevitable consequence of aging, a strong association exists between age and increasing incidence of OA. We hypothesized that this association is due to *in vivo* articular cartilage chondrocyte senescence which causes an age-related decline in the ability of the cells to maintain articular cartilage, that is, increasing age increases the risk of OA because chondrocytes lose their ability to replace their extracellular matrix. To test this hypothesis, we measured senescence markers in human articular cartilage chondrocytes from 27 donors ranging in age from one to 87 years. The markers included expression of the senescence-associated enzyme β-galactosidase, mitotic activity measured by ^{3}H-thymidine incorporation, and telomere length. β-galactosidase expression increased with age ($r = 0.84$, $p = 0.0001$) while mitotic activity and mean telomere length declined ($r = -0.774$, $p = 0.001$ and $r = -0.71$, $p = 0.0004$, respectively). Decreasing telomere length was strongly correlated with increasing expression of β-galactosidase and decreasing mitotic activity. These findings help explain the previously reported age related declines in chondrocyte synthetic activity and responsiveness to anabolic growth factors and indicate that *in vivo* articular cartilage chondrocyte senescence is responsible, at least in part, for the age related increased incidence of OA. The data also imply that people vary in their risk of developing OA because of differences in onset of chondrocyte senescence; and, the success of chondrocyte transplantation procedures performed to restore damaged articular surfaces in older patients could be limited by the inability of older chondrocytes to form new cartilage. New efforts to prevent the development or progression of OA might include strategies that delay the onset of chondrocyte senescence or replace senescent cells.

1. Introduction

Life long maintenance of articular cartilage depends on the ability of chondrocytes to continuously replace degraded matrix macromolecules. In most young people, synthesis of extracellular matrix macromolecules prevents the progressive loss of articular cartilage associated with the clinical syndrome of osteoarthritis (OA), however, the incidence and prevalence of synovial joint degeneration increase with increasing age, suggesting that age-related cartilage changes render the cells less capable of maintaining the tissue. This phenomenon has been attributed to the harmful effects of mechanical and chemical stresses which are thought to impair maintenance activity by killing chondrocytes outright or by inducing apoptosis [21,26,38]. While such environmental factors probably contribute to degeneration of articular cartilage in some individuals, they do not explain the seemingly irreversible age-dependent decline in chondrocyte growth factor responsiveness and matrix synthesis found by a number of investigators [1,6,13,15,16,30,33,35].

A recently formulated hypothesis suggests that cell aging is regulated by an intrinsic genetic "clock" associated with changes in telomeres (DNA sequences at the ends of chromosomes that are necessary for chromosomal replication [4]. Every cell division advances this "clock" until the cells can no longer replicate. Chromosomes from young, normal somatic cells have relatively long telomeres of >9 kilobase

*Address for correspondence: Joseph Buckwalter, Department of Orthopaedic Surgery, 01008 JPP, University of Iowa, 200 Hawkins Drive, Iowa City, IA 52242, USA. Tel.: 319 356 2595; Fax: 319 356 8999; E-mail: joseph-buckwalter@uiowa.edu.

pairs (kbp) but these are eroded at the rate of 100–200 base pairs (bp) with each cell division cycle [2,3]. Erosion beyond the minimum length necessary for DNA replication (5–7.6 kbp) results in cell cycle arrest, a condition referred to as replicative senescence [2–4,19].

Most cell types reach cell cycle arrest after a characteristic number of population doublings. This fundamental barrier to unbridled growth, termed the Hayflick limit, is common to somatic cells that lack telomerase, an enzyme responsible for replacing telomere sequences [4,9,17,24]. The Hayflick limit for human fibroblasts has been estimated at $\sim$60 population doublings [24] while the estimated limit for human chondrocytes is $\sim$35 doublings [5]. Telomerase activity is not always sufficient in and of itself to immortalize cells, however, germ cell lines and most cancer cell lines in which the "telomerase" enzyme is active are virtually immortal [5,23,25]. In telomerase-negative cells, telomere length can be viewed as a cumulative history of preceding cell division as well as a predictor of future capacity to divide [17,24].

Cell function may begin to deteriorate before cells reach cell cycle arrest. Declining protein synthesis, altered growth factor and cytokine responses, and longer population doubling times are senescence-associated phenotypic changes that begin to appear in continuously grown somatic cell cultures long before the cells reach the Hayflick limits [8,11,14,22,27,31,32,37]. This suggests that cell populations begin to drift toward senescence relatively early in their replicative lifespans, before telomeres have eroded to critical lengths.

The role of chondrocyte turnover in articular cartilage aging and disease has not been systematically studied due in part to the difficulty of assessing the *in vivo* replicative history of chondrocytes. Terminal restriction fragment length analysis of telomeres offers a simple means to overcome this problem as cell-turnover should be detectable as an age-related decline in average telomere length. If telomere erosion causes senescence, telomere length should correlate with phenotypic measures of senescence. Based on these rationales we hypothesized that telomere length in human articular cartilage chondrocytes declines as a function of donor age as phenotypic measures of senescence (declining DNA synthesis and increasing expression of senescence-associated β-galactosidase activity) increase.

2. Materials and methods

Human articular cartilage samples were harvested, chopped and digested overnight in Dulbecco's modified Eagle medium (DMEM) containing 10% fetal calf serum (FCS) (Gibco-BRL) and 0.5 mg/ml pronase E (Sigma), and 0.5 mg/ml collagenase type 1A (Sigma). Cells were then plated in monolayer culture and incubated for 1–5 days in DMEM/10%FCS without enzymes.

Genomic DNA was isolated from $\sim 1 \times 10^6$ to 5×10^6 cells using a DNEasy kit (Qiagen) according to the manufacture's directions. The DNA concentration of each sample was determined by UV spectrophotometry and 2 μg was digested to completion with 10 units each of Rsa I and Hinf I (New England Biolabs) in a 60 μl reaction. The reactions were electrophoresed on 0.5% SeaKem Gold agarose (FMC Bioproducts) in parallel with digoxigenin-labeled λ Hind III size standards (Roche). The gels were transferred by capillary action to Hybond-N+ (Amersham) nylon membranes in 20$\times$ standard sodium citrate (SSC) and baked for 2 hours at 80°C. Non-radioactive methods were used to detect telomere sequences according to Genius system directions published by the manufacturer (Roche). In brief, the membranes were pre-hybridized for 4–16 hours at 37°C in hybridization buffer (50% formamide, 5$\times$ SSC, 0.1% sodium lauryl sulfate, 0.02% sodium dodecyl sulfate [SDS], 2% block). A synthetic oligonucleotide complimentary to human telomeric repeat sequences [$(CCCTAA)_3$] and labeled at the 3′ end with digoxigenin (Sigma, Genosys) was diluted to 50 pM in hybridization buffer and the membrane probed for 16–24 hours at 37°C. Excess probe was removed by washing the membranes twice in 2× SSC with 0.1% SDS at ambient temperature (2 × 15 minutes), then in 0.5× SSC with 0.1 SDS at 37°C (2 × 15 minutes). A goat anti-digoxigenin alkaline phosphatase-conjugated antibody and a chemiluminescent substrate, CDP-Star (Roche), were used to detect the digoxigenin-labeled probe. Autoradiograms of the blots (15–120 minute exposures) were digitized using a flat bed scanner (Hewlett Packard ScanJet II CX). Optical density scans of each lane were performed using Scion Image (Scion Corporation) on a personal computer. The positions of the λ Hind III standard bands were plotted (log molecular weight versus relative migration distance) and the data fitted using linear regression analysis (Microcal Origin). Mean telomere terminal restriction fragment lengths (MTL) were derived as described [17]. In brief, the standard regression line was used to calculate the molecular weight at each pixel row from the origin and to demarcate the region corresponding to 3–17 kbp. MTL was then calculated for each lane as $\sum \mathrm{OD}_i / \sum \mathrm{OD}_i / Li$ where OD is the optical density at position i and L is the length in kilobase pairs at position i. All DNA samples were digested and analyzed on at least two gels.

Senescence-associated β-galactosidase activity assays were performed essentially as described [12]. In brief, chondrocytes were transfered to 4-well chamber slides (65,000 cells per well) and incubated overnight. The cell layers were washed twice using phosphate-buffered saline (PBS), then fixed for 2 minutes in 2% paraformaldehyde. After 3 PBS washes the cell layers were overlaid with assay solution (2.0 mg/ml X-Gal in 40 mM citric acid–sodium phosphate pH 6.0, 5 mM K ferricyanide, 150 mM NaCl, 2 mM $MgCl_2$) and were incubated in a sealed chamber at 37°C without CO_2 supplementation for -6-10 hours. The reactions were stopped by removal of the substrate and repeated washing in cold PBS. The slides were mounted and viewed on a Olympus BX60 microscope fitted with DIC optics. At least 4 images taken were recorded on color slide film using a 20× objective (25–50 cells/field) and the percentage of positively stained cells in the field was scored by an observer who was unaware of sample donor age.

^{3}H-thymidine incorporation was measured by establishing 3 replicate cultures in 24-well plates at a concentration of 130,000 cells/well. The cells were incubated overnight in DMEM/10% FCS before adding fresh medium containing 5 μCi/ml ^{3}H-thymidine (Amersham). After 24 hours the medium was removed and the wells were washed 3 times for 5 minutes in PBS at 4°C before trysinization with 0.25% trypsin, EDTA in Hanks balanced salt solution (Gibco-BRL). Cells in the trypsinized suspension were counted using a hemacytometer then pelleted by centrifugation at 200g. Cell pellets were extracted by boiling for 2 minutes in 7.7 M urea with 1% SDS. An aliquot of the extract was added to scintillation cocktail and counted on a Beckman LSII liquid scintillation counter. The total counts in the extracts were normalized to the cell number.

3. Results

Measurements of mean telomere length (MTL), β-galactosidase expression and mitotic activity showed strong evidence of increasing chondrocyte senescence with increasing age. Absolute MTL values varied from a maximum of 11,759 base pairs (13 year-old) to a minimum of 8,731 base pairs (87 year-old). These values fell in the range calculated for human fibroblasts and smooth muscle cells [2,3]. Mean telomere length (MTL) progressively declined with increasing age (Fig. 1) and regression analysis showed a significant linear correlation between MTL and age ($r = -0.71$, $p = 0.0004$). The frequency of cells expressing senescence-associated β-galactosidase (SA β-gal) increased with age (Fig. 2). The minimum value was 4.5% positive (1 year-old) and the maximum value was 55% (77 year-old). The

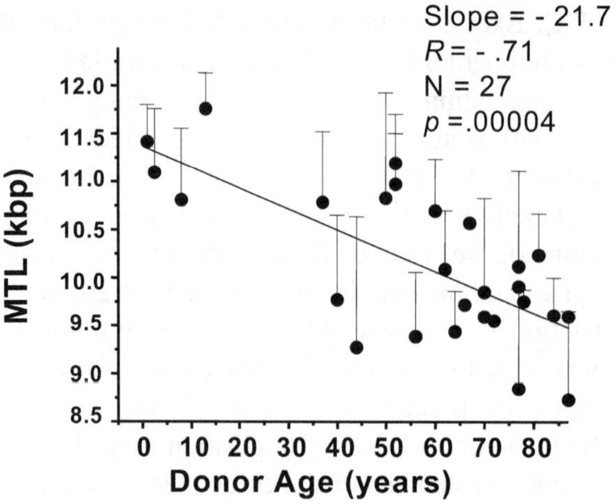

Fig. 1. Mean terminal restriction fragment lengths. Mean telomere lengths (MTL) for 27 chondrocyte strains are plotted as a function of donor age. Each data point shows the mean of at least 2 determinations. For clarity, error bars indicate only positive standard deviations (means from 3 or more determinations). The results of regression analysis are shown (inset) and the line of best fit is drawn.

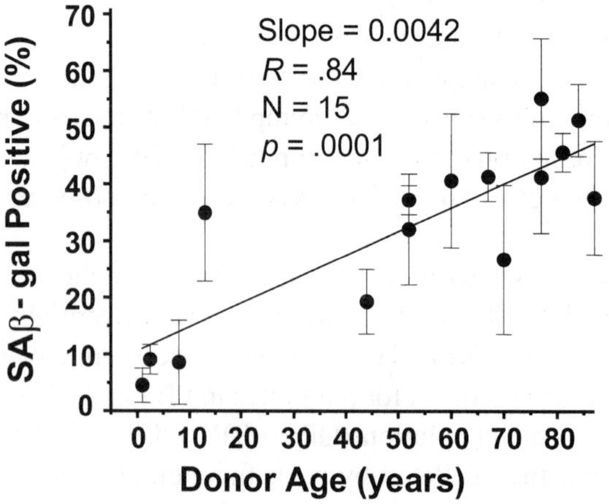

Fig. 2. Senescence-associated β-galactosidase (SAβ-Gal) scores. SAβ-Gal activity (SAβ-Gal % positive) is plotted as a function of donor age. Each point represents means and standard deviations (error bars) from 4 microscope fields. Linear regression parameters are shown (inset) and the line of best fit is drawn.

percentage of positive staining cells increased at an average rate of 4% per decade. Regression analysis of these data revealed a significant linear relationship between the β-galactosidase stain and donor age ($r = 0.80$, $p = 0.0001$). ^{3}H-thymidine incorporation assays used to measure DNA synthesis showed that incorporation values ranged from 2.93 CPM/cell (1 year-old donor) to 0.62 CPM/cell (87 year-old donor). Regression analysis of the plot showed a significant negative linear relationship between the age of the donor and DNA synthesis activity (Fig. 3) ($r = -0.77$, $p = 0.001$).

Telomere length was closely correlated with the two measures of phenotypic senscence. MTL data were plotted as a function of frequency of cells expressing senescence-associated β-galactosidase (Fig. 4A) and as a function of ^{3}H-thymidine incorporation (Fig. 4B) to determine if MTL correlated with these senescence markers. The plot for β-galactosidase expression revealed a negative linear relationship ($r = -0.62$) that was statistically significant ($p = 0.01$). These findings indicated that the

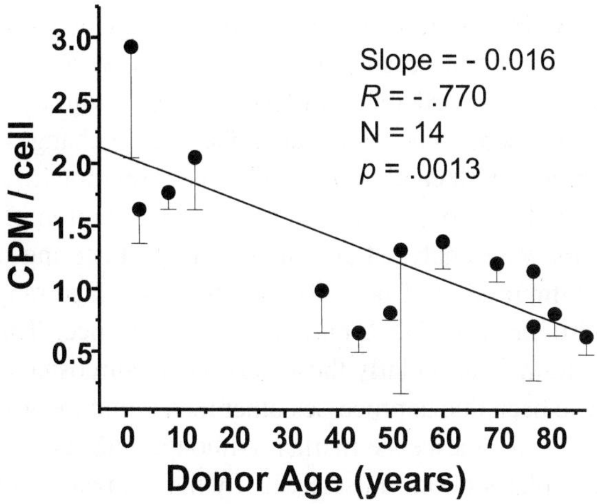

Fig. 3. ^{3}H-thymidine incorporation as a function of donor age. The graph shows ^{3}H-thymidine incorporation data plotted against donor age. The data (CPM/cell) are the means and standard deviations (error bars) from triplicate cultures. The best-fit line from linear regression analysis is shown together with regression parameters (inset).

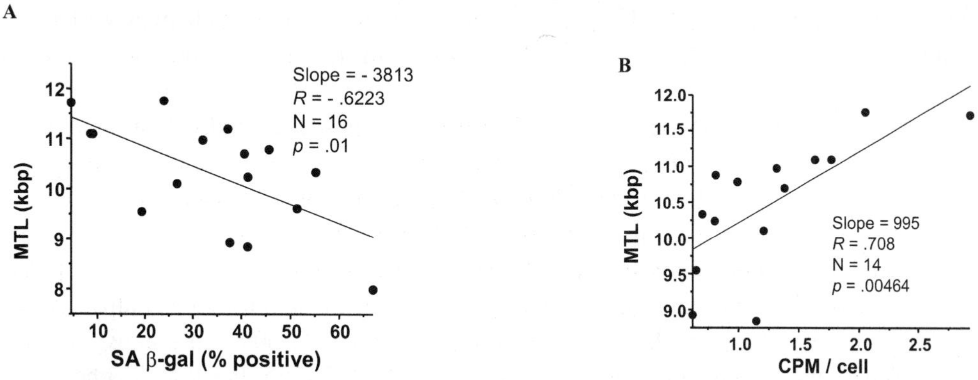

Fig. 4. Mean telomere length versus senescence markers. (A) Linear regression parameters (inset) for Mean Terminal restriction fragment Length (MTL) versus SAβ-Gal % activity. (B) Linear regression parameters (inset) for MTL versus ^{3}H-thymidine incorporation.

proportion of senescent chondrocytes increases as MTL declines. Similar results were found when MTL was plotted against ^{3}H-thymidine incorporation: regression analysis showed a significant positive, linear relationship ($r = 0.77$, $p = 0.001$) between the two variables indicating that incorporation declines with decreasing MTL.

4. Discussion

We hypothesized that the risk of OA increases with age because senescence decreases the ability of chondrocytes to maintain articular cartilage. Our results show that human articular cartilage chondrocytes become senescent with increasing age. In particular, we found that telomere length decreased with the chronological age of articular chondrocyte donors and that telomere changes were linked to two changes in phenotype associated with cell senescence, expression of SAβGal, a senescence marker enzyme, and mitotic activity, as measured by ^{3}H-thymidine incorporation. This evidence falls far short of proving that cell senescence is responsible for the increasing incidence of OA with age, but it strongly implies that the ability of chondrocytes to maintain articular cartilage decreases with age.

Although telomere erosion and senescence associated phenotypic changes were correlated with donor age, these data must be interpreted with caution. First, the linear relations we found between senescence markers and donor age may be due in part to the uneven age distribution of the donors. Relatively few young and middle aged donors were analyzed and the resulting clustering of points at very young and old ages could lead to a false impression of linearity over the entire age range. Second, other processes such as oxidative stress, and damage to DNA may induce senescence. Thus, some of the senescence we observed in chondrocyte strains, particularly those harvested from osteoarthritic donors following inflammatory episodes, may have been due to processes other than telomere erosion. Third, β-galactosidase activity at pH 6.0 is a controversial senescence marker. Although our results with articular chondrocytes are similar to findings for vascular smooth muscle cells, which appear to exhibit the same strong correlation between telomere length, age and activity, investigators studying human fibroblasts have been unable to observe a relationship between telomere length and donor age [7,10,34,36]. This suggests that senescent cells accumulate in different tissues at different rates. Moreover, although β-galactosidase activity is strongly associated with replicative senescence, it is also present in quiescent cells which may be common in some cartilage samples [36]. Lastly, the apparent age-related changes we observed might have been due to other ongoing disease processes in osteoarthritic donors which were clustered in the old age range.

The relationships between articular cartilage chondrocyte senescence and OA are undoubtedly complex. Potentially relevant disease processes include chondrocyte "cloning", a classic histologic feature of OA that refers to isolated clusters of chondrocytes formed by clonal expansion of a single cell. Cells within such clusters show accelerated mitotic activity [20] suggesting a cause for rapid telomere erosion. Thus the rapid decline in mitotic activity and ECM synthesis typical of end stage OA chondrocytes may reflect replicative senescence brought on by cloning, a hypothesis which might explain why OA samples typically showed shorter telomeres than non-osteoarthritic samples. Although this hypothesis indicates that replicative senescence is a result rather that a cause of OA, it also implies that the phenomenon plays an important role in the progression from early to end-stage articular cartilage degeneration.

The relevance of telomeres to cartilage aging and disease rests on proof that *in vivo* chondrocyte turnover rates are sufficient to cause telomere erosion. Short term DNA labeling studies indicate that chondrocyte mitoses are relatively rare in normal cartilage [18,20,28]. Although this apparent rate of turnover is too slow to result in significant telomere erosion over the short term, decades of turnover might well be sufficient. Furthermore, mitotic activity increases several-fold following cartilage injury, therefore repetitive joint injury could accelerate telomere erosion [29]. Increased mitotic activity during cartilage degeneration may also speed up the accumulation of senescent, growth-arrested chondrocytes in end-stage OA [20]. These observations suggest that, in many cases, *in vivo* chondrocyte turnover is sufficient to result in biologically significant telomere erosion.

The results summarized here demonstrate that chondrocyte telomeres erode *in vivo* in parallel with phenotypic changes associated with cell senescence. How these processes contribute to degenerative joint disease is not yet clear, however, our data strongly suggest that cell senescence contributes to either the development or progression of OA. Telomere length varies considerably among normal individuals as does the risk of developing OA. The findings of this study suggest that some of the variability in the risk of developing OA is due to genetically determined differences in onset of chondrocyte senescence.

Furthermore, *in vivo* chondrocyte senescence may adversely affect the results of chondrocyte transplantation procedures performed to restore damaged articular surfaces in older patients. Future efforts to prevent and treat OA might include strategies that delay the onset of chondrocyte senescence or replace senescent cells.

Acknowledgements

The authors thank Aaron Schroeder and Stacy Smith for technical assistance. This work was funded by the Veterans Administration (Merit Review) and the Department of Orthopaedic Surgery at the University of Iowa

References

[1] M. Adolphe, X. Ronot, P. Jaffray, C. Hecquet and J.L. Fontagne, Effects of donor's age on growth kinetics of rabbit articular chondrocytes in culture, *Mech. Ageing Dev.* **23** (1983), 191–198.
[2] R.C. Allsopp, H. Vaziri, M.A. Piatyszek, S. Goldstein, E.V. Younglai, A.B. Futcher, C.W. Greider and C.B. Harley, Telomere length predicts replicative capacity of human fibroblasts, *Proc. Natl. Acad. Sci. USA* **89** (1992), 10 114–10 118.
[3] R.C. Alsopp, E. Chang, M. Kashefi-Aazam, E.I. Rogaec, M.A. Piatyszek, J.W. Shay and C.B. Harley, Telomere shortening is associated with cell division *in vitro* and *in vivo*, *Exp. Cell. Res.* **220** (1995), 194–200.
[4] E.H. Blackburn, Structure and function of telomeres, *Nature* **350** (1991), 569–572.
[5] A.G. Bodnar, M. Ouellette, M. Frolkis, S.E. Holt, C.-P. Chiu, G.B. Morin, C.B. Harley, J.W. Shay, S. Lichtsteiner and W.E. Wright, Extension of life-span by introduction of telomerase into normal human cells, *Science* **279** (1998), 349–352.
[6] J.A. Buckwalter, S.L. Woo, V.M. Goldberg, E.C. Hadley, F. Booth, T.R. Oegema and D.R. Eyre, Soft-tissue aging and musculoskeletal function, *J. Bone Joint Surg. Am.* **75** (1993), 1533–1548.
[7] J. Campisi, Cancer, aging and cellular senescence, *In Vivo* **14** (2000), 183–188.
[8] E. Chang and C.B. Harley, Telomere length and replicative aging in human vascular tissues, *Proc. Natl. Acad. Sci. USA* **92** (1995), 11 190–11 194.
[9] C.M. Counter, A.A. Avilion, C.E. LeFeuvre, N.G. Stewart, C.W. Greider, C.B. Harley and S. Bacchetti, Telomere shortening associated with chromosome instability is arrested in immortal cells which express telomerase activity, *EMBO J.* **11** (1992), 1921–1929.
[10] V.J. Cristafalo, R.G. Allen, R.J. Pignolo, B.G. Martin and J.C. Beck, Relationship between donor age and the replicative lifespan of human cells in culture: A reevaluation, *Proc. Natl. Acad. Sci. USA* **95** (1998), 10 614–10 619.
[11] S. Decary, C.B. Hamida, V. Mouly, J.P. Barbet, F. Hentati, G.S. Butler-Browne, Shorter telomeres in dystrophic muscle consistent with extensive regeneration in young children, *Neuromuscul. Disord.* **10** (2000), 113–120.
[12] G.P. Dimri, X. Lee, G. Basile, M. Acosta, G. Scott, C. Roskelley, E.E. Medrano, M. Linskens, I. Rubelj, O. Pereira-Smith, M. Peacocke and J. Campesi, A biomarker that identifies senescent human cells in culture and in aging skin *in vivo*, *Proc. Natl. Acad. Sci. USA* **92** (1995), 9363–9367.
[13] J. Dominice, C. Levasseur, S. Larno, X. Ronot and M. Adolphe, Age-related changes in rabbit articular chondrocytes, *Mech. Aging Dev.* **37** (1987), 231–240.
[14] R.B. Efros, Loss of CD28 expression on T lymphocytes: a marker of replicative senescence, *Dev. Comp. Immunol.* **21** (1997), 471–478.
[15] P.A. Guerne, F. Blanco, A. Kaelin, A. Desgeorges and M. Lotz, Growth factor responsiveness of human articular chondrocytes in aging and development, *Arthritis Rheum.* **38** (1995), 960–968.
[16] T. Hardingham and M. Bayliss, Proteoglycans of articular cartilage: changes in aging and in joint disease, *Semin. Arthritis Rheum.* **20S1** (1990), 12–33.
[17] C.B. Harley, A.B. Futcher and C.W. Greider, Telomeres shorten during aging of human fibroblasts, *Nature* **345** (1990), 458–460.
[18] T. Havdrup and H. Telhag, Mitosis of chondrocytes in normal adult joint cartilage, *Clin. Orthop.* **153** (1980), 248–252.
[19] L. Hayflick and P.S. Moorehead, The serial cultivation of human diploid cell strains, *Exp. Cell. Res.* **25** (1961), 585–621.
[20] H. Hirotani and T. Ito, Chondrocyte mitosis in the articular cartilage of femoral heads with various diseases, *Acta Orthop. Scand.* **46** (1975), 979–986.
[21] W.E. Horton, L. Feng and C. Adams, Chondrocyte apoptosis in development, aging, and disease, *Matrix Biol.* **17** (1998), 107–115.

[22] M. Kassem, L. Ankersen, E.F. Ericksen, B.F.C. Clark and S.I.S. Ratten, Demonstration of aging and senescence in serially passaged long-term cultures of human trabecular osteoblasts, *Osteoporosis Int.* **7** (1997), 514–524.
[23] T. Kiyono, S.A. Foster, J.I. Koop, J.K. McDougall, D.A. Galloway and A.J. Klingelhutz, Both Rb/p16ink4a inactivation and telomerase activity are required to immortalize human epethelial cells, *Nature* **396** (1998), 84–88.
[24] A.J. Klingelhutz, Telomerase activation and cancer, *J. Mol. Med.* **75** (1997), 45–49.
[25] E. Kolettas, L. Buluwela, M.T. Bayliss and H.I. Muir, Expression of cartilage-specific molecules is retained in long-term cultures of human articular chondrocytes, *J. Cell Sci.* **108** (1995), 1991–1999.
[26] M. Lotz, S. Hashimoto and K. Kuhn, Mechanisms of chondrocyte apoptosis, *Osteoart. Cart.* **7** (1999), 389–391.
[27] A. Maciero-Coelho, Markers of cell senescence, *Mech. Aging Dev.* **103** (1998),105–109.
[28] H.J. Mankin, Mitosis in articular cartilage of immature rabbits. A histologic stathmokinetic (colchicine) and autoradiographic study, *Clin. Orthop.* **34** (1964), 170–183.
[29] H.J. Mankin, The response of articular cartilage to mechanical injury, *J. Bone Joint Surg. Am.* **64** (1982), 460–446.
[30] J.A. Martin, S.M. Ellerbroek and J.A. Buckwalter, Age-related decline in chondrocyte response to insulin-like growth factor-I: the role of growth factor binding proteins, *J. Orthop. Res.* **15** (1997), 491–498.
[31] A. Melk, V. Ramassar, L.M. Helms, R. Moore, D. Rayner, K. Solez and P.F. Halloran, Telomere shortening in kidneys with age, *J. Am. Soc. Nephrol.* **11** (2000), 444–453.
[32] R.R. Reddel, A reassessment of the telomere hypothesis of senescence, *Bioessays* **20** (1998), 977–984.
[33] F. Rosen, G. McCabe, J. Quach, J. Solan, R. Terkeltaub, J.E. Seegmiller and M. Lotz, Differential effects of aging on human chondrocyte responses to transforming growth factor beta: increased pyrophosphate production and decreased cell proliferation, *Arthritis Rheum.* **40** (1997), 1275–1281.
[34] I. Rubelj and Z. Vondracek, Stochastic mechanism of cellular aging-abrupt telomere shortening as a model for stochastic nature of cellular aging, *J. Theoret. Biol.* **197** (1999), 425–438.
[35] R.L. Sah, A.C. Chen, A.J. Grodzinsky and S.B. Trippel, Differential effects of bFGF and IGF-I on matrix metabolism in calf and adult bovine cartilage explants, *Arch. Biochem. Biophys.* **308** (1994), 137–147.
[36] J. Severino, R.G. Allen, S. Balin, A. Balin and V.J. Cristofalo, Is beta-galactosidase staining a marker of senescence in vitro and in vivo?, *Exp. Cell. Res.* **257** (2000), 162–171.
[37] C.M. Shanahan and P.L. Weissberg, Smooth muscle cell phenotypes in atherosclerotic lesions, *Curr. Opin. Lipidol.* **10** (1999), 507–513.
[38] S.R. Tew, A.P. Kwan, A. Hann, B.W. Thomson and C.W. Archer, The reactions of articular cartilage to experimental wounding: role of apoptosis, *Arthritis Rheum.* **43** (2000), 215–225.

Biorheology 39 (2002) 153–160
IOS Press

The functional expression of connexin 43 in articular chondrocytes is increased by interleukin 1β: Evidence for a Ca^{2+}-dependent mechanism

Rossana Tonon and Paola D'Andrea
Dipartimento di Biochimica, Biofisica e Chimica delle Macromolecole, Università di Trieste, via Licio Giorgieri 1, I-34127 Trieste, Italy
Tel.: +39 40 676 3690; Fax: +39 40 676 3691; E-mail: dandrea@bbcm.univ.trieste.it

Abstract. Cell-to-cell interactions and gap junctions-dependent communication are crucially involved in chondrogenic differentiation, while in adult articular cartilage direct intercellular communication occurs mainly among chondrocytes facing the outer cartilage layer. Chondrocytes extracted from adult articular cartilage and grown in primary culture express connexin 43 and form functional gap junctions capable of sustaining the propagation of intercellular Ca^{2+} waves. Degradation of articular cartilage is a characteristic feature of arthritic diseases and is associated to increased levels of interleukin-1 (IL-1) in the synovial fluid. We have examined the effects of IL-1 on gap junctional communication in cultured rabbit articular chondrocytes. Incubation with IL-1 potentiated the transmission of intercellular Ca^{2+} waves and the intercellular transfer of Lucifer yellow. The stimulatory effect was accompanied by a dose-dependent increase in the expression of connexin 43 and by an enhanced connexin 43 immunostaining at sites of cell-to-cell contact. IL-1 stimulation induced a dose-dependent increase of cytosolic Ca^{2+} and activates protein tyrosine phosphorylation. IL-1-dependent up-regulation of connexin 43 could be prevented by intracellular Ca^{2+} chelation, but not by inhibitors of protein tyrosine kinases, suggesting a crucial role of cytosolic Ca^{2+} in regulating the expression of connexin 43. IL-1 is one of the most potent cytokines that promotes cartilage catabolism: its modulation of intercellular communication represents a novel mechanism by which proinflammatory mediators regulate the activity of cartilage cells.

1. Introduction

Intercellular communication confers tissues the ability to co-ordinate many different cellular functions, such as the regulation of cell volume, the intracellular ionic composition and cell metabolism [3]. Cell-to-cell communication through gap junctions represents the pathway for direct intercellular exchange of small cytosolic constituents that diffuse through the intercellular channels constituted by the multimeric assembly of connexins, a family of closely related proteins [3].

Gap junction-mediated intercellular communication is critically involved in the development of cartilage during differentiation [6]; in adult articular cartilage, on the other hand, chondrocytes exist as individual cells embedded in the extracellular matrix, and gap junctions are mainly expressed by the flattened chondrocytes facing the outer cartilage layer [24]. Chondrocytes extracted from adult articular cartilage, however, retain the ability to form functional gap junctions in culture, as demonstrated by the expression of connexin 43 (Cx43), the 43 kD gap junction protein [4,8], and by the ability to respond to extracellular chemical or mechanical stimuli with co-ordinated patterns of intercellular Ca^{2+} oscillations and waves [7,12].

Degradation of articular cartilage is a characteristic feature of arthritic diseases and results from increased levels of matrix metalloproteinases such as collagenase and stromelysin [18]. Interleukin-1 (IL-1) is one of the most potent cytokines that promotes cartilage catabolism; its levels are elevated in arthritic joints and induce in chondrocytes the production of metalloproteinases [26], suppresses type II collagen and proteoglycan production [11], inhibits chondrocyte proliferation [2] and induces proinflammatory mediators such as prostaglandins and nitric oxide [18]. Interestingly, chondrocytes from the outer cartilage layer show a greater vulnerability to the harmful effects of IL-1 than cells in the deeper layers of the tissue [13].

Among the articular pathologies linked to increased IL-1 levels, rheumatoid arthritis induce an unscheduled remodelling of cartilage and the degradation of extracellular matrix [1]. As a part of synovial tissue reaction, proliferating synovial cells penetrate the superficial cartilage layer in the form of a pannus, causing the destruction of cartilage [5]. Synovial pannus formation is initiated with the recognition and adhesion of synovial cells to chondrocytes and to cartilage matrix [15,22,23,27]. Studies *in vitro* demonstrated that IL-1 stimulates dose-dependently the attachment of monocytes and synovial cells to cartilage through a mechanism involving an increased and altered expression of integrins and cadherins [16].

There is growing evidence that intercellular adhesion precedes, and may even be a prerequisite for, the establishment of gap junctions [17,20]. Moreover, evidence has been reported that inflammatory diseases in kidneys and lungs are associated with an increased expression of Cx43, resulting in a higher level of cell-to-cell coupling [9,14].

In this study we show that stimulation of rabbit articular chondrocytes in primary culture with IL-1β induces a dose-dependent up-regulation of Cx43. Increased Cx43 levels are associated with a higher level of intercellular dye coupling and with a potentiation of mechanically-induced intercellular Ca^{2+} waves, indicating that IL-1 enhances intercellular communication and signalling in articular cartilage. IL-1 induces dose-dependent increases of the cytosolic Ca^{2+} concentration and protein tyrosine phosphorylation. Intracellular Ca^{2+} chelation totally prevents IL-1-dependent Cx43 up-regulation, while tyrosine kinases inhibitors did not affect the level of IL-1β-induced Cx43 expression: the action of the cytokine on intracellular communication appears therefore mediated cytosolic Ca^{2+} increases.

2. Results

Chondrocytes extracted from rabbit articular cartilage and grown in primary cultures for 6–7 days form confluent layers of 40–60 cells; complete confluence is normally attained 1–2 days later. The intracellular Ca^{2+} concentration of chondrocytes under resting conditions was 92 ± 11 nM ($n = 368$), as measured by digital video imaging in cells loaded with fura-2.

We showed in previous studies [4,12] that mechanical stimulation of a single cell, obtained by briefly distorting the plasmamembrane with a fire-polished glass micropipette, induced a rapid increase of the cytosolic Ca^{2+} concentration to 1957 ± 270 nM ($n = 368$) followed by a subsequent decline to resting levels. Besides inducing a Ca^{2+} rise in the stimulated cell, mechanical stimulation gave rise to intercellular Ca^{2+} waves which propagated radially from the point of stimulation and involved 10 ± 2 chondrocytes ($n = 57$). We recently demonstrated that Ca^{2+} wave propagation depends on the intercellular diffusion of $InsP_3$ through gap junctions, followed by the intracellular Ca^{2+} release in neighbouring cells [12]. Given the pleiotropic effects induced by IL-1 on cartilage cells and its role in promoting cellular interactions at the chondro-synovial junction, we tested possibility that the cytokine could modulate also

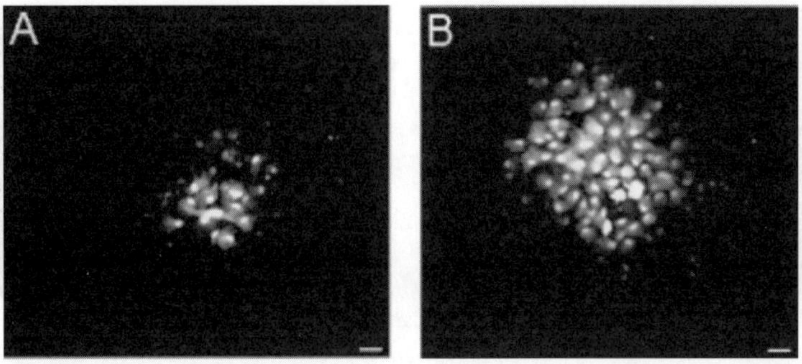

Fig. 1. IL-1β increases the extent of dye coupling. Intercellular dye transfer was measured in confluent chondrocytes cultures by micronjection of Lucifer yellow. Fluorescence micrographs were taken 8 min after microinjection of Lucifer yellow into one cell. Injections were performed before (A) or after incubation with 10^{-8} M IL-1β for 24 h (B). Bar: 40 μm.

the level of cell-to-cell coupling among articular chondrocytes. Incubation of cells for 24 h with 10^{-8} M IL-1β resulted in an increased propagation of the mechanically-induced intercellular Ca^{2+} wave which diffused to higher distance and involved 24 ± 3 cells ($n = 22$).

The extent of intercellular coupling was assessed in confluent cell layers by microinjecting individual cells with Lucifer yellow and measuring the intercellular dye transfer (Fig. 1). In control cultures the intercellular diffusion of Lucifer yellow was observed in 14 ± 1.5 cells ($n = 7$) (Fig. 1A), while in cultures treated with 10^{-8} M IL-1β (24 h) the number of fluorescence-labelled cells was 37 ± 3 ($n = 18$) (Fig. 1B), indicating that the cytokine increased the degree of cell-to-cell coupling.

Among the most widely connexins expressed, articular chondrocytes express elevated levels of Cx43, but not Cx32 or Cx26 [4,8]. Indirect immunofluorescence using a commercially available anti Cx43 antibody revealed a punctate distribution of Cx43 immunoreactivity particularly abundant at the interface of adjoining cells (Fig. 2). With respect to control cultures (Fig. 2A), cell treated for 24 h with increased concentrations of IL-1β (5×10^{-11} M, Fig. 2B; 5×10^{-10} M, Fig. 2C and 10^{-8} M, Fig. 2D) show a dose-dependent increase of Cx43 immunoreactivity. Higher cytokine doses (up to 10^{-7} M) failed to further increase the level of staining (not shown).

Immunoblotting experiments demonstrated the expression of four bands on chondrocyte cell lysates (Fig. 3) corresponding to the unphosphorylated (NP) and to differently phosphorylated isoforms of Cx43 (P1, P2, P3) [21]. Cell stimulation with increasing concentrations of IL-1β (5×10^{-11} M, lane 2; 5×10^{-10} M, lane 3, 5×10^{-9} M, lane 4 and 10^{-8} M, lane 5) for 6 h (Fig. 3, upper panel) or 24 h (Fig. 3, lower panel) induced a dose-dependent up-regulation of all Cx43 bands with respect to controls (lane 1). At both incubation times, the maximal stimulation of Cx43 expression was obtained with IL-1β 10^{-8} M, while higher cytokine doses (up to 10^{-7} M) did not stimulate further Cx43 increase (not shown). The concentration of IL-1β 10^{-8} M was therefore employed in all subsequent experiments.

In time–course experiments, an increase of Cx43 expression was detected after 6 h stimulation ($166 \pm 15\%$, $n = 3$), although a slight increase could sometimes be observed after 30 min (see below). Stimulation for 24 h maximally increased the level of Cx43 ($180 \pm 13\%$, $n = 3$) while prolonged stimulation times (up to 48 h) failed to induce a further up-regulation (not shown).

In a next series of experiments we investigated the possible mechanism responsible for the increased expression of Cx43 induced by IL-1β. In articular chondrocytes IL-1β has been shown to induce dose-dependent increases of the cytosolic Ca^{2+} concentration [19]. Although the transduction mechanism has been not fully elucidated, a necessary role of focal adhesions has been suggested.

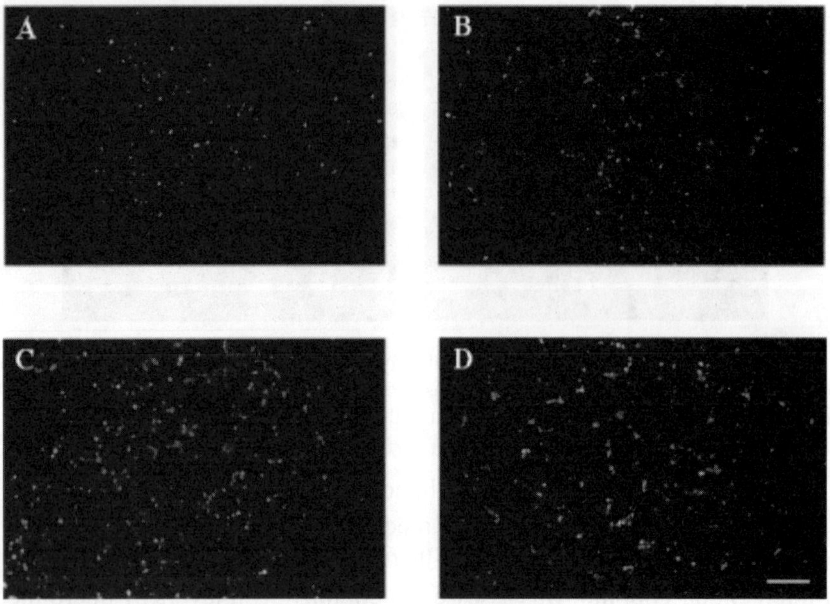

Fig. 2. IL-1β increases Cx43 immunoreactivity. Confluent chondrocytes cultures were stimulated, fixed, permeabilized and incubated with a polyclonal antibody to Cx43. A FITC-coniugated secondary antibody was used for fluorescence visualisation. Under control conditions (A) chondrocytes exhibit a punctate staining pattern that represents gap junctions composed of Cx43 located primarily in the plasma membrane between adjacent cells. Stimulation of the cells for 24 h with IL-1β 5×10^{-11} M (B), 5×10^{-10} M (C) and 10^{-8} M (D) increased dose-dependently the abundance of the punctuate staining. Bar: 40 μm.

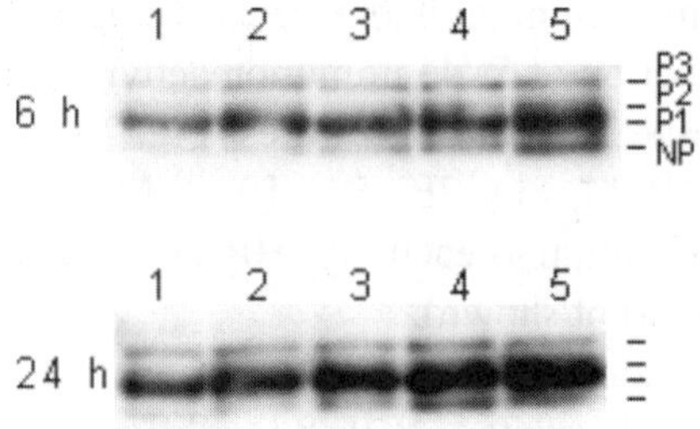

Fig. 3. IL-1β induces dose-dependent increases of Cx43 protein levels. Cells grown to confluence were incubated with increasing concentrations of IL-1β for 6 h (A) or 24 h (B). The protein extracts (10–15 μg for each sample) were separated by SDS polyacrylamide gel electrophoresis, transferred to nitro-cellulose and analysed using a polyclonal antibody to rat Cx43. Control (lane 1), IL-1β 5×10^{-11} M (lane 2), 5×10^{-10} M (lane 3), 5×10^{-9} (lane 4) and 10^{-8} M (lane 5). Multiple bands corresponding to phosphorylated (P1, P2, and P3) and unphosphorylated (NP) forms of Cx43 are present.

Stimulation of confluent cell monolayers with IL-1β induced dose-dependent increases of the cytosolic Ca^{2+} concentration (Fig. 4). At lower cytokine doses (5×10^{-11} M, Fig. 5A, 5×10^{-10} M, Fig. 4B) the increase was slow and long lasting (reaching 129 ± 18 nM, $n = 12$ and 133 ± 22 nM, $n = 9$, respectively), while at the maximal doses employed (10^{-8} M, Fig. 4C) the cell response appeared biphasic and composed by an initial spike (reaching 250 ± 30 nM, $n = 15$) followed by a sustained plateau (120 ± 5 nM, $n = 15$). The cytokine-induce Ca^{2+} response could be prevented by preincubating the cells (1 h) with the membrane-permeant Ca^{2+} chelator BAPTA-AM (25 μM, Fig. 4D).

Another important signal transduction pathway stimulated in chondrocytes by IL-1β is the activation of tyrosine kinases [10]. We, therefore, tested the ability of IL-1β to induce protein tyrosine phos-

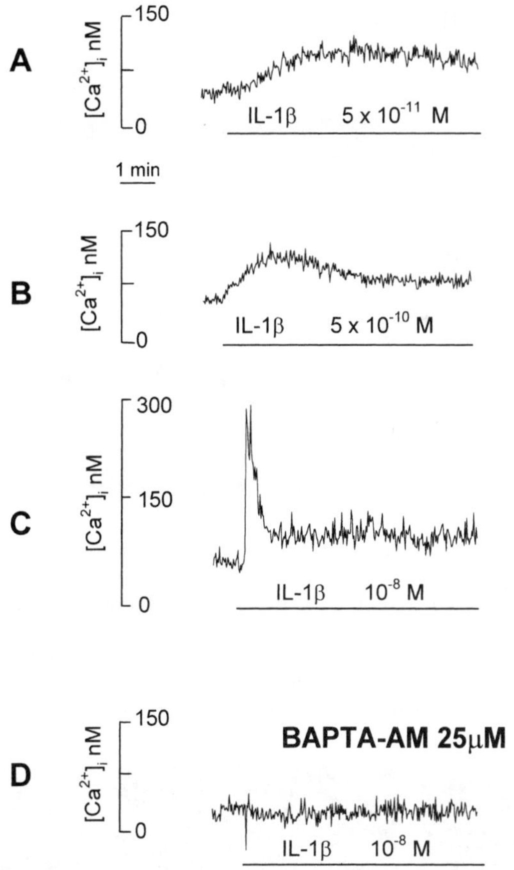

Fig. 4. IL-1β induces dose-dependent increases of the cytosolic Ca^{2+} concentration. Changes in intracellular Ca^{2+} were monitored by calcium imaging in cells loaded with fura-2. For each series of experiments is reported a representative tracing. (A–C) Cells were stimulated with indicated dosages of IL-1β. (D) Cells were preincubated with BAPTA-AM 25 μM for 1 h prior to stimulation with IL-1β 10^{-8} M.

phorylation in our chondrocyte cultures. Time-dependent activation of protein tyrosine phosphorylation was analysed in Western blots of whole cell lysates probed with an anti-phosphotyrosine antibody (not shown). IL-1β 10^{-8} M induced the tyrosine phosphorylation of several proteins migrating at 64, 53, 35 and 33 kDa. Tyrosine phosphorylation of these proteins was evident after 5 min stimulation and was maintained in cells stimulated for 24 h.

In order to verify whether the effects of IL-1β on Cx43 expression and protein tyrosine phosphorylation could derive from the increased Ca^{2+} levels induced by the cytokine, we assayed the expression of Cx43 in cells stimulated in the presence or in the absence of BAPTA-AM 25 μM, the concentration effective in preventing the rise of cytosolic Ca^{2+} induced by maximal IL-1β doses (10^{-8} M). Cell pre-treatment with the chelator alone (1 h) did not appreciably altered the level of Cx43 compared to controls (Fig. 5A) nor the basal tyrosine phosphorylation of cellular proteins (Fig. 5B). The chelator, however, consistently inhibited the up-regulation of Cx43 induced by IL-1β. The effect could be already detected in cells stimulated with the cytokine for 30 min, and fully evident in cells stimulated for 6 h (Fig. 5A). Cell pre-treatment with BAPTA-AM appeared effective also in preventing protein tyrosine phosphorylation induced by IL-1β (Fig. 5B).

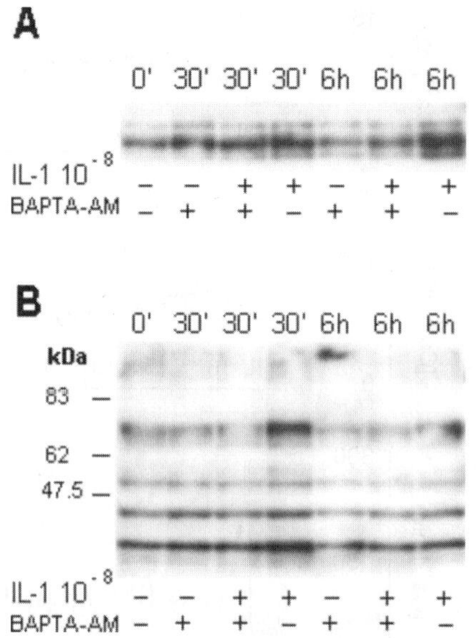

Fig. 5. BAPTA-AM prevents the effects of IL-1β on Cx43 expression and protein tyrosine phosphorylation. (A) Cells grown to confluence were incubated with IL-1β 10^{-8} M for the times indicated. In parallel experiments, cells were incubated for 1 h with 25 μM BAPTA-AM prior to cytokine stimulation. The protein extracts (10–15 μg for each sample) were separated by SDS polyacrylamide gel electrophoresis, transferred to nitro-cellulose and analysed using the polyclonal anti-Cx43. (B) 20 μg of the same protein extracts were analysed using a monoclonal antibody to phosphotyrosine (4G10).

Fig. 6. BAPTA-AM prevents the effects of IL-1β on Cx43 immunolocalization. Confluent chondrocytes cultures were fixed, permeabilized and incubated with a polyclonal antibody to Cx43 in resting conditions (A), after incubation either with BAPTA-AM 25 μM for 1 h prior to stimulation with IL-1β 10^{-8} M for 6 h (B), or after stimulation with IL-1β 10^{-8} M for 6 h (C). A FITC-coniugated secondary antibody was used for fluorescence visualisation. Bar: 40 μm.

In another set of experiments stimulation with IL-1β was carried out in the presence and in the absence of tyrosine kinases inhibitors, to test the effects of protein tyrosine phosphorylation on Cx43 expression. Cell pretreatment with herbimycin A (1 μM, 2 h) did not prevent the IL-1β-induced Cx43 up-regulation, and did not interfere with the inhibitory action of BAPTA-AM, suggesting that tyrosine phosphorylation is not directly involved in the signalling pathway leading to the increased expression of Cx43 (not shown). Similar results were obtained preincubating the cells in the presence of tyrphostin A 25 (100 μM, 15 min, not shown).

Immunofluorescence experiments were carried out to verify whether the treatment with BAPTA-AM induced major modifications in the cellular localisation of Cx43 (Fig. 6). The punctate distribution of Cx43 immunoreactivity, evident in control cultures (Fig. 6A) was not altered in cells pre-treated with BAPTA-AM and stimulated with 10^{-8} M IL-1β for 6 h (Fig. 6B). Moreover, the average level of staining was similar in both conditions. In the absence of BAPTA-AM, cell stimulation with 10^{-8} M IL-1β greatly enhanced the Cx43-specific staining at appositional membranes (Fig. 6C).

3. Conclusions

In this study, we examined the effect of the proinflammatory cytokine IL-1β on gap junction-mediated intercellular communication in articular chondrocytes. We found that the cytokine increases dose-dependently the expression of Cx43, its localisation at the plasmamembrane, the extent of dye transfer and the propagation of mechanically-induced intercellular Ca^{2+} waves, indicating a positive modulation of cell-to-cell coupling. Among the intracellular signals activated by IL-1β in these cells, an increase of cytosolic Ca^{2+} appears necessarily required, since preventing Ca^{2+} rises with BAPTA-AM completely abolished the cytokine-induced Cx43 up-regulation.

Modulation of intercellular coupling represents a novel aspect of cartilage response to proinflammatory signals and is likely to have important consequences in cartilage pathology. Inflammatory and degenerative joint diseases are invariably associated with elevated levels of IL-1, one of the most important mediators responsible for cartilage matrix catabolism [18]. IL-1-induced chondrocytes responses represent central pathogenic events in rheumatoid arthritis and osteoarthritis and include the induction of metalloproteinases, of other proinflammatory cytokines and the inhibition of both extracellular matrix synthesis and chondrocyte proliferation [18]. The resulting dysregulation of cartilage homeostasis leads to progressive destruction of the joints.

The IL-1-dependent up-regulation of Cx43 expression, reported in this study, appears to be associated to the activation of Ca^{2+} signalling, since it is prevented by intracellular Ca^{2+} chelation. Interestingly, this condition prevents also part of the cytokine-induced protein tyrosine phosphorylation, suggesting that IL-1 activates multiple tyrosine kinases in cartilage cells. The activation of Ca^{2+} signalling by IL-1 is strictly dependent on focal adhesions [19]: a likely candidate for a direct linkage between Ca^{2+} rises and tyrosine phosphorylation could be the Ca^{2+}-dependent tyrosine kinase Pyk2, a member of the focal adhesion kinase family which can be activated by stress signals [25]. In this context, chondrocyte response to proinflammatory stimuli would depend on the adhesive status of the cells, which is deeply modified in pathological cartilage: the erosion of the matrix embedding individual cells leads to the direct exposition of chondrocytes to novel cell–matrix and cell-to-cell interactions, and in rheumatoid arthritis, to the direct interaction with the infiltrating inflammatory cells [5].

Gap junctions-mediated intercellular communication provide a pathway for a bi-directional transfer of ions and small molecules between the cells; their permeability to intracellular second messengers confers tissues the ability to respond uniformly to localised stimuli [3]. Enhanced communication competence among chondrocytes and, possibly, between chondrocytes and synovial cells, may facilitate the diffusion of locally generated signals, such as mechanical strain, and thus regulate the sensitivity of cartilage and synovial pannus to hormonal and physical factors.

Acknowledgements

We would like to thank Franco Vittur for support and helpful discussions. The work was supported by MURST-PRIN2000 and by University of Trieste, Italy.

References

[1] W.P. Arend, The pathophysiology and treatment of rheumatoid arthritis, *Arthritis Rheum.* **40** (1997), 595–597.

[2] F. Blanco and M. Lotz, IL-1 induced nitric oxide inhibits chondrocyte proliferation via PGE2, *Exp. Cell Res.* **218** (1995), 319–325.

[3] R. Bruzzone, T.W. White and D.L. Paul, Connections with connexins: the molecular basis of direct intercellular signalling, *Eur. J. Biochem.* **238** (1996), 1–27.
[4] I. Capozzi, R. Tonon and P.D. Andrea, Ca^{2+}-sensitive phosphoinositide hydrolysis is activated in synovial cells but not in articular chondrocytes, *Biochem. J.* **344** (1999), 545–553.
[5] M.W.K. Chew, B. Henderson and J.C.W. Edwards, Antigen-induced arthritis in rabbit: ultrastructural changes at the chondrosynovial junction, *Int. J. Exp. Path.* **71** (1990), 879–894.
[6] C.N.D. Coelho and R.A. Kosher, Gap junctional communication during limb cartilage differentiation, *Dev. Biol.* **144** (1991), 47–53.
[7] P. D'Andrea and F. Vittur, Gap junctions modulate intercellular Ca^{2+} signalling in cultured articular chondrocytes, *Cell Calcium* **20** (1996), 389–397.
[8] H. Donahue, F. Guilak, M.A. Vander Moolen, K.J. McLeod, C.T. Rubin, D.A. Grande and P.R. Brink, Chondrocytes isolated from mature articular cartilage retain the capacity to form functional gap junctions, *J. Bone Min. Res.* **10** (1995), 1359–1364.
[9] M. Fernandez-Cobo, C. Gingalewski and A. Di Maio, Expression of the connexin43 gene is increased in the kidney and the lungs of rats injected with bacterial lipopolysaccharide, *Shock* **10** (1998), 97–102.
[10] Y. Geng, J. Valbracht and M. Lotz, Selective activation of the mitogen-activated protein kinase subgroups c-Jun NH sub 2 terminal kinase and p38 by IL-1 and TNF in human articular chondrocytes, *J. Clin. Invest.* **98** (1996), 2425–2430.
[11] M.B. Goldring, J.R. Brikhead, L.F. Suen, R. Yamin, S. Mizuno, J. Glowacki, J.L. Arbiser and J.F. Apperley, Interleukin-1β-modulated gene expression in immortalised human chondrocytes, *J. Clin. Invest.* **94** (1994), 2307–2316.
[12] M. Grandolfo, A. Calabrese and P. D'Andrea, Mechanism of mechanically-induced intercellular calcium waves in rabbit articular chondrocytes and in HIG-82 synovial cells, *J. Bone Min. Res.* **13** (1998), 443–453.
[13] H.J. Häuselmann, J. Flechtenmacher, L. Michal, E. Thonar, M. Shinmei, K.E. Kuettne and M.B. Aydelotte, The superficial layer of human articular cartilage is more susceptible to interleukin-1-induced damage than deep layers, *Arthritis Rheum.* **39** (1996), 478–488.
[14] G.S. Hillis, L.A. Duthie, P.A.J. Brown, J.G. Simpson, A.M. Macleod and N.E. Haites, Upregulation and co-localization of connexin43 and cellular adhesion molecules in inflammatory renal diseases, *J. Pathol.* **182** (1997), 373–379.
[15] H. Ishikawa, S. Hirata, Y. Nishibayashi, S. Imura, H. Kubo and O. Ohno, The role of adhesion molecules in synovial pannus formation in rheumatoid arthritis, *Clin. Orthop.* **300** (1994), 297–303.
[16] H. Ishikawa, O. Ohno, R. Sura, T. Matsubara and K. Hirohata, Cytokine enhancement of monocyte/synovial cell attachment to the surface of cartilage: a possible trigger of pannus formation in arthritis, *Rheumatol. Int.* **11** (1991), 31–36.
[17] W.M.F. Jongen, D.J. Fitzgerald, M. Asamoto et al., Regulation of connexin43-mediated gap junctional intercellular communication by Ca^{2+} in mouse epidermal cells is controlled by E-cadherin, *J. Cell Biol.* **114** (1991), 545–556.
[18] M. Lotz, F.J. Blanco, J. von Kempis, J. Dudler, R. Maier, P.M. Villiger and Y. Geng, Cytokine regulation of chondrocyte functions, *J. Rheumatol.* **22** (1995), 104–108.
[19] L. Luo, T. Cruz and C. McCulloch, Interleukin 1-induced calcium signalling in chondrocytes requires focal adhesions, *Biochem. J.* **324** (1997), 653–658.
[20] R.-M. Mege, F. Matsuzaki, W.J. Gallin, J.I. Goldberg, B.A. Cunningham and G.M. Edelman, Construction of epithelioid sheets by transfection of mouse sarcoma cells with cDNAs for chicken adhesion molecules, *Proc. Natl. Acad. Sci. USA* **85** (1988), 7274–7278.
[21] A.P. Moreno, J.C. Sàez, G.I. Fishman and D.C. Spray, Human connexin43 gap junction channels: regulation of unitary conductances by phosphorylation, *Circ. Res.* **74** (1994), 1050–1057.
[22] A. Ramachandrula, K. Tiku and M.L. Tiku, Tripeptide RGD-dependent adhesion of articular chondrocytes to synovial fibroblasts, *J. Cell Sci.* **101** (1992), 859–871.
[23] N. Rinaldi, M. Schwarz-Eywill, D. Weis, P. Leppelmann-Jansen, M. Lukoschek, U. Keilholz and T.F.E. Barth, Increased expression of integrins on fibroblast-like synoviocytes from rheumatoid arthritis in vitro correlates with enhanced binding to extracellular matrix proteins, *Ann. Rheum. Dis.* **56** (1997), 45–51.
[24] W. Schwab, A. Hofer and M. Kasper, Immunohistochemical distribution of connexin 43 in the cartilage of rats and mice, *Histochem. J.* **30** (1998), 413–419.
[25] G. Tokiwa, I. Dikinc, S. Lev and J. Schlessinger, Activation of Pyk2 by stress signals and coupling with JNK signalling pathway, *Science* **273** (1996), 792–794.
[26] M.P. Vincenti, C.I. Coon, O. Lee and C.E. Brinckerhoff, Regulation of collagenase geneexpression by IL-1β requires transcriptional and post-transcriptional mechanisms, *Nucleic Acid Res.* **22** (1994), 4818–4827.
[27] A.Z. Wang, J.C. Wang, G.W. Fisher and H. Diamond, Interleukin-1β-stimulated invasion of articular cartilage by rheumatoid synovial fibroblasts is inhibited by antibodies to specific integrin receptors and by collagenase inhibitors, *Arthritis Rheum.* **40** (1997), 1298–1307.

Biorheology 39 (2002) 161–169
IOS Press

Ultrasonic characterization of articular cartilage

J. Töyräs [a], H.J. Nieminen [a], M.S. Laasanen [a], M.T. Nieminen [b], R.K. Korhonen [b], J. Rieppo [b], J. Hirvonen [a], H.J. Helminen [b] and J.S. Jurvelin [a,*]
[a] *Department of Clinical Physiology and Nuclear Medicine, Kuopio University Hospital and University of Kuopio, Kuopio, Finland*
[b] *Department of Anatomy, University of Kuopio, Kuopio, Finland*

Abstract. Osteoarthrosis is the most important joint disease that threatens health of the musculoskeletal system of elderly people. Today, there is a need for sensitive, quantitative diagnostic methods for successful and early diagnosis of the disorder. In the present study, we aimed at evaluating the applicability of ultrasound for quantitative assessment of cartilage structure and properties. Bovine articular cartilage was investigated both *in vitro* and *in situ* using high frequency ultrasound. Cartilage samples were also tested mechanically *in vitro* to reveal relationships between acoustic and mechanical parameters of the tissue. The collagen organization and proteoglycan content of cartilage samples were mapped, using quantitative polarized light microscopy and digital densitometry, respectively, to reveal their effect on the acoustic properties of tissue. The high frequency pulse-echo ultrasound (20–30 MHz) technique proved to be sensitive in detecting the degeneration of the superficial collagen-rich cartilage zone. In addition, ultrasound was found to be a potential tool for measuring cartilage thickness. When the results from biomechanical indentation measurements and ultrasound measurements of normal and enzymatically degraded articular cartilage were combined, collagen or proteoglycan degradation in the tissue could be sensitively and specifically differentiated from each other. To conclude, high frequency ultrasound is a useful tool for evaluation of the quality of superficial articular cartilage as well as for the measurement of cartilage thickness. Therefore, ultrasound appears to be a valuable supplement to the mechanical measurements of articular cartilage stiffness.

1. Introduction

Osteoarthrosis (OA) is a common musculoskeletal disorder affecting millions of individuals worldwide. The social and economical burden of OA is enormous. Obviously, a diagnostic tool that could help in making reliably and early diagnosis of the degenerative joint disease would have significant economical impact and would alleviate pain and disability.

One of the first signs of cartilage degeneration is tissue softening, which can occur before any histological changes are observed [19]. With time, softening can lead to cartilage fibrillation and OA. Before any disruption of the collagen meshwork occurs, the changes are believed to be reversible [4,11]. For preventive actions, it would therefore be desirable to detect the tissue changes in the earliest stages of the disease process. During arthroscopy, cartilage quality is typically evaluated visually and by palpating the stiffness of the articular surface manually with a blunt probe. Improved, preferably quantitative techniques, however, are required for early, sensitive and reliable evaluation of cartilage status. At present there are a few quantitative *in vivo* devices available for the estimation of cartilage condition [3,9,20,21, 29,34]. Physical theories of acoustics suggest that it is also possible to use ultrasound for structural characterization of the cartilage surface and even for the determination of viscoelastic properties of tissue.

*Address for correspondence: J.S. Jurvelin, Ph.D., Dept. of Clinical Physiology and Nuclear Medicine, P.O. Box 1777, FIN-70211 Kuopio, Finland. Tel.: +358 17 173 261; Fax: +358 17 173 244; E-mail: Jukka.Jurvelin@uku.fi.

In addition, cartilage thickness, which is an important diagnostic parameter *per se*, may be determined from the ultrasound measurements.

In this study, we have gathered new information on the applicability of quantitative mechano-acoustic evaluation of articular cartilage. For articular cartilage, enzymatic degradation *in vitro* was utilized to selectively degrade structural constituents that are crucial for cartilage function. Subsequently, ultrasound was used to reveal tissue changes. Finally, the results from ultrasound measurements and mechanical indentation tests were combined for the mechano-acoustic evaluation of cartilage status.

1.1. Acoustic properties of articular cartilage

Ultrasound has been utilized for the measurement of articular cartilage thickness by several authors [14,22,23,25,34]. Exact knowledge of ultrasound velocity in cartilage, especially when examining different joints [33], is crucial for an accurate determination of cartilage thickness. Ultrasound microscopy of articular cartilage has indicated that the speed of sound in cartilage may be related to collagen fibril orientation [2]. No significant correlations were found between the sound velocity and biochemical composition of human cartilage tissue, even though ultrasound velocity in OA cartilage was lower than in healthy cartilage [23]. A recent study reported that ultrasound velocity was significantly reduced after digestion of proteoglycans (PGs) by trypsin [34]. The typical mean values for speed of sound in healthy cartilage are within the range from 1650 to 1770 m/s [18,23,34], while in degenerated cartilage they vary between 1580 and 1600 m/s [23,34]. Recently, ultrasound velocities were found to vary widely in cartilage obtained from different sites of the human ankle and hip joints (1419–2428 m/s, mean: 1892 m/s) [33].

Only a few studies have focused attention on ultrasound attenuation in articular cartilage. Integrated attenuation values of 3.2–7.5 Np/cm (~2.8–6.5 dB/mm) for bovine cartilage have been measured in the frequency range 10–40 MHz [27], while values of 101–169 Np/cm (~88–147 dB/mm) were revealed at 100 MHz [2]. These results suggest that, just as observed in most soft tissues, the attenuation is significantly frequency dependent in cartilage. A previous study suggested that cartilage from the areas exposed to high stresses during joint motion shows a higher capability to attenuate ultrasound than the cartilage from the areas exposed to light stresses [27]. A positive correlation between ultrasound attenuation and compressive stiffness was demonstrated in bovine knee articular cartilage [18].

Human cartilage fibrillation, as well as differences between different weight-bearing areas of porcine articular cartilage have been detected by altered echogenicity [17,23]. In contrast, no changes were observed in the echogenicity in samples taken from different areas of normal human femoral condyles [23]. In experimental OA of rat knees, specific morphological changes were observed in the echo pattern of cartilage and subchondral bone [6,26]. Also, acoustic properties of rat cartilage were affected by maturation-related changes in tissue composition and morphology [5]. Angular ultrasound scattering from the articular surface demonstrated differences between the normal and degenerated cartilage, revealing an increase of surface roughness in OA [1,7,8]. In an *in vitro* study, ultrasound was found to be an accurate and reliable tool for the detection and grading of knee OA [10]. Recently, an ultrasound indentation technique, which combines ultrasonic and mechanical measurements has been presented [16, 34,35]. The combination of ultrasound and mechanical measurement may offer certain advantages, such as information of both cartilage thickness and stiffness.

Earlier measurements of cartilage density, approximately 1050 kg/m^3 [13], and ultrasound velocity, yield an acoustic impedance between 1.7–1.8 $\times$ 10^6 kg/(m^2 s). Due to the heterogeneous and anisotropic nature of cartilage structure, it is also possible that there is a gradient of acoustic impedance as a function of cartilage depth.

In this study, capability of ultrasound to measure cartilage mechanical and structural properties was studied. For this aim cartilage acoustic, structural and mechanical properties were experimentally determined, *in vitro* and *in situ,* using a variety of cartilage samples taken from several anatomical locations.

2. Materials and methods

2.1. Sample preparation

Sterile cylindrical osteochondral plugs (∅ = 6 mm, $n = 60$) were prepared for ultrasound, mechanical and microscopic analyses from patellae of 1- to 3-year-old bulls. The joints were delivered intact from the local slaughterhouse within 2 hours post mortem. Ultrasound and mechanical measurements were performed for 0 hour control samples ($n = 24$) immediately after preparation. Other samples were initially incubated under physiological conditions (37°C, 5% CO_2), for 44 hours in cell culture medium with antibiotics and a specific enzyme. The enzyme addition was used for experimental degradation of the tissue matrix using collagenase (Sigma Chemical Co., St. Louis, MO, USA) to degrade collagen [28] and chondroitinase ABC (Seikagaku, Co., Tokyo, Japan) to cleave PGs [32]. Also 44 hour controls ($n = 12$) were prepared. In addition, 12 samples (∅ = 9 mm) were degraded either with trypsin (200 μg/ml) to degrade primarily PGs [12], or collagenase (30 U/ml), for 4 or 6 hours, respectively. Also 6 hour controls ($n = 6$) were prepared. In addition, bovine humeral head cartilage ($n = 7$) was used for ultrasound measurements. After the ultrasound measurements, a cylindrical full thickness cartilage disk (∅ = 3 mm, $n = 29$) was cut from the site of measurement, using a biopsy punch and a razor blade [15].

2.2. Acoustic measurements

A-mode pulse-echo measurements (Fig. 1) were performed with a high frequency, broadband (mean frequency 22 MHz) ultrasound instrument (Minhorst Osteoson DCIII, Minhorst GmbH&Co., Meudt, Germany) [30]. The PBS–cartilage interface reflection coefficient was determined as a ratio between the reflected amplitude and the total ultrasound amplitude reaching the cartilage surface. Ultrasound

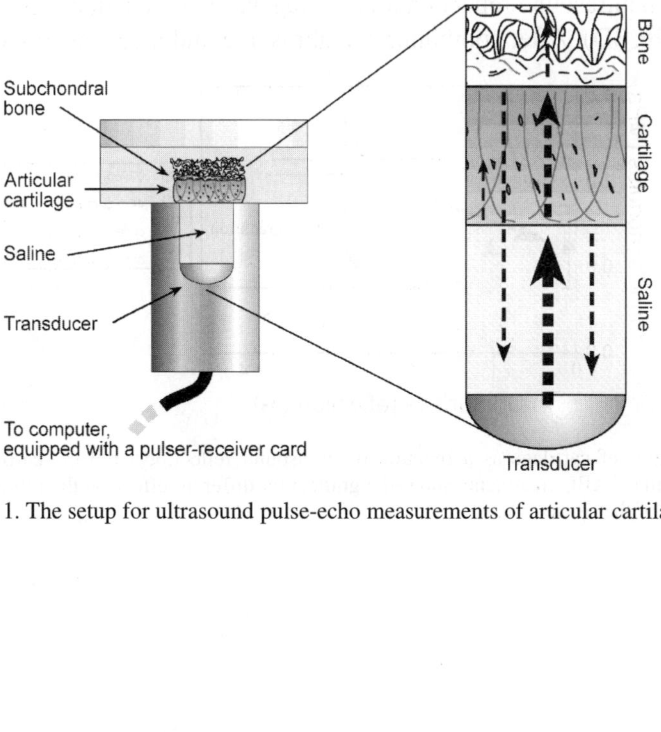

Fig. 1. The setup for ultrasound pulse-echo measurements of articular cartilage.

velocity was determined using the time of flight method [30]. Ultrasound attenuation coefficient (α) was determined using the following equation

$$\alpha = \frac{1}{2h}\ln\left(\frac{A_1 R_2(1 - R_1^2)}{A_2 R_1}\right), \tag{1}$$

where A_1 is the amplitude of the cartilage surface echo, A_2 is the amplitude of the echo from the cartilage–bone interface, R_1 is the cartilage surface reflection coefficient, R_2 is the reflection coefficient of cartilage–bone interface and h is the cartilage thickness. R_2 was approximated to have the same constant value for all samples ($R_2 = 0.374$) [30].

For the *in situ* A-mode ultrasonic measurements, the transducer was manually positioned perpendicular to the articular surface at the measurement site and ultrasound reflections from the cartilage surface and cartilage–bone interface were registered. Also cartilage thickness, based on the constant sound velocity of 1654 m/s, was determined.

The ultrasound measurements as a function of time during enzyme degradation were performed using ULTRAPAC system (Physical Acoustics Corporation, Princeton, NJ, USA), equipped with a high frequency broadband probe (Panametrics PZ25-0.25"-SU-R1.00"). The C-, D- and F-mode measurements of healthy articular cartilage were done using Panametrics Application Lab's MULTISCAN system with a 5900PR pulser receiver, 5627RPP remote pulser preamplifier and PI50-F1.00" transducer (frequency range 25–75 MHz, −6 dB) (Panametrics Inc., Waltham, MA, USA). The B-mode *in situ* ultrasound measurements were made using Dermascan C, 20 MHz ultrasound probe (Cortex Technology, Hadsund, Denmark).

3. Results

Collagenase and chondroitinase ABC selectively degraded the collagen fibril network and digested proteoglycans, respectively, as revealed by the quantitative microscopic methods [30]. The study revealed that the ultrasound reflection coefficient of the physiological saline–cartilage interface (R_1) decreased significantly (-96.4%, $p < 0.01$, Fig. 2) in the collagenase digested cartilage compared to controls. The first signs of decrease in ultrasound reflection after collagenase degradation were found after fifteen minutes of degradation (Fig. 3D). After combining the ultrasonic and mechanical measurements, PG deple-

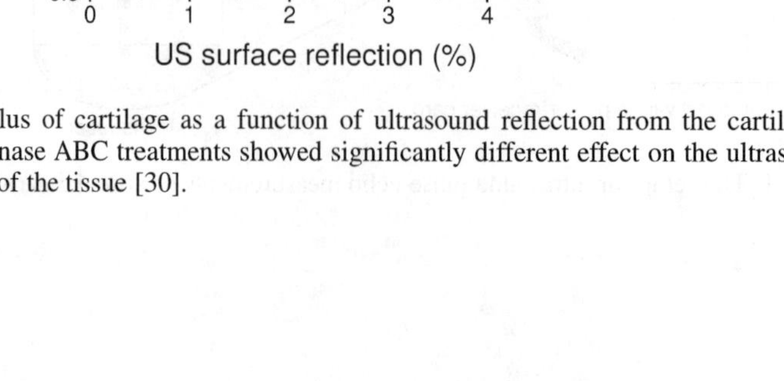

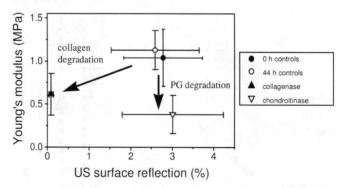

Fig. 2. The Young's modulus of cartilage as a function of ultrasound reflection from the cartilage surface (mean ± SD). Collagenase and chondroitinase ABC treatments showed significantly different effect on the ultrasound reflection, but similar effect on Young's modulus of the tissue [30].

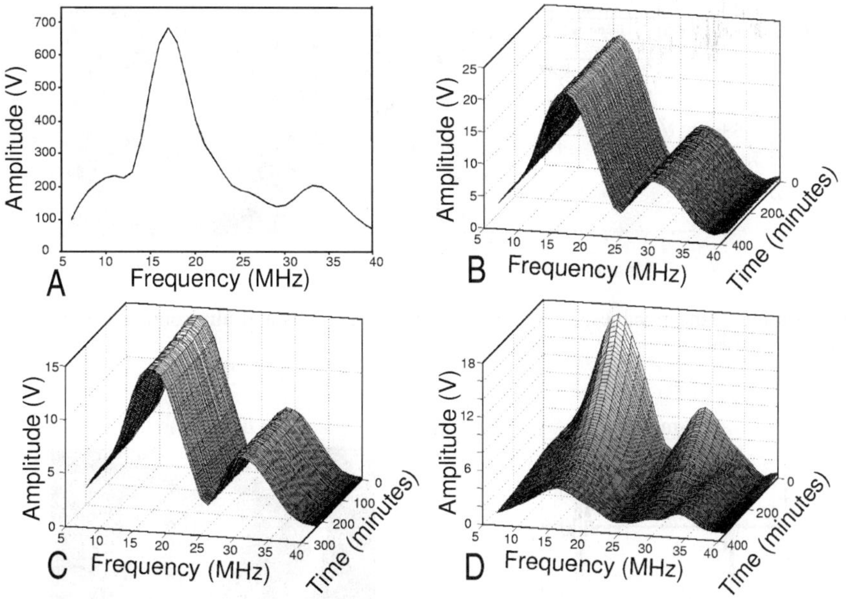

Fig. 3. The ultrasound amplitude spectrum from a perfect reflector at the distance of articular surface (A). Typical amplitude spectra of the ultrasound reflections from saline–cartilage interface as a function of incubation time: Control (B), trypsin digestion (C), collagenase digestion (D). Note significant and rapid changes in reflection after collagenase treatment [24].

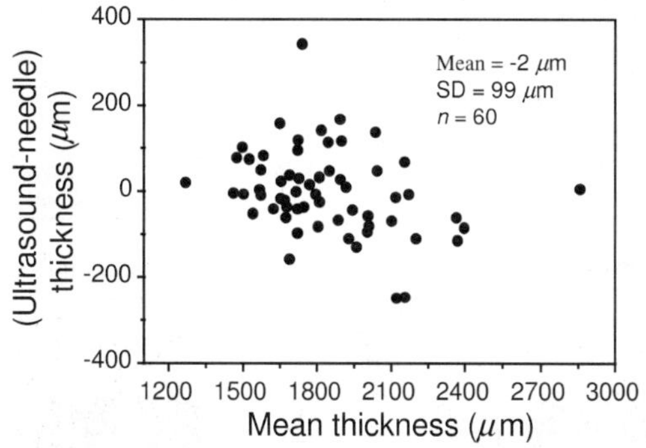

Fig. 4. Difference in cartilage thickness determined with ultrasound and reference techniques vs. mean thickness measured with the ultrasound and needle-probe techniques.

tion (after chondroitinase digestion) could be separated from the degradation of collagen meshwork (after collagenase digestion) (Fig. 2). The changes detected in ultrasound velocity and attenuation after enzymatic degradation were minor and depended on the magnitude of cleavage so that attenuation increased only after trypsin treatment and velocity decreased only after trypsin treatment or 44 hour collagenase treatment [24,30]. Cartilage thickness, as determined with ultrasound, revealed a high, linear correlation ($r = 0.943$, $n = 60$) with the reference thickness obtained with a needle–probe technique [30]. The accuracy of ultrasound technique to measure cartilage thickness was studied by a Bland–Altman plot (Fig. 4). The mean difference between ultrasound and needle-probe measurements was 2 μm with SD of 99 μm. A similar pattern of topographical variation in ultrasound surface reflection and cartilage Young's modulus of bovine humerus was recorded (Fig. 5).

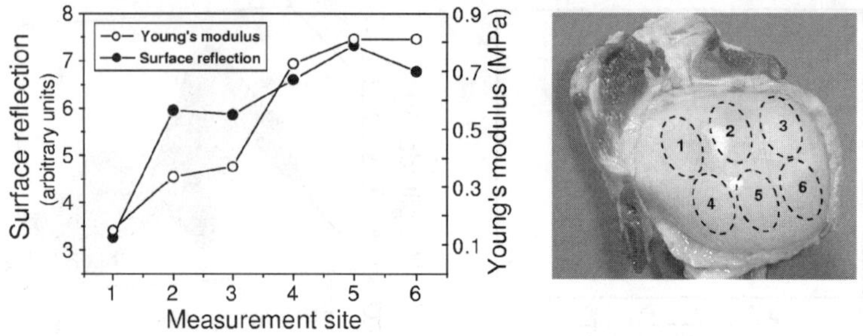

Fig. 5. Topographical variation (mean values, $n = 2$–9) of Young's modulus and ultrasound surface reflection in numbered sites of bovine humeral cartilage.

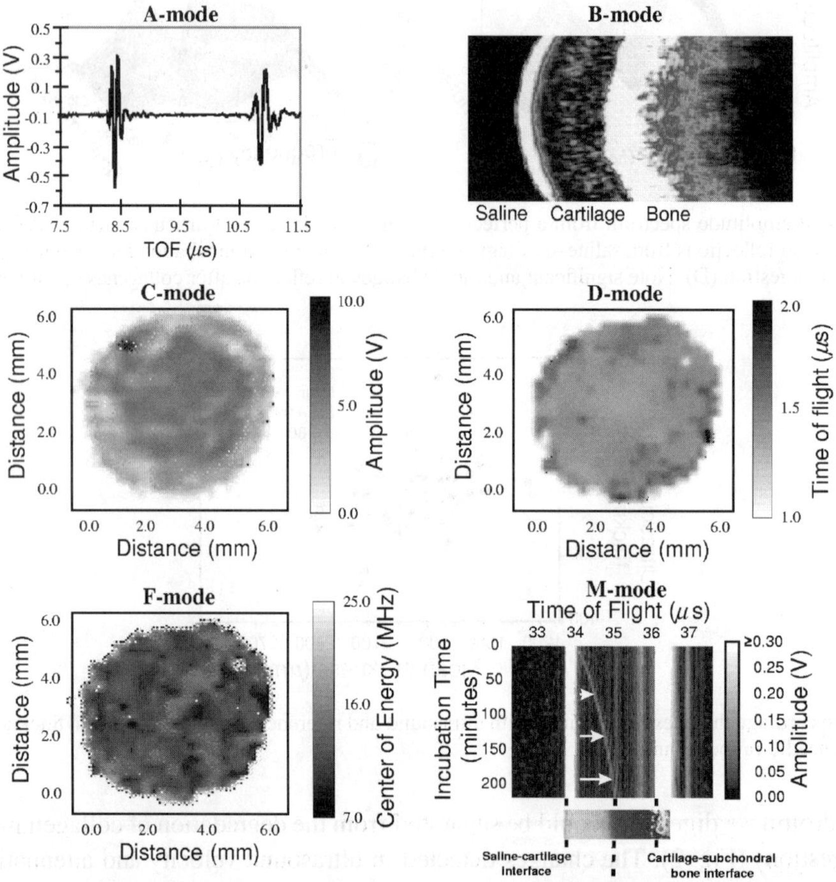

Fig. 6. The A-, B-, C-, D-, F- and M-mode ultrasonic images of articular cartilage. The C- and F-mode images represent only the cartilage surface, while others represent the full thickness sample [24].

An extra echo after PG depletion was found after trypsin treatment (Fig. 6, M-mode image). The scanning ultrasound systems enabled generation of A-, B-, C-, D-, F- and M-mode images (Fig. 6) of bovine articular cartilage. These high resolution images suggested that ultrasound can be utilized in evaluation of spatial changes in cartilage properties.

4. Discussion

High frequency ultrasound was used to characterize the properties of articular cartilage *in vitro* and *in situ* to reveal if ultrasound could provide a novel tool for the diagnosis of cartilage degradation. After collagenase digestion, the structural changes in the superficial collagen network were sensitively detected by ultrasound reflection. PG degradation in the matrix was detected by the extra echo at the depth of the chondroitinase ABC or trypsin penetration front [24,30]. By combining mechanical indentation and ultrasound measurements, cartilage degeneration related to collagen and PG depletion could be separated from each other [30]. Based on these results, a combination of ultrasound and mechanical measurements not only differentiates normal articular cartilage from degenerated one, but may even discern a superficial collagen fibril network injury from PG depletion. This information is certainly of clinical relevance when selecting the best patient treatment and rehabilitation, as well as when evaluating progress of the disease.

The echo amplitude from the articular surface is related to the ratio of acoustic impedances of cartilage and the immersion medium, as well as to the surface roughness, which controls scattering of ultrasound [1,7,8]. Both collagenase and chondroitinase ABC digested samples showed no macroscopical signs of degradation. However, only collagenase induced a significant reduction of the ultrasound surface reflection coefficient. This result indicates that the tangential collagen fibrils act as ultrasound reflectors at the cartilage surface. Therefore, ultrasound provides a sensitive tool for characterizing the quality of the collagenous superficial cartilage layer. This finding is important, since the injury of the collagen network is thought to represent the "point of no return" in the process of OA progression [4,11].

It has been suggested that ultrasound velocity is greater in normal cartilage than in OA cartilage [23] and that the digestion of PGs reduces the ultrasound velocity in cartilage [34]. In this study, only a minor decrease of ultrasound velocity in PG or collagenase-digested cartilage was detected. The small difference in ultrasound velocity between healthy and degraded cartilage suggests that a predefined, constant ultrasound velocity can be used in ultrasonic measurements in order to obtain a diagnostically acceptable accuracy for cartilage thickness. Possibly, different ultrasound velocities, based on normative data, may be used at different articular sites. Ultrasonic measurement of cartilage thickness can be utilized to correct mechanically measured cartilage stiffness values, as reported in our earlier study [31].

In contrast to trypsin induced PG loss, we found no statistically significant changes in the ultrasound attenuation after chondroitinase ABC or collagenase digestions. PG removal after chondroitinase ABC digestion was typically limited to the upper 30% of the cartilage thickness. The superficial tissue is poor in PGs and, therefore, only a small portion of all PGs was removed from the tissue by chondroitinase ABC. After trypsin digestion, PGs were more effectively cleaved from the tissue, and the attenuation increased as the trypsin penetration increased during digestion. Recently, an ultrasound indentation technique [35] was utilized to characterize changes in the acoustic and mechanical properties of articular cartilage after enzymatic removal of PGs using trypsin [34]. In that study, a significant decrease in ultrasound velocity as well as in cartilage stiffness was revealed.

5. Conclusions

In a series of experiments, structural and mechanical properties of articular cartilage were investigated using several *in vitro* and *in situ* techniques. The quantitative information obtained from the analyses was then compared with the acoustic properties recorded for the same samples. The most important results can be summarized as follows:

- High frequency ultrasound provides a potential method for the measurement of cartilage thickness.
- Ultrasound behavior in PG-depleted cartilage is different from that in cartilage with a degraded collagen network. Collagen was found to act as a reflector of ultrasound at the articular surface.
- By combining the indentation and ultrasound measurements, PG depletion and collagen degradation in articular cartilage can be discerned. This finding may have a major diagnostic significance, since degradation of the collagen fibril network is believed to be irreversible while PG depletion can be reversible.
- Ultrasound can be used to correct mathematically the effect of cartilage thickness on the measured indentation forces. Combination of ultrasound and indentation measurements would likely help making early clinical diagnosis of OA and monitoring of it.

Acknowledgements

Financial support from the Technology Development Center (TEKES), Finland, Kuopio University Hospital, Kuopio, Finland (EVO 5103), and the Graduate School for Musculoskeletal Diseases, Finland is acknowledged.

References

[1] R.S. Adler et al., Quantitative assessment of cartilage surface roughness in osteoarthritis using high frequency ultrasound, *Ultrasound Med. Biol.* **18** (1992), 51–58.
[2] D.H. Agemura et al., Ultrasonic propagation properties of articular cartilage at 100 MHz, *J. Acoust. Soc. Am.* **87** (1990), 1786–1791.
[3] S.I. Berkenblit et al., Nondestructive detection of cartilage degeneration using electromechanical surface spectroscopy, *J. Biomech. Eng.* **116** (1994), 384–392.
[4] J.A. Buckwalter and H.J. Mankin, Articular cartilage, Part II: Degeneration and osteoarthritis, repair, regeneration, and transplantation, *J. Bone Joint. Surg. Am.* **79** (1997), 612–632.
[5] E. Cherin et al., Assessment of rat articular cartilage maturation using 50 MHz quantitative ultrasonography, *Osteoarthritis Cartilage* **9** (2001), 178–186.
[6] E. Cherin et al., Evaluation of acoustical parameter sensitivity to age-related and osteoarthritic changes in articular cartilage using 50 MHz ultrasound, *Ultrasound Med. Biol.* **24** (1998), 341–354.
[7] E.H. Chiang et al., Quantitative assessment of surface roughness using backscattered ultrasound: the effects of finite surface curvature, *Ultrasound Med. Biol.* **20** (1994), 123–135.
[8] E.H. Chiang et al., Ultrasonic characterization of in vitro osteoarthritic articular cartilage with validation by confocal microscopy, *Ultrasound Med. Biol.* **23** (1997), 205–213.
[9] J.H. Dashefsky, Arthroscopic measurement of chondromalacia of patella cartilage using a microminiature pressure transducer, *Arthroscopy* **3** (1987), 80–85.
[10] D.G. Disler et al., Articular cartilage defects: in vitro evaluation of accuracy and interobserver reliability for detection and grading with US, *Radiology* **215** (2000), 846–851.
[11] M.A. Freeman, Is collagen fatigue failure a cause of osteoarthrosis and prosthetic component migration?, A hypothesis, *J. Orthop. Res.* **17** (1999), 3–8.
[12] E.D. Harris, Jr. et al., Effects of proteolytic enzymes on structural and mechanical properties of cartilage, *Arthritis Rheum.* **15** (1972), 497–503.
[13] D. Joseph et al., True density of normal and enzymatically treated bovine articular cartilage, *Transact. Orthop. Res. Soc.* **24** (1999), 642.
[14] J.S. Jurvelin et al., Comparison of optical, needle probe and ultrasonic techniques for the measurement of articular cartilage thickness, *J. Biomech.* **28** (1995), 231–235.
[15] J.S. Jurvelin et al., Optical and mechanical determination of Poisson's ratio of adult bovine humeral articular cartilage, *J. Biomech.* **30** (1997), 235–241.
[16] G.N. Kawchuk and P.D. Elliott, Validation of displacement measurements obtained from ultrasonic images during indentation testing, *Ultrasound Med. Biol.* **24** (1998), 105–111.

[17] H.K. Kim et al., Imaging of immature articular cartilage using ultrasound backscatter microscopy at 50 MHz, *J. Orthop. Res.* **13** (1995), 963–970.

[18] P. Kolmonen et al., Experimental comparison of acoustic and mechanical properties of bovine knee articular cartilage, *Transact. Orthop. Res. Soc.* **20** (1995), 513.

[19] J.M. Lane et al., Experimental knee instability: early mechanical property changes in articular cartilage in a rabbit model, *Clin. Orthop.* (1979), 262–265.

[20] A. Legare et al., Evaluation of cartilage quality using streaming potential maps, *Transact. ICRS* **3** (2000), 89.

[21] T. Lyyra et al., Indentation instrument for the measurement of cartilage stiffness under arthroscopic control, *Med. Eng. Phys.* **17** (1995), 395–399.

[22] V.E. Modest et al., Optical verification of a technique for in situ ultrasonic measurement of articular cartilage thickness, *J. Biomech.* **22** (1989), 171–176.

[23] S.L. Myers et al., Experimental assessment by high frequency ultrasound of articular cartilage thickness and osteoarthritic changes, *J. Rheumatol.* **22** (1995), 109–116.

[24] H.J. Nieminen et al., Real-time ultrasound analysis of articular cartilage degradation *in vitro*, *Ultrasound Med. Biol.* (2002) (in press).

[25] H. Nötzli et al., Automated ultrasound measurements can map contour and thickness of articular cartilage, *Transact. Orthop. Res. Soc.* **19** (1994), 409.

[26] A. Saied et al., Assessment of articular cartilage and subchondral bone: subtle and progressive changes in experimental osteoarthritis using 50 MHz echography in vitro, *J. Bone Miner. Res.* **12** (1997), 1378–1386.

[27] D.A. Senzig et al., Ultrasonic attenuation in articular cartilage, *J. Acoust. Soc. Am.* **92** (1992), 676–681.

[28] W.D. Shingleton et al., Collagenase: a key enzyme in collagen turnover, *Biochem. Cell Biol.* **74** (1996), 759–775.

[29] H. Tkaczuk, Human cartilage stiffness. In vivo studies, *Clin. Orthop.* **206** (1986), 301–312.

[30] J. Töyräs et al., Characterization of enzymatically induced degradation of articular cartilage using high frequency ultrasound, *Phys. Med. Biol.* **44** (1999), 2723–2733.

[31] J. Töyräs et al., Estimation of the Young's modulus of articular cartilage using an arthroscopic indentation instrument and ultrasonic measurement of tissue thickness, *J. Biomech.* **34** (2001), 251–256.

[32] T. Yamagata et al., Purification and properties of bacterial chondroitinases and chondrosulfatases, *J. Biol. Chem.* **243** (1968), 1523–1535.

[33] J.Q. Yao and B.B. Seedhom, Ultrasonic measurement of the thickness of human articular cartilage in situ, *Rheumatology (Oxford)* **38** (1999), 1269–1271.

[34] I. Youn et al., Determination of the mechanical properties of articular cartilage using a high-frequency ultrasonic indentation technique, *Transact. Orthop. Res. Soc.* **24** (1999), 162.

[35] Y.P. Zheng and A.F. Mak, An ultrasound indentation system for biomechanical properties assessment of soft tissues in-vivo, *IEEE Trans. Biomed. Eng.* **43** (1996), 912–918.

Biorheology 39 (2002) 171–179
IOS Press

COX-1, COX-2 and articular joint disease: A role for chondroprotective agents

Paul R. Colville-Nash * and Derek A. Willoughby
Department of Experimental Pathology, William Harvey Research Institute, Barts and The London, Queen Mary's School of Medicine and Dentistry, University of London, London, UK

Abstract. It is widely accepted that whilst exhibiting clinically useful anti-inflammatory and analgesic activity, the application of non-steroidal anti-inflammatory drugs (NSAIDs) does not affect the underlying pathogenesis of articular diseases such as rheumatoid arthritis. The demonstration of a role for COX-2 in the resolution of inflammation may partly underly the lack of disease modifying activity seen with NSAIDs in long term use in these inflammatory joint diseases. This has led to the suggestion that the anti-arthritic efficacy of these agents may be improved by altering prescribing practice such that they are not given during periods of disease remission, which may be difficult to achieve in the clinic. Alternatively, they may benefit from concomitant administration of chondroprotective agents, such as diacetylrhein, which may protect against the deleterious effects of traditional NSAIDs on cartilage degradation and, further, inhibit additional pathways such as cytokine elaboration which are important in joint destruction.

1. Key concepts in the pathogenesis of chronic inflammatory diseases

Inflammation is a complex and highly regulated process initiated in response to tissue injury. Once activated there is a sequential release of mediators which results in the development of the classical signs of inflammation, with vasodilation, increased blood flow, exudation and activation of sensory pain pathways. Inflammatory cells are attracted to the site by the activation of adhesion molecule expression on themselves and the endothelium in the affected tissue and the local production of chemoattractants which guide their migration. The process of cell migration is also highly regulated and during the acute phases of the inflammatory response the lesion is dominated by the presence of polymorphonuclear leukocytes (PMNs). Later there is migration into the area by mononuclear cells (MNs; predominantly macrophages but also other cell types such as lymphocytes depending upon circumstance). These typify the chronic phases of inflammatory reactions and these gradually disappear during resolution of the response which may also be accompanied by activation of wound healing pathways as damaged tissue is repaired or replaced.

In many chronic inflammatory diseases in man such as rheumatoid arthritis, the pathogenesis of which is the focus of this review, a relapsing and remitting disease course is seen. Thus there is an underlying chronic inflammatory response, thought to be propagated by the presence of an underlying immune-driven stimulus, perhaps to endogenous antigens such as collagen II in the case of rheumatoid arthritis. This is overlaid by the presence of acute inflammatory processes during periods of relapse as well as evidence of attempts by the host to remodel and repair damaged tissue. It is therefore possible to see all the phases of the inflammatory response throughout the lifetime of the disease and it is important to

*Address for correspondence: Department of Experimental Pathology, William Harvey Research Institute, Barts and The London, Queen Mary's School of Medicine and Dentistry, University of London, Charterhouse Square, London EC1M 6BQ, UK. Tel.: +44 20 7882 6160; Fax: +44 20 7882 6095; E-mail: p.r.colville-nash@qmul.ac.uk.

recognise this in order to develop an understanding of the potential ramifications of inhibiting various reactions which contribute to the development of the disease as well as perhaps also participating in attempts to resolve the lesion.

2. Prostaglandins and the development of the non-steroidal anti-inflammatory drugs

The prostaglandins (PGs) are amongst the earliest of the inflammatory mediators which are released during the initiation of inflammation. They have been the subject of much research and only a few major points will be raised here; the reader is directed to other reviews for further information regarding their role in inflammation [64]. Much of the work in this area has historically concentrated on PGE_2, the major PG produced in acute inflammation, and prostacyclin or PGI_2. PGE_2 is a powerful vasodilator and acts synergistically with other early mediators such as histamine and the kinins to promote increased vascular permeability [69]. Similarly PGI_2 plays a prominent role in the development of acute inflammatory reactions [46]. Both of these are hyperalgesic and can generate a pyrexic response. They are thus capable of mimicking the cardinal signs of inflammation. The interest in them as a therapeutic target was greatly increased by the demonstration by Vane in 1971 [65] that the therapeutic mechanism of action of aspirin was via the inhibition of the production of these lipid mediators by the enzyme cyclooxygenase (COX), which converts arachidonic acid into PGH_2 which is then subsequently further metabolised to the major prostanoids by a variety of downstream synthase enzymes [59]. This finding resulted in the development of a wide variety of agents targeting this enzyme system, collectively termed the non-steroidal anti-inflammatory drugs (NSAIDs). These are amongst the most widely used drugs in the control of inflammation.

However, there was evidence that the prostaglandins were important in a variety of other pathways. In normal physiology, it became recognised that they play a key role in the maintenance of gastric mucosal integrity and renal function, amongst a variety of other processes also, and that inhibition of these processes was the cause of the major side effects of NSAIDs such as gastric ulceration, which is the major cause of morbidity associated with the use of these agents [3]. Further, in the immune system, PGE_2 was shown to exert a predominantly immunosuppressive role, inhibiting immune driven reactions *in vivo* [68] and also *in vitro* where, for example, lymphocyte function [29,40,63] and cytokine production [7] could be inhibited with this mediator. Certain cancers which produce high levels of PGE_2 were thought to use this as a means of avoiding some of the host defence mechanisms. There was also the suggestion early on that there may be an anti-inflammatory role for other prostaglandins such as $PGF_{2\alpha}$ [14]. During the development of a variety of inflammatory reactions of different aetiologies, it was shown that whilst PGE_2 dominated the early stages of inflammation, during the resolving phases $PGF_{2\alpha}$ levels increased and it was suggested that there was a switch in the PG synthetic machinery from pro-inflammatory to anti-inflammatory PGs as part of the control mechanism in inflammation.

More recently, the story was further expanded by the recognition that there were in fact two isoforms of COX, a constitutive COX-1 and an inducible isoform COX-2 [60]. The demonstration that COX-2 was highly induced in inflammation, producing high levels of PGs, whilst COX-1 seemed to be the isoform in normal tissue responsible for the maintenance of routine homeostasis led to the search for new generations of NSAIDs which would selectively target COX-2, thus avoiding the major side effects of traditional NSAIDs [66]. This idea was strengthened by studies which showed that the side effect profile of the existing NSAIDs could be related by their selectivity for COX-1 or COX-2, with COX-1 selective agents such as piroxicam being the most likely to cause adverse reactions [6,25,33,39,44,48,

54]. The development of the new generation of "super aspirins" and their use *in vivo* showed that they had little effect on the gastric mucosa even at high doses [22], and the first of these, celecoxib [49] and rofecoxib [50], are now in clinical use. However, the development of these agents has also allowed the examination of the role of the two isoforms in inflammation and other diseases and has highlighted a number of potential pitfalls associated with NSAID therapy.

3. The pro-inflammatory role of COX-2

The earliest publications in this new field examined the induction of COX-2 in various acute inflammatory reactions. Thus in the rat carrageenan pleurisy, expression of COX-2 correlated with the production of PGE_2 [62], both reaching maximal levels at 2 hours and waning up to the 24 hour end point of the study. Similarly, in a carrageenan air pouch [37] and paw oedema models [58] there was also good correlation between the expression of COX-2 and PG synthesis. These latter studies showed that SC-58125, a novel COX-2 selective agent could suppress the development of the inflammatory reaction and inhibit the associated hyperalgesia without causing inhibition of gastrointestinal PG synthesis and associated gastric irritation, unlike the comparator NSAID, indomethacin. Similarly, using an acute adjuvant arthritis model [1], COX-2 induction in the synovium was shown to correlate with disease signs, and SC-58125 was again shown to be efficacious, reducing paw swelling and PG production in the inflamed joints.

However, there was the beginnings of the recognition that COX-2 may be playing a further role in inflammation. Thus early studies examining the distribution of COX isoforms in chronic inflammatory reactions led to the suggestion that there may be an association between the expression of COX-2 and the phase of wound healing in a murine chronic granulomatous air pouch reaction [2]. Loss of COX-2 also did not seem to affect the development of inflammatory reactions in genetically modified animals [19], although it was noted that in COX-2 deficient animals, there seemed to be a persistence of lymphocyte infiltrates in the inflamed site. Concern was also raised that effective anti-inflammatory effects were only seen with the COX-2 selective agents during the early phases of inflammation [28], whilst traditional NSAIDs were effective over a longer time course, or at concentrations where inhibition of COX-1 was also possible [67].

4. The anti-inflammatory role of COX-2

Many of these aforementioned studies concentrated on the acute phases of inflammatory reactions and did not examine the expression of COX-2 and the effects of NSAIDs during the resolution phases of an inflammatory reaction. That there was a potential role in wound healing was suggested by early studies showing expression during the wound healing phase of a granulomatous reaction in mice [2], which was strengthened by the demonstration that loss of COX-2 by application of selective NSAIDs or use of knockout mice could delay healing of gastric lesions in models of ulceration [56] and also exacerbate colitis [52] and allergic airway hypersensitivity reactions [23].

A potential mechanism for an anti-inflammatory role of COX-2 in inflammation has now been described [26]. Examination of a self-resolving acute inflammatory reaction, the rat carrageenan pleurisy, throughout the duration of the reaction through to resolution, revealed two phases of COX-2 expression. Thus whilst there was an early peak associated with the onset of inflammation and the production of PGE_2, there was a second peak in expression associated with resolution and in the absence of production of PGE_2. Inhibition of COX-2 with both COX-2 selective and non-selective agents during this latter phase resulted in the prolongation of the inflammatory response. Further examination of the production of PGs during this second phase of COX-2 expression revealed a switch to production of PGD_2 and it's cyclopentenone PGJ series metabolites. It was the loss of these during resolution that resulted in the delay in terminating the reaction, which could be restored in the presence of the NSAIDs by exogenously replacing these. It was hypothesised that COX-2 may therefore contribute to the initiation of inflammation by generating pro-inflammatory PGs such as PGE_2 in the early stages of an inflammatory reaction, but perform an anti-inflammatory role in resolution by helping to generate an alternative set of anti-inflammatory prostaglandins such as the PGJs.

A variety of pathways have now been described whereby mediators such as the cyclopentenone PGs may lead to resolution of inflammation, and it is perhaps likely that these may be acting in concert. These are natural endogenous ligands for peroxisome proliferator-activated receptor γ (PPAR-γ) [21] and studies have revealed a number of actions *in vitro* which contribute to possible anti-inflammatory effects of activating this system. PPAR-γ activation has been shown to be able to induce apoptosis of a wide variety of cell types including macrophages [16], synoviocytes [35] and endothelial cells [8], which could contribute to failure of pannus development in RA. It has been demonstrated that PPAR-γ is induced upon differentiation of monocytes to macrophages [16] and that activation of this pathway in macrophages can reduce the production of pro-inflammatory mediators shown to be of import to the pathogenesis of RA and other chronic inflammatory diseases. Thus production of IL-1, IL-6, TNF-α and induction of inducible nitric oxide synthase (iNOS) in macrophages are all inhibited by these agents [17,32,53]. In T cells, application of selective agents for PPAR-γ as well as $15dPGJ_2$ has been shown to inhibit T cell proliferation in response to phytohaemaglutinin and to reduce their production of IL-2 [71]. Induction of PPAR-γ receptors has further been shown to be important in the inhibition of iNOS expression as a result of IL-4 treatment in macrophages [31].

Recent publications have suggested inhibition of the activation of NF-κB may be another target for the anti-inflammatory properties of cyclopentenone prostaglandins. Treatment with $15dPGJ_2$ was shown to inhibit TPA stimulated NF-κB activation in Hela cells [61]. Rossi et al. [55] showed that cyclopentenone prostaglandins $15dPGJ_2$ and PGA_1 specifically inhibit the IKKβ subunit of the IKK complex in the μM range *in vitro*. These studies clearly showed that cyclopentenone prostaglandins target IKKβ activation by pro-inflammatory stimuli such as TNF-α, IL-1β and TPA without affecting the activation of other kinases such as JNK and p38. $15dPGJ_2$ has also been demonstrated to alkylate directly the p65/p50 dimer which also reduces their DNA binding activity [61].

The transcription factor NF-κB plays an important role in the regulation of pro-inflammatory mediator production and hyperplasia in rheumatoid synovium [41]. NF-κB is a generic term for a dimeric transcription factor consisting of hetero- or homodimers of the Rel family of protein [34]. NF-κB is an ubiquitously expressed transcription factor normally held inactive in the cytoplasm by association with an inhibitory molecule of the IκB (inhibitor of κB) family, IκBα. Pro-inflammatory cytokines such as TNF-α and IL-1β act through distinct receptors and signalling pathways which culminate in the activation of an IκB kinase (IKK) complex [18]. The subsequent phosphorylation, ubiquitination and degradation of IκBα by the 26S proteosome complex allows NF-κB to translocate to the nucleus where it induces transcription of target genes with κB response elements in their promoter. The recent description of the components of this IKK complex has led to the isolation of the IKKβ subunit responsible for the phosphorylation of IκBα in response to pro-inflammatory stimuli [36]. This pathway has now been shown to important in the activation of NF-κB in T and B lymphocytes *in vivo* [10,15] and primary synovial fibroblast-like cells [4,30].

The target genes for NF-κB include those that encode many of the pro-inflammatory mediators described above [34] and the enzymes involved in cartilage degradation [9]. NF-κB activation has been demonstrated in samples of human rheumatoid synovium [4] and also in joint tissue from murine collagen-induced arthritis where it preceded development of clinical signs [30]. It has therefore been suggested that the NF-κB pathway may be a target for the development of novel immunosuppressive and anti-inflammatory therapies [5] and inhibition of the NF-κB pathway leads to the suppression of pro-inflammatory cytokine release both *in vitro* and *in vivo* [24]. Equally, inhibition of natural anti-inflammatory pathways such as the production of cyclopentenone PGs by the application of NSAIDs may lead to a failure to control this pathway and possible reactivation of important pro-inflammatory cascades.

5. NSAIDs and arthritis

It is generally accepted that NSAIDs, whilst considerably improving the quality of life for patients by inhibiting many of the signs of inflammation such as pain and swelling [65]. However, they have little impact upon the progression of the disease, and in fact it is now recognised that they may actually accelerate the loss of cartilage in patients suffering from chronic inflammatory joint diseases such as osteoarthritis and rheumatoid arthritis [47,51]. As well as evidence from patients, there is a wide variety of evidence from animal models of arthritis which also suggest the same. Thus for example, proteoglycan synthesis by cartilage from RA patients is reduced following NSAID treatment [20]. NSAIDs *in vitro* will also suppress chondrocyte proteoglycan synthesis and increase cartilage degeneration [12, 13]. The intra-articular injection of *Mycobacterium tuberculosis* in the stifle joint of a rat results in an inflammatory reaction with synovial hyperplasia, loss of patella cartilage and bone erosion [57]. In this *in vivo* model of irritant induced monoarticular arthritis in the rat, the NSAIDs naproxen, piroxicam and indomethacin all accelerate the loss of patellar cartilage, although they all reduced the clinical signs of inflammation. Studies using the experimental COX-2 selective agent NS-398 has revealed that whilst it does not accelerate cartilage loss in the same way as these traditional NSAIDs, but it nevertheless fails to protect against cartilage loss [27]. Further investigations into the effect of treatment with NSAIDs have been performed in a model of cotton-induced granuloma-mediated cartilage destruction in mice. These studies further confirmed that NSAID treatment with indomethacin resulted in greater loss of cartilage matrix [11]. Examination of the effect of diclofenac, another commonly used NSAID for the treatment of arthritis, also showed that this NSAID could not affect the destruction of cartilage [45].

6. Implications for future treatment of chronic inflammatory diseases

The above findings have led to the suggestion that the failure of NSAIDs to show disease modifying activity in diseases such as rheumatoid arthritis may result from their inhibition of some of the beneficial functions of PG synthesis [70], allowing reactivation of pro-inflammatory pathways which may otherwise have been inhibited by the presence of anti-inflammatory mediators such as the cyclopentenone prostaglandins. Thus in rheumatoid arthritis, patients will tend to be prescribed and take NSAIDs routinely on a daily basis. Whilst they do improve quality of life and will be efficacious during those periods of relapse where there is reactivation of the inflammatory reaction, it has been suggested that during periods of remission, they may in fact be hampering the attempts by the host to resolve the inflammatory reaction by inhibiting the production of anti-inflammatory PGs. It has therefore been suggested that during periods of remission, there is a need to reduce NSAID therapy to avoid inhibiting such endogenous anti-inflammatory pathways or indeed potentially reducing the duration of periods of remission. It is however recognised that there is a need to develop adequate markers of disease activity which would make such an alteration in prescribing practice viable. At present these are not available.

Alternative approaches are therefore being sought. One such approach has been the development of chondroprotective agents such as diacerhein [38]. This is a drug used to treat osteoarthritis and with anti-arthritic and chondroprotective effects in animal models of arthritis. Importantly, in relation to the ideas presented here, it has no impact on the biosynthesis of prostaglandins. As well as reducing the release of free radicals from activated neutrophils [43] and reducing macrophage accumulation in inflammatory sites [42], diacerhein has been shown to inhibit the loss of both glycosaminoglycans and hydroxyproline in the cotton-induced granuloma-mediated cartilage degradation model in mice [45]. Additionally, the inflammatory tissue was analysed for the production of key pro-inflammatory mediators such as TNFα, IL-1 and IL-6. Diacerhein treatment reduced levels of these mediators which could further explain it's ability to inhibit cartilage destruction. In contrast, the comparator drug in this study, diclofenac, tended to enhance cytokine production.

Whilst it is unknown whether diacerhein specifically targets any of the pathways which may be affected for better or worse by the NSAIDs mentioned above, it is suggested that perhaps some of the potential deleterious effects of NSAID therapy may be mitigated by the concomitant administration of agents such as diacerhein.

Acknowledgements

Dr Paul Colville-Nash is a Post-Doctoral Research Fellow funded by the Arthritis Research Campaign, UK (Grants C0594, C0635). The author also gratefully acknowledges the financial support of Laboratories NEGMA-LERADS.

References

[1] G.D. Anderson, S.D. Hauser, K.L. McGarity, M.E. Bremer, P.C. Isakson and S.A. Gregory, Selective inhibition of cyclooxygenase (COX)-2 reverses inflammation and expression of COX-2 and interleukin 6 in rat adjuvant arthritis, *J. Clin. Invest.* **97** (1996), 2672–2679.

[2] I. Appleton, A. Tomlinson, J.A. Mitchell and D.A. Willoughby, Distribution of cyclooxygenase isoforms in murine chronic granulomatous inflammation. Implications for future anti-inflammatory therapy, *J. Pathol.* **176** (1995), 413–420.

[3] C.P. Armstrong and A.L. Blower, Non-steroidal anti-inflammatory drugs and life threatening complications of peptic ulceration, *Gut* **28** (1987), 527–532.

[4] K.R. Aupperle, B.L. Bennett, D.L. Boyle, P.P. Tak, A.M. Manning and G.S. Firestein, NF-κB regulation by IκB kinase in primary fibroblast-like synoviocytes, *J. Immunol.* **163** (1999), 427–433.

[5] P.A. Baeurle and V.R. Baichwal, NF-κB as a target for immunosuppressive and anti-inflammatory molecules, *Adv. Immunol.* **65** (1997), 111–118.

[6] J. Barnett, J. Chow, D. Ives, M. Chiou, R. Mackenzie, E. Ösen, B. Nguyen, S. Tsing, C. Bach, J. Freire et al., Purification, characterization and selective inhibition of human prostaglandin G/H synthase 1 and 2 expressed in the baculovirus system, *Biochim. Biophys. Acta* **1209** (1994), 130–139.

[7] M. Betz and B.S. Fox, PGE2 inhibits production of Thl lymphokines but not of Th2 lymphokines, *J. Immunol.* **146** (1991), 108–113.

[8] D. Bishop-Bailey and T. Hla, Endothelial cell apoptosis induced by the peroxisome proliferator-activated receptor (PPAR) ligand 15-deoxy-Delta-12,14-prostaglandin J2, *J. Biol. Chem.* **274**(24) (1999), 17 042–17 048.

[9] J. Bondeson, B. Foxwell, F. Brennan and M. Feldmann, Defining therapeutic targets by using adenovirus: Blocking NF-κB inhibits both inflammatory and destructive mechanisms in rheumatoid synovium but spares anti-inflammatory mediators, *Proc. Natl. Acad. Sci. USA* **96** (1999), 5668–5673.

[10] M.R. Boothby, A.L. Mora, D.C. Scherer, J.A. Brockman and D.W. Ballard, Perturbation of the T lymphocyte lineage in transgenic mice expressing a constitutive repressor of nuclear factor (NF)-κB, *J. Exp. Med.* **185** (1997), 1897–1907.
[11] K.M.K. Bottomley, R.J. Griffiths, T.J. Rising and A. Steward, A modified mouse air pouch model for evaluating the effects of compounds on granuloma-induced cartilage degradation, *Br. J. Pharmacol.* **93** (1986), 627–635.
[12] K.D. Brandt, Effects of non-steroidal anti-inflammatory drugs on chondrocyte metabolism in vitro and in vivo, *Am. J. Med.* **83** (1987), 29–34.
[13] K.D. Brandt, Non-steroidal anti-inflammatory drugs and articular cartilage, *J. Rheumatol.* **14** (1987), 132–133.
[14] F. Capasso, C.J. Dunn, S. Yamamoto, D.A. Deporter, J.P. Giroud and D.A. Willoughby, Pharmacological mediators of various immunological and non-immunological inflammatory reactions produced in the pleural cavity, *Agents and Actions* **5**(5) (1975), 528–533.
[15] C.L. Chen, N. Singh, F.E. Yull, D. Strayhorn, L. Van Kaer and L.D. Kerr, Lymphocytes lacking IκB-α develop normally, but have selective defects in proliferation and function, *J. Immunol.* **165** (2000), 5418–5427.
[16] G. Chinetti, S. Griglio, M. Antonucci, I.P. Torra, P. Delerive, Z. Majd, J.C. Fruchart, J. Chapman, J. Najib and B. Staels, Activation of proliferator-activated receptors alpha and gamma induces apoptosis of human monocyte-derived macrophages, *J. Biol. Chem.* **273**(40) (1998), 25 573–25 580.
[17] P.R. Colville-Nash, S.S. Qureshi, D. Willis and D.A. Willoughby, Inhibition of inducible nitric oxide synthase by peroxisome proliferator-activated receptor agonist: Correlation with induction of heme oxygenase-1, *J. Immunol.* **161** (1998), 978–984.
[18] J.A. DiDonato, M. Hayakawa, D.M. Rothwarf, E. Zandi and M. Karin, A cytokine-responsive IκB kinase that activates the transcription factor NF-κB, *Nature* **388** (1997), 548–554.
[19] J.E. Dinchuk, B.D. Car, R.J. Focht, J.J. Johnston, B.D. Jaffee, M.B. Covington, N.R. Contel, V.M. Eng, R.J. Collins, P.M. Czerniak et al., Renal abnormalities and an altered inflammatory response in mice lacking cyclooxygenase II, *Nature* **378** (1995), 406–409.
[20] J.T. Dingle and M.J. Shield, The interactions of cytokines, NSAIDs and prostaglandins in cartilage destruction and repair, *Adv. Prost. Thromb. Leukotr. Res.* **21** (1990), 955–966.
[21] B.M. Forman, P. Tontonoz, J. Chen, R.P. Brun, B.M. Spiegelman and R.M. Evans, 15-Deoxy-Delta-12,14-prostaglandin J2 is a ligand for the adipocyte determination factor PPAR gamma, *Cell* **83** (1995), 803–812.
[22] N. Futaki, K. Yoshikawa, Y. Hamasaka, I. Arai, S. Higuchi, H. Iizuka and S. Otomo, NS-398, a novel non-steroidal anti-inflammatory drug with potent analgesic and antipyretic effects, which causes minimal stomach lesions, *Gen. Pharmacol.* **24** (1993), 105–110.
[23] S.H. Gavett, S.L. Madison, P.C. Chulada, P.E. Scarborough, W. Qu, J.E. Boyle, H.F. Tiano, C.A. Lee, R. Langenbach, V.L. Roggli and D.C. Zeldin, Allergic lung responses are increased in prostaglandin H synthase-deficient mice, *J. Clin. Invest.* **104** (1999), 721–732.
[24] D.M. Gerlag, L. Ransone, P.P. Tak, Z. Han, M. Palanki, M.S. Barbosa, D. Boyle, A.M. Manning and O.S. Firestein, The effect of a T cell-specific NF-κB inhibitor or in vitro cytokine production and collagen-induced arthritis, *J. Immunol.* **165** (2000), 1652–1658.
[25] J.K. Gierse, S.D. Hauser, O.P. Creely, C. Koboldt, S.H. Rangwala, P.C. Isakson and K. Seibert, Expression and selective inhibition of the constitutive and inducible forms of human cyclooxygenase, *Biochem. J.* **305** (1995), 479–484.
[26] D.W. Gilroy, P.R. Colville-Nash, D. Willis, J. Chivers, M.J. Paul-Clark and D.A. Willoughby, Inducible cyclooxygenase may have anti-inflammatory properties, *Nat. Med.* **5**(6) (1999), 598–701.
[27] D.W. Gilroy, A. Tomlinson, K. Greenslade, M.P. Seed and D.A. Willoughby, The effects of cyclooxygenase 2 inhibitors on cartilage erosion and bone loss in a model of Mycobacterium tuberculosis induced monoarticular arthritis in the rat, *Inflammation* **22**(5) (1998), 509–519.
[28] D.W. Gilroy, A. Tomlinson and D.A. Willoughby, Differential effects of inhibitors of cyclooxygenase (cyclooxygenase 1 and cyclooxygenase 2) in acute inflammation, *Eur. J. Pharmacol.* **355** (1998), 211–217.
[29] J.S. Goodwin, R.P. Messner and G.T. Peake, Prostaglandin suppression of mitogen-stimulated lymphocytes in vitro. Changes with mitogen dose and preincubation, *J. Clin. Invest.* **62** (1978), 753–760.
[30] Z. Han, D.L. Boyle, A.M. Manning and G.S. Firestein, AP-1 and NF-κB regulation in rheumatoid arthritis and murine collagen-induced arthritis, *Autoimmunity* **28** (1998), 197–208.
[31] J.T. Huang, J.S.Welch, M. Ricote, C.J. Binder, T.M. Willson, C. Kelly, J.L. Witztum, C.D. Funk, D. Conrad and C.K. Glass, Interleukin-4-dependent production of PPAR-gamma ligands in macrophages by 12/15-lipoxygenase, *Nature* **400**(6742) (1999), 378–382.
[32] C. Jiang, A.T. Ting and B. Seed, PPAR-gamma agonists inhibit production of monocyte inflammatory cytokines, *Nature* **391** (1998), 82–86.
[33] S. Kargman, E. Wong, G.M. Greig, J.P. Falgueyret, W. Cromlish, D. Ethier, J.A. Yergey, D. Riendeau, J.F. Evans, B. Kennedy et al., Mechanism of selective inhibition of human prostaglandin G/H synthase-1 and -2 in intact cells, *Biochem. Pharmacol.* **52** (1996), 1113–1125.

[34] M. Karin and Y. Ben-Neriah, Phosphorylation meets ubiquitination: The control of NF-κB activity, *Annu. Rev. Immunol.* **18** (2000), 621–663.

[35] Y. Kawahito, M. Kondo, Y. Tsubouchi, A. Hashiramoto, D. Bishop-Bailey, K. Inoue, M. Kohno, R. Yamada, T. Hla and H. Sano, 15-deoxy-$\Delta^{12,14}$-PGJ_2 induces synoviocyte apoptosis and suppresses adjuvant-induced arthritis in rats, *J. Clin. Invest.* **106** (2000), 189–197.

[36] Z.W. Li, W. Chu, Y. Hu, M. Delhase, T. Deerinck, M. Ellisman, R. Johnson and M. Karin, The IKKβ subunit of IκB kinase (IKK) is essential for nuclear factor κB activation and prevention of apoptosis, *J. Exp. Med.* **189** (1999), 1839–1845.

[37] J.L. Masferrer, B.S. Zweifel, P.T. Manning, S.D. Hauser, K.M. Leahy, W.G. Smith, P.C. Isakson and K. Seibert, Selective inhibition of inducible cyclooxygenase 2 in vivo is anti-inflammatory and nonulcerogenic, *Proc. Natl. Acad. Sci. USA* **91** (1994), 3228–3232.

[38] B. Mazieres, L. Berdah, M. Thiechart and G. Viguler, Diacetylrhein on a post-contusion model of experimental osteoarthritis in the rabbit, *Rev. Rheumatol.* **60**(6pt2) (1993) s77–s81.

[39] E.A. Meade, W.L. Smith and D.L. DeWitt, Differential inhibition of prostaglandin endoperoxide synthase (cyclooxygenase) isozymes by aspirin and other non-steroidal anti-inflammatory drugs, *J. Biol. Chem.* **268** (1993), 6610–6614.

[40] H.G. Meerphol and T. Bauknecht, Role of prostaglandin on the regulation of macrophage proliferation and cytotoxic functions, *Prostaglandins* **B31B** (1986), 961–972.

[41] A.V. Miagov, D.V. Kovalenko, C.E. Brown, J.R. Didsbury, J.P. Cogswell, S.A. Stimpson, A.S. Balswin and S.S. Makarov, NF-κB activation provides the potential link between inflammation and hyperplasia in the arthritic joint, *Proc. Natl. Acad. Sci. USA* **95** (1998), 13 859–13 864.

[42] M. Mian, D. Benetti, S. Rosini and R. Fantozzi, Effects of diacerein on the quantity and phagocytic activity of thioglycollate-elicited mouse peritoneal macrophages, *Pharmacology* **39**(6) (1989), 362–366.

[43] M. Mian, S. Brunelleschi, S. Tarli, A. Rubino, D. Benetti, R. Fantozzi and L. Zilletti, Rhein: an anthraquinone that modulates superoxide anion production from human neutrophils, *J. Pharm. Pharmacol.* **39**(10) (1987), 845–847.

[44] J.A. Mitchell, P. Akarasereenont, C. Thiemermann, R.J. Flower and J.R. Vane, Selectivity of nonsteroidal antiinflammatory drugs as inhibitors of constitutive and inducible cyclooxygenase, *Proc. Natl. Acad. Sci. USA* **90** (1993), 11 693–11 697.

[45] A.R. Moore, K.J. Greenslade, C.A.S. Alam and D.A. Willoughby, Effects of diacerhein on granuloma induced cartilage breakdown in the mouse, *Osteoarthritis and Cartilage* **6**(1) (1998), 19–23.

[46] T. Murata, F. Ushikubi, T. Matsuoka, M. Hirata, A. Yamasaki, Y. Sugimoto, A. Ichikawa, Y. Aze, T. Tanaka, N. Yoshida, A. Ueno, S. Oh-ishi and S. Narumiya, Altered pain perception and inflammatory response in mice lacking prostacyclin receptor, *Nature* **388**(6643) (1997), 678–682.

[47] N.M. Newman and R.S. Ling, Acetabular bone destruction related to non-steroidal anti-inflammatory drugs, *Lancet* **2** (1985), 11–14.

[48] G.P. O'Neill, J.A. Mancini, S. Kargman, J. Yergey, M.Y. Kwan, J.P. Falgueyret, M. Abramovitz, B.P. Kennedy, M. Quellet, W. Cromlish et al., Overexpression of human prostaglandin G/H synthase-1 and -2 by recombinant vaccinia virus: inhibition by nonsteroidal anti-inflammatory drugs and biosynthesis of 15-hydroxyeicosatetraenoic acid, *Mol. Pharmacol.* **45** (1994), 245–254.

[49] T.D. Penning, J.J. Talley, S.R. Bertenshaw, J.S. Carter, P.W. Collins, S. Docter, M.J. Graneto, L.F. Lee, J.W. Malecha, J.M. Miyashiro et al., Synthesis and biological evaluation of the 1,5-diarylpyrazole class of cyclooxygenase-2 inhibitors: identification of 4-[5-(4-methylphenyl)-3-(trifluoromethyl)-1H-pyrazol-1-yl]benzenesulfonamide (SC-58635, celecoxib), *J. Med. Chem.* **40** (1997), 1347–1365.

[50] P. Prasit, Z. Wang, C. Brideau, C.C. Chan, S. Charleson, W. Cromlish, D. Ethier, J.F. Evans, A.W. Ford-Hutchinson, J.Y. Gauthier et al., The discovery of rofecoxib, [MK 966, Vioxx, 4-(4′-methylsulfonylphenyl)-3-phenyl-2(5H)-furanone], an orally active cyclooxygenase-2-inhibitor, *Bioorg. Med. Chem. Lett.* **9** (1999), 1773–1778.

[51] S. Rashad, P. Revell, A. Hemingway, F. Low, K. Rainsford and F. Walker, Effect of non-steroidal anti-inflammatory drugs on the course of osteoarthritis *Lancet* **ii** (1989), 519–522.

[52] B.K. Reuter, S. Asfaha, A. Buret, K.A. Sharkey and J.L. Wallace, Exacerbation of inflammation-associated colonic injury in rat through inhibition of cyclooxygenase-2, *J. Clin. Invest.* **98** (1996), 2076–2085.

[53] M. Ricote, A.C. Li, T.M. Willson, C.J. Kelly and C.K. Glass, The peroxisome proliferator-activated receptor gamma is a negative regulator of macrophage activation, *Nature* **391** (1998), 79–82.

[54] D. Riendeau, M.D. Percival, S. Boyce, C. Brideau, S. Charleson, W. Cromlish, D. Ethier, J. Evans, J.P. Falgueyret, A.W. Ford-Hutchinson et al., Biochemical and pharmacological profile of a tetrasubstituted furanone as a highly selective COX-2 inhibitor, *Br. J. Pharmacol.* **121** (1997), 105–117.

[55] A. Rossi, P. Kapahl, G. Natoli, T. Takahashi, Y. Chen, M. Karin and M.G. Santoro, Anti-inflammatory cyclopentenone prostaglandins are direct inhibitors of IκB kinase, *Nature* **403** (2000), 103–108.

[56] A. Schmassmann, B.M. Peskar, C. Stettler, P. Netzer, B. Flogerzi and F. Halter, Effects of inhibition of prostaglandin endoperoxide synthase-2 in chronic gastro-intestinal ulcer models in rats, *Br. J. Pharmacol.* **123** (1998), 795–804.

[57] M.P. Seed, A.R. Bowden, F.P. Parker, S. Johns, A. Curnock and C.R. Gardner, Rapid quantitation of joint destruction in

irritant monoarticular arthritis induced by heat killed Mycobacterium tuberculosis, *Eur. J. Rheumatol. Inflamm.* **13** (1993), 27–33.
[58] K. Seibert, Y. Zhang, K. Leahy, S. Huser, J. Masferrer, W. Perkins, L. Lee and P. Isakson, Pharmacological and biochemical demonstration of the role of cyclooxygenase 2 in inflammation and pain, *Proc. Natl. Acad. Sci. USA* **91** (1994), 12 013–12 017.
[59] W.L. Smith and D.L. Dewitt, Prostaglandin endoperoxide H synthases-1 and -2, *Adv. Immunol.* **62** (1996), 167–215.
[60] W.L. Smith, R.M. Garavito and D.L. DeWitt, Prostaglandin endoperoxide H synthases (cyclooxygenases)-1 and -2, *J. Biol. Chem.* **271** (1996), 33 157–33 160.
[61] D.S. Straus, G. Pascual, M. Li, J.S. Welch, M. Ricote, C.H. Hsiang, L.L. Sengchanthalangsy, G. Ghosh and C.K. Glass, 15-Deoxy-$\Delta^{12,14}$prostaglandinJ_2 inhibits multiple steps in the NF-κB signaling pathway, *Proc. Natl. Acad. Sci. USA* **97** (2000), 4844–4849.
[62] A. Tomlinson, I. Appleton, A.R. Moore, D.W. Gilroy, D. Willis, J.A. Mitchell and D.A. Willoughby, Cyclooxygenase and nitric oxide synthase isoforms in rat carrageenin-induced pleurisy, *Br. J. Pharmacol.* **113** (1994), 693–698.
[63] D.E. Van Epps, Suppression of human lymphocyte migration by PGE2, *Inflammation* **5** (1981), 81–87.
[64] J.R. Vane and R.M. Botting, The prostaglandins, in: *Aspirin and Other Salicylates*, J.R. Vane and R.M. Botting, eds, Chapman and Hall Medical, London, 1992, pp. 17–34.
[65] J.R. Vane, Inhibition of prostaglandin synthesis as a mechanism of action of aspirin like drugs, *Nat. New Biol.* **231** (1971), 232–235.
[66] J.R. Vane, Towards a better aspirin, *Nature* **367** (1994), 215–216.
[67] J.L. Wallace, A. Bak, W. McKnight, S. Asfaha, K. Sharkey and W.K. MacNaughton, Cyclooxygenase 1 contributes to inflammatory responses in rats and mice: implications for gastrointestinal toxicity, *Gastroenterology* **115**(1) (1998), 101–109.
[68] D.R. Webb and P.L. Osheroff, Antigen stimulation of prostaglandin synthesis and control of immune responses, *PNAS* **73** (1976), 1300–1304.
[69] K.I. Williams and G.A. Higgs, Eicosanoids and inflammation, *J. Pathol.* **156** (1998), 101–110.
[70] D.A. Willoughby, A.R. Moore and P.R. Colville-Nash, COX-1, COX-2 and COX-3 and the future treatment of chronic inflammatory disease, *Lancet* **355**(9204) (2000), 646–648.
[71] X.Y. Yang, L.H. Wang, T. Chen, D.R. Hodge, J.H. Resau, L. DaSilva and W.L. Farrar, Activation of human T lymphocytes is inhibited by peroxisome proliferator-activated receptor gamma (PPARgamma) agonists. PPARgamma co-association with transcription factor NFAT, *J. Biol. Chem.* **275**(7) (2000), 4541–4544.

PART III

Matrix Interactions – Clinical and Pharmacological Applications

Biorheology 39 (2002) 183–191
IOS Press

The influence of exercise on the composition of developing equine joints

Chris H.A. van de Lest[a,b,*], Pieter A.J. Brama[b] and P. René van Weeren[b]

[a] *Department of Biochemistry Cell Biology & Histology, Division of Biochemistry, P.O. Box 80.176, 3508 TD Utrecht, The Netherlands*

[b] *Department of Equine Sciences, Faculty of Veterinary Medicine, Utrecht University, Utrecht, The Netherlands*

Abstract. An overview is given of the direct and long-term effects of exercise on the biochemical characteristics of cartilage and subchondral bone, and on the metabolic activity of chondrocytes in the juvenile horse. In the experimental setup 43 foals were reared until weaning at 5 months of age under similar conditions, except for the type and amount of exercise. Fifteen foals remained at pasture (Pasture group and also control group), 14 foals were kept in box stalls (Box group), and 14 foals were kept in the same box stalls but were subjected daily to an increasing number of gallop sprints (Training group). After weaning 8 foals from each group were euthanised. All remaining 19 animals were housed together in a loose box with access to a small paddock to study a possible reversibility of exercise-induced effects. *Post mortem* subchondral bone and cartilage samples were collected and analysed for bone morphogenic enzymes, matrix composition, chondrocyte metabolic activity, and bone mineral density.

It resulted that lack of exercise leads to a retardation of the normal development of the joint. This is largely compensated for when afterwards a more normal exercise regimen is followed. Most parameters in the Training group approximated those of the pastured foals at age 5 months. However, at age 11 months there were indications for a reduced performance of the investigated tissues in this group.

It is concluded that regular, sub-maximal loading, as occurred in the Pasture group, seems best for an optimal development of the musculoskeletal tissues. The combination of short bouts of heavy exercise superimposed on a basic box rest regimen appears to have adverse effects on long-term viability of the tissues and may hence lead to an impaired resistance to injury.

1. Introduction

There is a world-wide tendency that intensive sports training starts at an ever earlier age. This is the case with both human athletes, and with the only domesticated animal that is kept primarily for its athletic potential: the horse. Although many reports have demonstrated the influence of mechanical stress on chondrocyte metabolism, and on cartilage and bone composition, only little is known about the effects of intensive training on the still developing musculoskeletal tissues in the juvenile individual.

Joints form a particularly vulnerable part of the musculoskeletal system. The composition of the extracellular matrix determines to a great extent the biomechanical properties of the cartilage and the underlying subchondral bone. Both tissue types are subjected to high mechanical stresses that have to be resisted properly. A less than optimal molecular composition of the extracellular matrix may impair the biomechanical properties of the joint tissue. This, in turn, may lead to failure in coping with the strong compressive, shear and tensile forces to which the joint is continuously subjected. Matrix damage may

*Address for correspondence: C.H.A. van de Lest (PhD), Laboratory of Veterinary Biochemistry, Utrecht University, P.O.B. 80.176, 3508 TD Utrecht, The Netherlands. Tel.: +31 30 2535396; Fax: +31 30 2535492; E-mail: C.vandeLest@vet.uu.nl.

ensue and will finally lead to overt pathology of the joint. Therefore, a high-quality matrix, which is able to resist day-to-day wear and tear and the occasional peak load is of great importance.

Physical training has often been considered to be beneficial for the properties of articular cartilage and of bone, although catabolic responses to heavy exercise have also been reported [7]. In experiments with dogs, the effects of running on cartilage were shown to depend on the intensity of training. Positive responses, like increased glycosaminoglycan levels and cartilage thickness, were observed with light or moderate training [17,20], but strenuous training produced degenerative changes [24] and a concomitant decrease in glycosaminoglycan content [1,18]. Also, a reduced loading due to splint immobilization of the joint decreased aggrecan content in articular cartilage of young dogs [13].

In bone the positive effect of exercise on joint quality has been documented extensively in several species, including the horse [6,15,16]. As expected, the effect was most pronounced in young individuals [10,11]. So far, most efforts have been directed at the determination of bone mineral density (BMD), reports explicitly dealing with the effect of exercise on the organic part of bone are limited. Recently, the authors investigated the influence of maturation and exercise on the collagen characteristics [3,4], and on [3,4] bone morphogenic enzymes [21] of equine subchondral bone [21].

In the present paper the various studies on the direct and long-term effects of exercise on the biochemical and metabolic properties of articular cartilage and (subchondral) bone in the horse are reviewed in order to obtain a comprehensive view of what various exercise regimens in young growing individuals might mean for the eventual quality, and hence for the injury resistance, of musculoskeletal tissues.

2. Materials and methods

2.1. *Origin of the foals and experimental set-up*

This report is the result of a comprehensive study into the effects of exercise at young age on the development of the equine musculoskeletal system that has been described in detail elsewhere [2,23]. The study was approved by the Utrecht University Ethical Committee.

In brief, 43 Dutch warmblood foals were kept at pasture with their dams during the first week after birth, and then randomly allotted to one of three exercise groups. The Box group ($n = 14$) stayed for 24 h per day in a 3×3.5 m box stall. The Training group ($n = 14$) was housed in the same kind of box stalls but was given an increasing training load, and the Pasture group ($n = 15$) was kept at pasture for 24 hours a day. At the age of 5 months all foals were weaned. Twenty-four (8 from each group) were randomly selected and sacrificed directly after weaning. The remaining 19 foals were housed together in a loose house with access to a small paddock for an additional 6 months before being sacrificed at the age of 11 months.

2.2. *Cartilage and subchondral bone samples*

Directly after euthanasia of the foals the right femoropatellar and the right metacarpophalangeal joints were dissected. Full thickness articular cartilage samples, and subchondral bone samples of approximately 2 mm in depth and 4 mm in diameter were taken from predefined areas of these joints [3–5,21,22].

2.3. Origin of data used

All data were obtained from the same experimental group of foals. Details of the analysis techniques that were used can be found in the original papers. Data on cartilage metabolic properties, glycosaminoglycan and DNA content were reported by van den Hoogen et al. [22] analysed metabolic levels of cartilage explants directly after harvesting, and after 4 days of serum stimulation by ^{35}S-incorporation. Data on bone morphogenic enzymes activities in subchondral bone; bone-specific alkaline phosphatase (ALP), tartrate resistant acid phosphatase (TRAP) and lysyl oxidase (LO) come from a study by van de Lest et al. [21]. The data on collagen composition of cartilage and on collagen composition, water-, DNA- and calcium content of subchondral bone come from studies by Brama et al. [3–5]. Finally, data on cross-sectional area and bone mineral density of the 3rd metacarpal bone (measured by peripheral quantitative computer tomography) stem from the study by Cornelissen et al. [10].

2.4. Statistical analysis

Statistical analysis for multiple group comparison was performed using two-way ANOVA; post multiple group comparison was performed using Dunnett's multiple comparison test, in which the Pasture group was assumed to be the control group, since pasture raising can be seen as the normal situation in equine husbandry. All data are represented as mean ± SD. Levels of statistically significant differences are indicated by P-values ($P < 0.05$; $P < 0.01$; $P < 0.001$).

3. Results

3.1. Metabolic properties of cartilage explants [22]

The metabolic properties of the cartilage explants were expressed in 4 parameters: (a) the release of endogenous glycosaminoglycans into the medium during culture; (b) the proteoglycan synthesis rate directly after sampling of the cartilage explants (*ex vivo* proteoglycan synthesis); (c) proteoglycan synthesis rate after a 4 days of serum stimulation, and (d) the release of newly synthesised proteoglycans into the culture medium.

The release of glycosaminoglycans from the explants during 4 days of serum stimulation was significantly higher in the Training group, both directly after the training was stopped, and after a 6 months period of equal exercise. Box-rest during the first 5 months of age had no influence on this parameter (Table 1).

In the 5 months old foals, the *ex vivo* proteoglycan synthesis of the explants was considerably higher in the Training group compared to the other groups ($P < 0.01$; Fig. 1). After 4 days of serum stimulation, however, the proteoglycan synthesis was equal in both the training and Pasture group, but in the Box group it was still significantly lower ($P < 0.05$; Fig. 1). The serum stimulation rates (degree to which the base level of synthesis could be raised by incubation with serum) of proteoglycan synthesis in the 5 months old groups were comparable in all three exercise groups. After an additional 6 months in which all foals were subjected to the same light exercise regimen, *ex vivo* proteoglycan synthesis in the former Training group was at the same level as that of the other groups. However, now the rate of serum stimulation was virtually zero in the former Training group and had become significantly less than in both other groups ($P < 0.001$; Fig. 1). No age effect was observed in the *ex vivo* proteoglycan synthesis levels, but

Table 1

Influence of different exercise regimens on the measured cartilage parameters (mean ± SD) in the 5 and 11 months old foal groups. GAG, sulphated glycosaminoglycan; LO, lysyl oxidase; TRAP, tartrate resistant acid phosphatase; HP, hydroxylysylpyridinoline cross-links; ww = wet weight

	Box rest group		Training group		Pasture group		Age effect	Source
	5 ($n = 8$)	11 ($n = 6$)	5 ($n = 8$)	11 ($n = 6$)	5 ($n = 8$)	11 ($n = 7$)		
GAG (mg/g ww)	39.7±4.2§	25.3±1.9	28.7±1.8	24.1±1.7	29.8±2.0	27.4±1.2	$P < 0.001$	[22]
DNA (ng/g ww)	270±10§§¶¶	206±6§§	204±12	179±10	206±11	166±11	$P < 0.0001$	[22]
Collagen (mg/g dw)	550±80	590±70	500±50	560±40	540±30	570±40	$P < 0.01$	[5]
HP (mmol/mol triple helix)	880±130	1000±90	920±100	1030±60	910±40	1000±80	$P < 0.01$	[5]
GAG-release (%)	9.2±1.4	15.4±1.6	13.2±2‡	23±4.1¶‡	8.3±0.4	13.1±1.2	$P < 0.0001$	[22]
Release of newly synthesised GAG's (%)	24.7±1.8§	20.7±0.9	20.9±1	20.8±0.7	18.9±1.1	22.5±0.9	–	

Legend for statistical symbols: § = significantly different between Groups Box and Pasture ($^{§}P < 0.05$; $^{§§}P < 0.01$); ¶ = significantly different between Groups Box and Training ($^{¶}P < 0.05$; $^{¶¶}P < 0.01$); ‡ = significantly different between Groups Training and Pasture ($P < 0.05$).

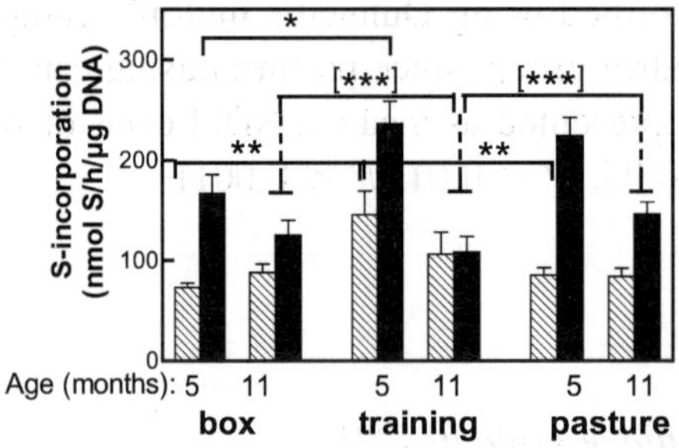

Fig. 1. Sulphate incorporation as a measure of proteoglycan synthesis in cartilage explants of foals that were subjected to different exercise regimens during the first 5 months of life. Cartilage was examined directly after these five months, and after 6 months of a similar exercise programme (11 months of age). The dashed bars represent the *ex vivo* S-incorporation rates ± SEM, and the solid bars represent the S-incorporation after 4 days of serum stimulation. Statistically significant differences are denoted by asterisks; $^{*}P < 0.05$, $^{**}P < 0.01$, $^{***}P < 0.001$. The notations between square brackets "[*]" refer to statistical significancy in the rate of stimulation. Data originate from van den Hoogen et al. [22].

the rate of stimulation and proteoglycan synthesis after 4 days were significantly lower in cartilage of 11 months old foals (both $P < 0.001$).

Even though the proteoglycan synthesis rate of cartilage in 5 months old foals after 4 days of serum stimulation was lowest in the Box group, the release of these newly synthesized proteoglycans, expressed as the relative amount of newly produced proteoglycans, was highest in this group ($P < 0.02$; Table 1). In the 11 months old foals, the relative release of newly synthesized proteoglycans was comparable for all former exercise groups. The level of release of newly synthesized proteoglycans was age independent.

The influence of exercise on the relative release of endogenous proteoglycans, as measured by their relative contribution to the total amount of glycosaminoglycans released into the culture medium during the first 4 days in culture, was significant in both 5 and 11 month old foals ($P < 0.05$). In both the 5 and 11 months old foals glycosaminoglycan release was highest in the (former) Training groups (Table 1). Proteoglycan release was significantly increased in 11 months old foals compared to 5 months old foals ($P < 0.001$).

Table 2

Influence of different exercise regimens on the measured subchondral bone parameters (mean ± SD) in the 5 and 11 months old foal groups. ALP, alkaline phosphatase; LO, lysyl oxidase; TRAP, tartrate resistant acid phosphatase; HP, hydroxylysylpyridinoline cross-links; LP, lysylpyridinoline cross-links; dw = dry weight; NA, not available

	Box rest group		Training group		Pasture group		Age effect	Source
	5 ($n = 8$)	11 ($n = 6$)	5 ($n = 8$)	11 ($n = 6$)	5 ($n = 8$)	11 ($n = 7$)		
ALP (U/mg protein)	14.9±4.5§	42.9±9.8	14.9±8.7§	34.5±4.7	23.5±4.9	40.9±10.8	$P < 0.001$	[21]
LO (U/mg protein)	0.40±0.10§¶	0.59±0.23§	0.99±0.37	1.13±0.74	0.73±0.28	1.42±1.01	$P < 0.001$	
TRAP (U/mg protein)	19.9±5.8§	22.7±7.6§¶	18.8±3.9§	12.9±4.4	12.0±1.6	16.7±4.1	$P < 0.001$	
Calcium (mg/g dw)	252±21§§¶	NA	280±13	NA	280±10	300±20	–	[3,4]
Collagen (mg/g demin.dw)	477±69	NA	527±61	NA	510±70	600±50	$P < 0.05$	
HP (mmol/mol triple helix)	109±16§§¶¶	NA	148±26	NA	155±33	149±30	–	
LP (mmol/mol triple helix)	59±13§§¶¶	NA	81±11	NA	87±26	83±18	–	

Legend for statistical symbols: § = significantly different between Groups Box and Pasture ($^{\S}P < 0.05$, $^{\S\S}P < 0.01$); ¶ = significantly different between Groups Box and Training ($^{\P}P < 0.05$, $^{\P\P}P < 0.01$); ‡ = significantly different between Groups Training and Pasture ($P < 0.05$).

3.2. *Biochemical composition of cartilage* [5,22]

Both glycosaminoglycan- and DNA-contents of the cartilage explants were considerably higher in the Box group (Table 1). At 11 months, DNA content in the Box group was still significantly influenced, although less prominent ($P < 0.01$), while the effect of box rest on glycosaminoglycan content had ceased completely. No effects of exercise level were observed on collagen or water content, nor on collagen cross-link levels. Glycosaminoglycan, DNA and water contents were significantly lower, and collagen and collagen cross-link levels were significantly higher in the older foals (Table 1).

3.3. *Biochemical composition of subchondral bone* [3,4,21]

At 5 months of age there were large differences between the Pasture (control) group and both other groups with respect to all enzymes that were measured. ALP was significantly decreased in both the Box and the Training group, whereas TRAP was increased. There were no differences between the Training and the Box groups for these two enzymes (Table 2). The ratio between ALP and TRAP, which was about 2 : 1 in the Pasture group, was about 1 : 1.3 in the two other groups. Lysyl oxidase was significantly lower in the Box group than in the Pasture group. In this case, levels in the Training and Pasture groups were comparable.

At 11 months of age, ALP levels were comparable in all former exercise groups. TRAP levels were still higher in the former Box group, but in the former Training group they had dropped to values comparable to those of the former Pasture group. The ALP : TRAP ratios varied from 2 to 2.5 : 1. Lysyl oxidase levels had increased from 5 to 11 months, but were still significantly lower in the former Pasture group.

3.4. *Bone mineral density* [10]

Bone mineral density (BMD) at 5 months was lowest in the box-rest group. The same applies to cross-sectional area (CSA) (Table 3). At 11 months CSA was comparable in all groups and BMD in the formerly box-rested foals had made up completely for lost ground. However, it is interesting to note that in the former Training group BMD did not increase at all, which led to a significant difference with the former Pasture group (Table 3).

Table 3

Influence of different exercise regimens on the bone density and cross-section area (mean ± SD) in the 5 and 11 months old foal groups. CSA, cross-sectional area; McIII, 3rd metacarpal bone; BMD, bone mineral density; PSB, proximal sesamoid bone. Data originate from Cornelissen et al. [10]

	Box rest group		Training group		Pasture group		Age effect
	5 ($n = 8$)	11 ($n = 6$)	5 ($n = 8$)	11 ($n = 6$)	5 ($n = 8$)	11 ($n = 7$)	
CSA McIII (mm^2)	509±41§	686±43	530±38	709±54	562±42	714±55	$P < 0.001$
BMD (mg/cm^3) apex PSB	457±50§§¶¶	582±30¶	521±33	528±29‡	506±30	577±43	$P < 0.001$

Legend for statistical symbols: § = significantly different between Groups Box and Pasture ($^{§}P < 0.05$, $^{§§}P < 0.01$); ¶ = significantly different between Groups Box and Training ($^{¶}P < 0.05$, $^{¶¶}P < 0.01$); ‡ = significantly different between Groups Training and Pasture ($P < 0.05$).

4. Discussion

Juvenile joints consist of dynamic tissues that are capable of continuous remodelling, which makes them apt to respond adequately to the changes in biomechanical loading they are subjected to. This is in contrast to mature joints, where remodelling activity has decreased to a low level and hence the capacity for functional adaptation is limited. This is especially true for articular cartilage. Juvenile cartilage is a very dynamic tissue. Apart from changes caused by growth, the deeper cartilage zones undergo mineralisation as well. This implicates a high metabolic activity of juvenile cartilage and requires a high nutritional flow through the tissue. The higher nutritional needs of juvenile cartilage can also be deduced from the fact that blood vessels are still present in the deeper zones of the cartilage in young animals. In the horse, it has been reported that such vessels exist up to 7 months of age [8]. Exercise may have contradictory effects on the nutritional flow in juvenile cartilage. On the one hand, exercise may stimulate flow of nutrients by the dynamic compression of the cartilage, however, on the other hand excessive exercise may lead to micro-trauma and the obstruction of blood vessels.

At 5 months, proteoglycan synthesis seemed favourably influenced by training, since the cartilage samples from the Training group were the most active *ex vivo*, and could also be stimulated well by serum *in vitro*. The *ex vivo* proteoglycan synthesis in the Pasture group was lower than in the Training group, but reached the same level after 4 days of serum stimulation. Hence, the overall proteoglycan synthesizing capacity was similar in these two groups. Box rest did not seem to be beneficial to proteoglycan synthesis. Proteoglycan synthesis *ex vivo* in the Box group was lower than in the Training group, and was still so after a 4 day serum culture period. At the long term, this pattern had substantially changed. At 11 months the *ex vivo* proteoglycan synthesis was still highest in the former Training group, yet these chondrocytes appeared to completely have lost their capacity to be stimulated while the cartilage of the Pasture and Box groups of 11 months old foals could still be stimulated. This should be considered an extremely important finding, as the unresponsiveness to stimulation will virtually nullify the capacity for repair and hence severely impair the response to injury, rendering the cartilage highly vulnerable to the development of chronic degenerative ailments such as osteoarthritis. Another strong indication of the negative effect of the exercise regimen that was used in the Training group is the fact that the release of endogenous proteoglycans at 5 months was highest in the cartilage of the Training group, which was even more evident 6 months after the training program had ceased.

A low proteoglycan content or an altered proteoglycan composition may result in cartilage that is less resistant to biomechanical loading, which will enhance wear and increase the risk of cartilage injury. Since proteoglycan synthesis is one of the major tasks of chondrocytes, the rate of proteoglycan synthesis and the capacity to properly react to stimulation can be considered as indications for the vitality of chondrocytes. The reduced ability to increase proteoglycan synthesis during stimulation in the Training group suggests a metabolic exhaustion of the chondrocytes, which was most probably caused by prolonged over stimulation.

In the hypertrophic zone of the cartilage in growing individuals chondrocytes show signs of differentiation such as matrix vesicle budding and the production of ALP and type X collagen. Later in this process chondrocytes move into apoptosis, and are replaced by osteoblasts, which complete the process of matrix mineralisation and bone maturation. ALP, which is located mainly on the outer surface of the cell membranes of cells and matrix vesicles [14], is responsible for the liberation of inorganic phosphate from organic- or pyrophosphate. Subsequently, inorganic phosphate associates with calcium ions into hydroxyapatite, which is the major mineral in calcified cartilage and bone. While ALP is involved mainly in the formation of bone, TRAP is an important enzyme in the process of bone resorption by osteoclasts [12]. An increased activity of this enzyme therefore most likely indicates an increase of bone resorption. The ALP : TRAP ratio may thus to a certain extent be seen as one of the indicators of bone metabolic activity. In the present study this ratio was found to be different in both the Training and the Box group in comparison with the control (Pasture) group. This coincided with a significantly larger cross-sectional area (thus a larger increase in bone mass) in the 3rd metacarpal bone in the Pasture group at 5 months. At 11 months, after the 6 months of similar exercise, the difference in CSA had levelled out as had the difference in ALP. Although TRAP activity still remained high at that age, the ALP : TRAP ratio had levelled out too and had attained a value comparable to the value in the Pasture group at 5 months.

Whereas the increase in bone mass seems to be governed by the influence of continuous loading rather than peak loads, this appears not the case with respect to calcium content, bone mineral density and collagen cross-linking. In all cases, differences existed at 5 months between the Box group and the other two groups, but not between the Training and the Pasture groups. Apparently, the short, but heavy bouts of exercise the Training group was subjected to were sufficient to bring all these parameters to levels comparable with the Pasture group. This is important from a viewpoint of injury resistance. The responsiveness of the HP and LP cross-links to exercise is in agreement with the findings of Mansell and Bailey [19], who demonstrated that collagen metabolism is most active in the subchondral bone, making this zone highly susceptible to external influences. The lower degree of cross-linking in the Box group corresponded very nicely to the decreased activity of lysyl oxidase activity, making this enzyme into the most likely mediator of the exercise effect. In cartilage no effect of exercise on collagen cross-linking was observed. However, here also HP cross-linking levels tended to be lower in the Box group at 5 months. The difference is probably explained by the high turnover rate of juvenile mineralising cartilage at the bone–cartilage interface in which type II collagen from cartilage is replaced by type I collagen in bone, making subchondral bone more likely to demonstrate exercise-induced effects in cross-linking than cartilage.

The BMD values used in this study were not taken from subchondral bone, but just distal to the apical level of the proximal sesamoid bone which is the site where the suspensory ligament attaches. It can be expected that also at sites near the insertion of heavily loaded ligaments bone metabolism is very active, similar to frequently loaded subchondral areas. Indeed, calcium levels and the related BMD were both significantly lower in the group that was least biomechanically challenged: the Box group. At 11 months the former Box group seemed to have regained full metabolic activity. The ALP level was at the same height as in the former Pasture group, and TRAP activity was even higher. BMD had also attained the level of the former Pasture group, which meant a huge increase from 5 to 11 months. In the Pasture group a less extensive, age-related increase in BMD was seen, but in the former Training group BMD levels remained virtually the same. In the study by Cornelissen et al. [10] it was suggested that this phenomenon might be due to an (over)stimulation of the osteoblasts by the exercise regimen from which they had not recovered yet 6 months later. This corresponds well with the negative long-term effect on chondrocyte metabolism as discussed earlier. In this respect it should be mentioned that a similar effect was found in tendon tissue in this same group of foals [9].

It may be concluded from this overview that there is an evident and rather profound effect of the exercise regimen during the first months of life on the metabolic and biochemical status of the chondrocytes in articular cartilage and subchondral bone. If exercise is not given in a natural, more or less continuous and evenly distributed way, the biochemical status is disturbed at various levels. For example, the levels of ALP, TRAP, and LO differ significantly from the normal situation, which in turn may lead to a less than normal increase in total bone mass. If exercise is artificially provided in the form of short bouts of heavy exercise superimposed on a basic box rest regimen, levels of collagen cross-linking, calcium and bone mineral density glycosaminoglycan content attain normal values. However, although the basic activity of chondrocytes in the *ex vivo* situation does not seem impaired or is even stimulated, these chondrocytes lose their ability to properly respond to stimuli, which may have serious consequences for the composition and hence for the quality of the extracellular matrix. Also, BMD data and the enzyme levels of ALP and TRAP in the subchondral bone samples at age 11 months from the Training group suggest such a negative long-term effect of the specific training regimen too. In the case of the pyridine HP and LP cross-links, variations in levels correspond nicely with variations in the level of lysyl oxidase, making this enzyme into the most probable mediator of the effect of exercise.

It can be resumed that natural pasture exercise comes out best and both other exercise regimens seemed to have deleterious effects. This leaves the intriguing question open if well-dosed exercise on top of natural exercise could have a beneficial effect on the composition of the extracellular matrix, and thus on injury resistance.

References

[1] J. Arokoski, I. Kiviranta, J. Jurvelin, M. Tammi and H.J. Helminen, Long-distance running causes site-dependent decrease of cartilage glycosaminoglycan content in the knee joints of beagle dogs, *Arthritis Rheum.* **36** (1993), 1451–1459.

[2] A. Barneveld and P.R. van Weeren, Conclusions regarding the influence of exercise on the development of the equine musculoskeletal system with special reference to osteochondrosis, *Equine Vet. J. Suppl.* (1999), 112–119.

[3] P.A. Brama, J.M. TeKoppele, R.A. Bank, A. Barneveld and P.R. van Weeren, Maturation of subchondral bone in the foal and the influence of physical activity, *Equine Vet. J.* (2001) (in press).

[4] P.A. Brama, J.M. TeKoppele, R.A. Bank, A. Barneveld and P.R. van Weeren, Training alters the collagen framework of subchondral bone in foals, *Vet. J.* (2001) (in press).

[5] P.A. Brama, J.M. TeKoppele, R.A. Bank, P.R. van Weeren and A. Barneveld, Influence of different exercise levels and age on the biochemical characteristics of immature equine articular cartilage, *Equine Vet. J. Suppl.* **31** (1999), 55–61.

[6] S.H. Buckingham, R.N. McCarthy, G.A. Anderson, R.N. McCartney and L.B. Jeffcott, Ultrasound speed in the metacarpal cortex – a survey of 347 thoroughbreds in training, *Equine Vet. J.* **24** (1992), 191–195.

[7] J.A. Buckwalter and N.E. Lane, Athletics and osteoarthritis, *Am. J. Sports Med.* **25** (1997), 873–881.

[8] C.S. Carlson, L.D. Cullins and D.J. Meuten, Osteochondrosis of the articular–epiphyseal cartilage complex in young horses: evidence for a defect in cartilage canal blood supply, *Vet. Pathol.* **32** (1995), 641–647.

[9] W. Cherdchutham, C. Becker, R.K. Smith, A. Barneveld and P.R. van Weeren, Age-related changes and effect of exercise on the molecular composition of immature equine superficial digital flexor tendons, *Equine Vet. J. Suppl.* (1999), 86–94.

[10] B.P. Cornelissen, P.R. van Weeren, A.G. Ederveen and A. Barneveld, Influence of exercise on bone mineral density of immature cortical and trabecular bone of the equine metacarpus and proximal sesamoid bone, *Equine Vet. J. Suppl.* **31** (1999), 79–85.

[11] E.C. Firth, P.R. van Weeren, D.U. Pfeiffer, J. Delahunt and A. Barneveld, Effect of age, exercise and growth rate on bone mineral density (BMD) in third carpal bone and distal radius of Dutch Warmblood foals with osteochondrosis, *Equine Vet. J. Suppl.* **31** (1999), 74–78.

[12] Y. Fujikawa, M. Shingu, T. Torisu, I. Itonaga and S. Masumi, Bone resorption by tartrate-resistant acid phosphatase-positive multinuclear cells isolated from rheumatoid synovium, *Br. J. Rheumatol.* **35** (1996), 213–217.

[13] J. Haapala, M.J. Lammi, R. Inkinen, J.J. Parkkinen, U.M. Agren, J. Arokoski, I. Kiviranta, H.J. Helminen and M.I. Tammi, Coordinated regulation of hyaluronan and aggrecan content in the articular cartilage of immobilized and exercised dogs, *J. Rheumatol.* **23** (1996), 1586–1593.

[14] F.M.D. Henson, M.E. Davies, J.N. Skepper and L.B. Jeffcott, Localisation of alkaline phosphatase in equine growth cartilage, *J. Anat.* **187** (1995), 151–159.

[15] L.B. Jeffcott, Bone quality in horses, *Br. Vet. J.* **144** (1988), 1–3.

[16] L.B. Jeffcott, J.H. Hyland, A.A. MacLean, T. Dyke and G. Robertson-Smith, Changes in maternal hormone concentrations associated with induction of fetal death at day 45 of gestation in mares, *J. Reprod. Fertil. Suppl.* **35** (1987), 461–467.

[17] I. Kiviranta, M. Tammi, J. Jurvelin, A.M. Saamanen and H.J. Helminen, Moderate running exercise augments glycosaminoglycans and thickness of articular cartilage in the knee joint of young beagle dogs, *J. Orthop. Res.* **6** (1988), 188–195.

[18] M.J. Lammi, T.P. Hakkinen, J.J. Parkkinen, M.M. Hyttinen, M. Jortikka, H.J. Helminen and M.I. Tammi, Adaptation of canine femoral head articular cartilage to long distance running exercise in young beagles, *Ann. Rheum. Dis.* **52** (1993), 369–377.

[19] J.P. Mansell and A.J. Bailey, Abnormal cancellous bone collagen metabolism in osteoarthritis, *J. Clin. Invest.* **101** (1998), 1596–1603.

[20] A.M. Säämänen, M. Tammi, I. Kiviranta, J. Jurvelin and H.J. Helminen, Levels of chondroitin-6-sulfate and nonaggregating proteoglycans at articular cartilage contact sites in the knees of young dogs subjected to moderate running exercise, *Arthritis Rheum.* **32** (1989), 1282–1292.

[21] C.H.A. van de Lest, P.A. Brama and P.R. van Weeren, The influence of exercise on bone morphogenic enzyme activity of immature equine subchondral bone, *Biorheology* (2001) (submitted).

[22] B.M. van den Hoogen, C.H.A. van de Lest, P.R. van Weeren, L.M.G. van Golde and A. Barneveld, Effect of exercise on the proteoglycan metabolism of articular cartilage in growing foals, *Equine Vet. J. Suppl.* **31** (1999), 62–66.

[23] P.R. van Weeren and A. Barneveld, Study design to evaluate the influence of exercise on the development of the musculoskeletal system of foals up to age 11 months, *Equine Vet. J. Suppl.* (1999), 4–8.

[24] N. Vasan, Effects of physical stress on the synthesis and degradation of cartilage matrix, *Connect. Tissue Res.* **12** (1983), 49–58.

Biorheology 39 (2002) 193–199
IOS Press

Tensile fatigue behaviour of articular cartilage

Giordano Bellucci and Bahaa B. Seedhom

Rheumatology and Rehabilitation Research Unit, University of Leeds, 36 Clarendon Road, Leeds LS2 9NZ, UK
Tel.: +44 0113 233 49 42; E-mail: {b.b.seedhom, mrpgb}@leeds.ac.uk

Abstract. Although fatigue has been implicated in cartilage failure there are only two studies by the same author, and in both of which cartilage was tested in the direction parallel to the collagen orientation in the surface layer. In the present work articular cartilage was tested *also* along the perpendicular direction, being the direction in which cartilage possesses lower tensile strength.

Specimens were tested under cyclic tensile load. Number of cycles at failure was recorded as well as elongation of the specimen. To date 72 specimens have been tested all from one knee joint.

The number of cycles to failure ranged between two and 1.5 million. The surface and deep layers have better fatigue properties whether tested in the parallel or the perpendicular direction, while the middle layer was far weaker. Better fatigue behaviour was observed with specimens tested in parallel than in perpendicular direction to the fibres.

1. Introduction

Although fatigue was suggested as a failure mechanism of cartilage [1], only two studies have thus far investigated fatigue behaviour of cartilage, both by Weightman [3,4]. In both studies cyclic tensile load was applied on cartilage specimens from the surface layer of femoral heads, along the direction of the collagen fibrils. The load was applied for 1 second followed by 20 seconds of resting with a rise time of approximately 0.1 second. The main conclusions drawn from both studies were: that cartilage exhibited a typical fatigue behaviour, that is, as the magnitude of the applied cyclic stress decreased an increasing number of load cycles was required to produce fracture. Further, fatigue resistance of cartilage varied widely with age and from one femoral head to another.

Cartilage fatigue is being revisited here on methodological grounds. Firstly, in the studies by Weightman and colleagues, specimens from the superficial layer were subjected to tensile load along the direction of collagen fibrils. Since cartilage possess a much lower tensile strength in the perpendicular direction [2]; its resistance to fatigue in that direction is also likely to be weaker than it is along the direction of the collagen fibrils. Also the protocol of loading the specimen for one second and the allowing it to recover for twenty greatly differs from the physiological.

The present study aimed to investigate cartilage response to prolonged cyclic tensile loading, applied under physiological conditions, and to monitor the elongation of the test specimens, through the entire test to specimen failure.

0006-355X/02/$8.00

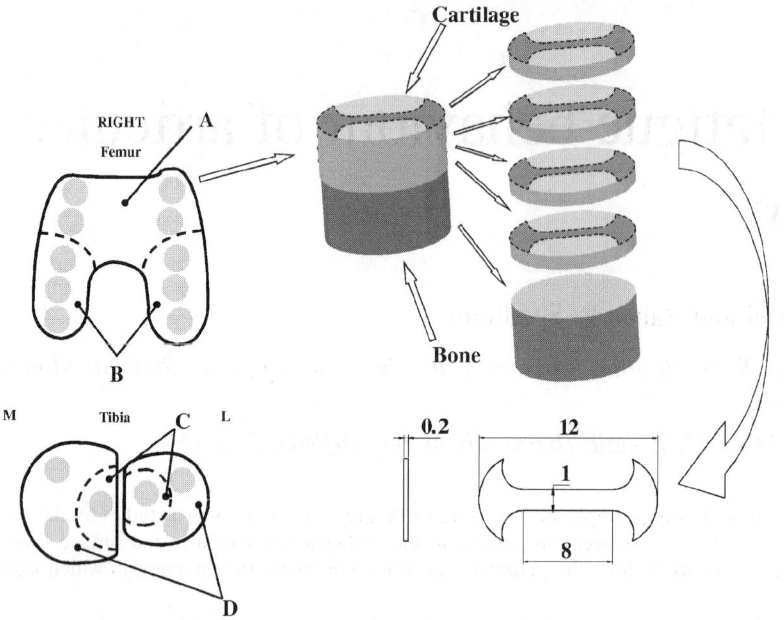

Fig. 1. Sites from which osteochondral plugs were harvested. (A) Patellar surface; (B) femoral condyles; (C) tibia plateau in direct contact with femur condyles; (D) tibia plateau covered with menisci. Also shown are the different layers tested and dimensions of the test specimen (mm).

2. Materials and methods

Only one human knee, 48 years old, has been used so far in this study. The degree of cartilage degeneration was visually assessed to single out fibrillated regions. A total of 16 osteochondral plugs were harvested from different regions and slices 200 μm thick were prepared from the different zones at each cartilage layer (Fig. 1). Dumbbell specimens were cut from these layers and tested in a specially constructed apparatus, which allowed the application of a cyclic tensile load onto the specimens (Fig. 2). The rise time and frequency of load application were possible to control and the system provided a near square wave load cycle with a rise time of approximately 150 ms.

Nominal tensile stress ranging 1.5 to 3 MPa was applied on the test specimens. Approximately half of the specimens were tested in the parallel direction to the collagen in the superficial layer and the other half in the perpendicular direction.

The data collection system together with software stored the various data, these being: (a) the number of load cycles to failure and (b) load and extension for each specimen. The number of load cycles to failure was determined using an electronic counter. After each 15 minutes (900 cycles at 1 Hz), the counter sent an impulse to the Microlink. When failure occurred, an electric switch automatically stopped the apparatus and the number of load cycles counted after the last impulse was recorded. The load and extension of the specimen were monitored through the test till specimen failure.

3. Preliminary results

The applied tensile stress versus the number of cycles to failure recorded for each cartilage specimen were plotted on a semi-logarithmic scale, in groups according to the orientation of load application with respect to that of the collagen in the superficial layer (Fig. 3).

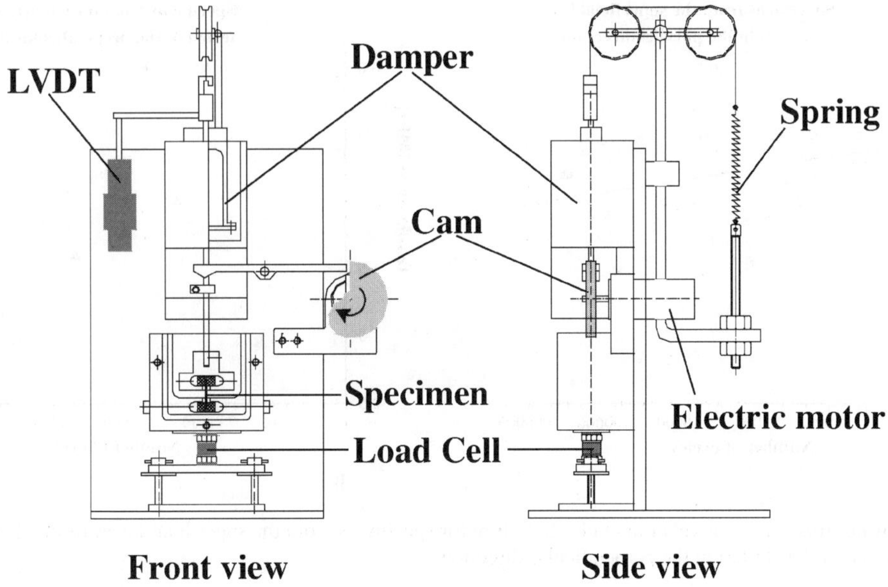

Fig. 2. Apparatus for applying cyclic tensile load.

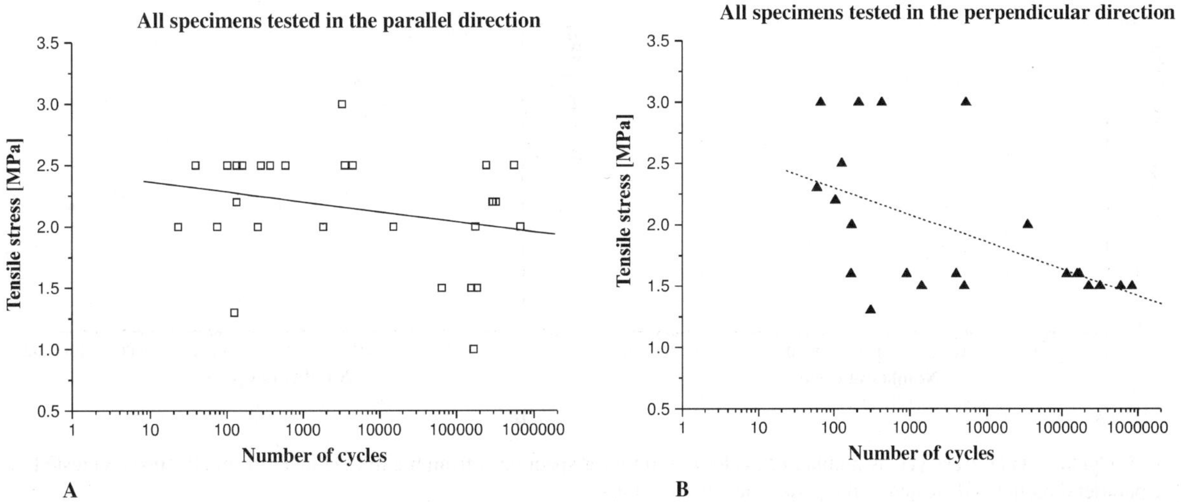

Fig. 3. Cyclic tensile stress versus number of cycles to failure for all specimens from all sites: (A) tested in the parallel direction; (B) tested in the perpendicular direction.

Next the data were divided in three groups from the superficial, the middle and the deep layers (Figs 4–6). With this division it was possible to compare the fatigue properties of each specimen with the different cartilage structure through the thickness. For each graph a linear relation representative of the fatigue behaviour was considered.

The tensile load applied was plotted against the elongation of every specimen to evaluate the changes of the tissue structure during the test (Fig. 7). The same elongation was plotted against the number of load cycles to failure (Fig. 8).

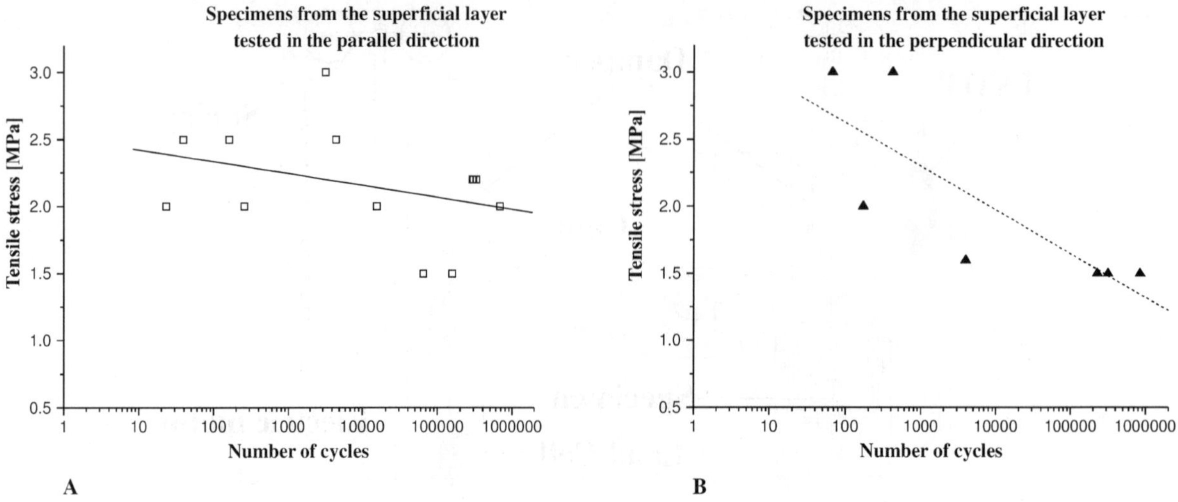

Fig. 4. Cyclic tensile stress versus number of cycles to failure for specimens from the superficial layer, from all sites: (A) tested in the parallel direction; (B) tested in the perpendicular direction.

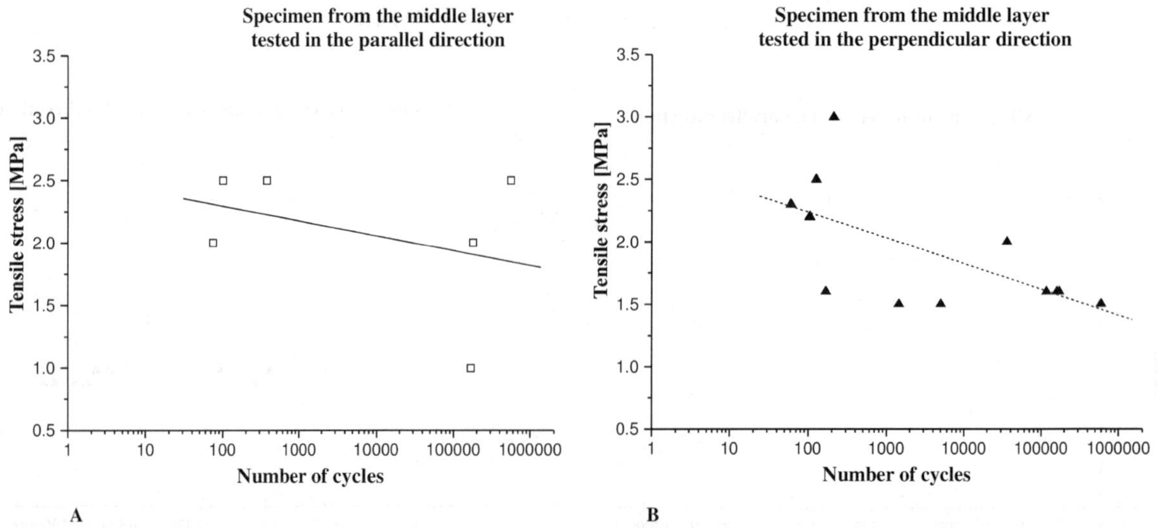

Fig. 5. Cyclic tensile stress versus number of cycles to failure for specimens from the middle layer, from all sites: (A) tested in the parallel direction; (B) tested in the perpendicular direction.

4. Comments on the results

The most striking finding of this investigation was the extremely wide scatter of the number of cycles to failure recorded for different specimens tested under the same tensile stress. The number of load cycles to failure thus varied from 20 to 1.5×10^6 in a range of stress applied varying between 1 and 3 MPa.

Considering the entire data collected to date, without regarding the zone or the region where the specimens were harvested from, a statistically significant linear correlation was found between applied stress and the number of cycles to failure, for the specimens tested in the perpendicular direction to that of the collagen in the surface layer ($r = -0.569$; $P = 0.004$). For the specimens tested in the parallel direction, the correlation was statistically much weaker ($r = -0.273$; $P = 0.159$).

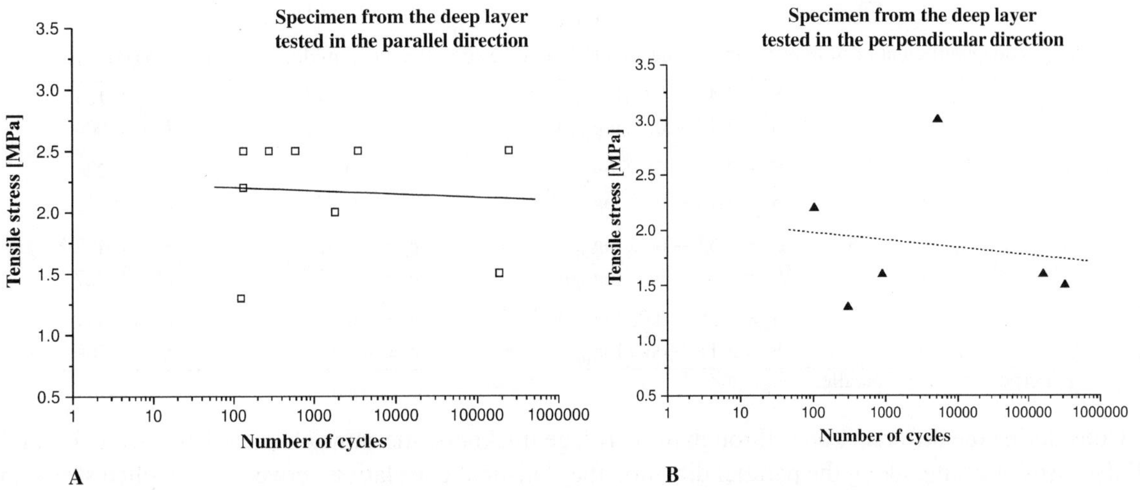

Fig. 6. Cyclic tensile stress versus number of cycles to failure for specimens from the deep layer, for all sites: (A) tested in the parallel direction; (B) tested in the perpendicular direction.

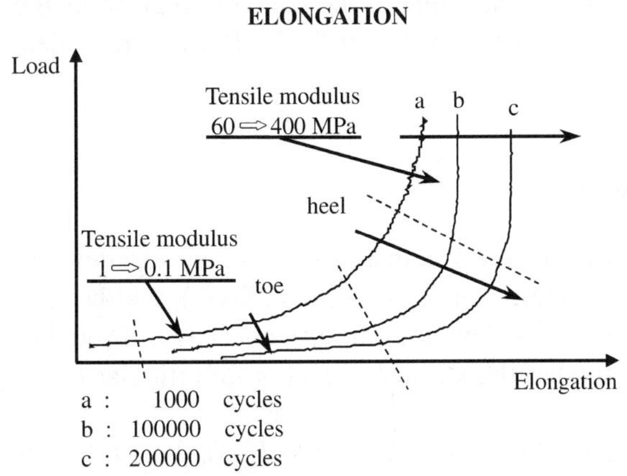

Fig. 7. Qualitative load-elongation trend during the fatigue test.

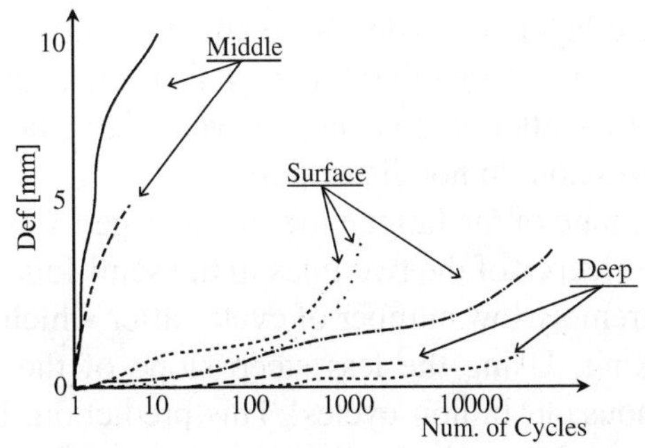

Fig. 8. Qualitative specimen elongation from different zones through the cartilage thickness during the fatigue test.

Table 1
Regression coefficient of fit lines for the whole data (W) and the superficial (S), middle (M) and deep (D) layer

W	=	$S = 2.44 - 0.08 \log_{10} N$	$r = -0.273$	$P = 0.159$
	+	$S = 2.74 - 0.22 \log_{10} N$	$r = -0.569$	$P = 0.004$
S	=	$S = 2.51 - 0.09 \log_{10} N$	$r = -0.346$	$P = 0.246$
	+	$S = 3.28 - 0.33 \log_{10} N$	$r = -0.823$	$P = 0.012$
M	=	$S = 2.53 - 0.12 \log_{10} N$	$r = -0.369$	$P = 0.472$
	+	$S = 2.65 - 0.21 \log_{10} N$	$r = -0.632$	$P = 0.027$
D	=	$S = 2.25 - 0.03 \log_{10} N$	$r = -0.075$	$P = 0.847$
	+	$S = 2.11 - 0.07 \log_{10} N$	$r = -0.155$	$P - 0.769$

+, Perpendicular; =, parallel.

Considering the different zones through the cartilage thickness, the same characteristics were found. In all the cases of testing along the parallel direction the statistical correlation between the applied stress and the number of cycles to failure was very weak (r ranging from -0.37 to -0.10 and P from 0.25 to 0.84). The weakest correlation was found among data from the deep layer, whether tested in the parallel or the perpendicular direction ($r = -0.075$, $P = 0.85$; $r = -0.155$, $P = 0.77$, respectively).

The fracture load of each specimen tested was determined as the intercept of the regression line with the vertical axis, i.e., the load axis. The value of the intercept represents the necessary load to break the specimen in one application. An average value of 2.55 MPa was recorded. This was much lower than published data [2,3].

5. Discussion and conclusions

The data obtained so far show that the specimens tested in the parallel direction to the split pattern (i.e., along the direction of fibre orientation in the superficial layer), had better fatigue properties than those tested in the perpendicular direction. This finding was consistent throughout the depth of cartilage and lends support to the hypothesis that the weakest direction of articular cartilage is the one perpendicular to the direction of collagen.

Comparing the present data with Weightman's results [3,4], the fatigue properties of cartilage revealed in this study seem to be better than shown by Weightman's data, whether the cartilage was tested in the parallel or perpendicular direction. This is evidenced by the smaller slope of the best line fitting the data on stress versus number of cycles to failure (on a semi-logarithmic scale), which indicates slower deterioration of the properties of the specimens.

The specimens from the middle layer seemed to have slightly worse fatigue resistance, breaking at a lower number of cycles than specimens from other layers. This is explicable by the much looser structure of collagen in the middle zone. The scatter of the data however is large between specimens from all zones and the slopes of the lines of regression do not differ significantly.

An attempt can be made to estimate of the fatigue life of cartilage from the data obtained in the current study. Basing the estimate on the steeper of the two lines in the semi-logarithmic expressions of cartilage fatigue behaviour predicts an extremely low number of cycles after which failure occurs; 100,000 cycles, barely one month of level walking. Using the less steep slope of the second expression predicts an extremely long life; some ten thousand billion cycles! This prediction, however seemingly unrealistic, confirms that cartilage resistance to fatigue in the direction parallel to the collagen fibres orientation is much greater than it is in the perpendicular direction.

Acknowledgements

The authors would like to express their thanks to their technicians Brian Whitham and Michael Pullan for their technical assistance during the course of this work. This work was supported by a University of Leeds scholarship.

References

[1] M.A.R. Freeman, *Modern Trends in Orthopaedics*, Butterworths, 1972.
[2] G.E. Kempson, Mechanical properties of articular cartilage, in: *Adult Articular Cartilage*, M.A.R. Freeman ed., Pitman Medical Ltd., London, 1979, pp. 333–414.
[3] B. Weightman, Tensile fatigue of human articular cartilage, *Journal of Biomechanics* **9** (1976), 193–200.
[4] B. Weightman, D.J. Chappell and E.A. Jenkins, A second study of tensile fatigue properties of human articular cartilage, *Annals of the Rheumatic Diseases* **37** (1978), 58–63.

Biorheology 39 (2002) 201–214
IOS Press

Nitric oxide (NO) and cartilage metabolism: NO effects are modulated by superoxide in response to IL-1

Jean-Yves Jouzeau *, Sandrine Pacquelet, Christelle Boileau, Emmanuelle Nedelec, Nathalie Presle, Patrick Netter and Bernard Terlain
Laboratoire de Pharmacologie et UMR 7561 CNRS-UHP, Faculté de Médecine, Avenue de la forêt de Haye, BP 184, 54505 Vandœuvre-lès-Nancy, France

Abstract. Nitric oxide (NO) is thought to mediate most effects of interleukin-1 (IL-1) on cartilage. *In vitro* evidence includes the decreased synthesis of extracellular matrix components, the abnormal cell renewal, the decreased production of IL-1 receptor antagonist, the induction of apoptosis and the enhanced sensitivity of chondrocytes to oxidative stress. Studies in $NOS2^{-/-}$ mice or administration of NO synthase inhibitors in animal models of joint disorders have confirmed its potent pathophysiological role in cartilage. Using L-NMMA (1 mM), as a NO synthase inhibitor, and CuDips (10 μM), as a SOD mimetic, we provide evidence that the inhibitory potency of IL-1β on proteoglycan synthesis and its stimulating effect on COX-2 activity depend both on NO and $O_2^{-\cdot}$ production. Peroxynitrite formation is further demonstrated by the occurrence of 3-nitrotyrosines in chondrocytes stimulated *in vitro* with 2.5 ng/ml IL-1 and in femoral condyles of rats injected locally with 1 μg IL-1. Preliminary data suggest that such contribution of reactive oxygen species is not shared in common by IL-17, another NO-producing cytokine. We conclude that superoxide is a key modulator of NO-mediated effects in chondrocyte stimulated with IL-1 and that a combined therapy with NO synthase inhibitors and antioxidants may be promising for a full cartilage protection.

Keywords: Interleukin-1, chondrocytes, proteoglycans, superoxide, nitric oxide, NO synthase inhibitors

1. Background

1.1. Moving from pro-inflammatory cytokines to nitric oxide

Joint disorders are characterized by a progressive loss of articular cartilage resulting from an imbalance between pro-inflammatory cytokines, as interleukin-1, -17 and -18 (IL-1, -17, -18), and regulatory or inhibitory systems including IL-1 receptor antagonist (IL-1ra), soluble receptors (IL-1 decoy) and immunomodulatory cytokines (IL-4, -10, -13) [2]. It is now widely accepted that IL-1, rather than TNFα, is a driving mediator of cartilage lesions in rheumatoid arthritis (RA) and osteoarthritis (OA) due to its ability to inhibit the synthesis of specific matrix components or tissue proteases inhibitors, and to favour their degradation by metalloproteases having a wide range of substrate specificity [31]. The major consequences are the disruption of the collagen network and the blurring of cell matrix interactions leading to a progressive decrease of the biomechanical properties of cartilage. Animals models have confirmed the relevance of IL-1 as a major target since its neutralisation either with specific antibodies [53] or with IL-1ra, administered systemically [9] or produced locally by genetically engineered synovial cells [32], provided cartilage protection. In this last decade, most of the effects of IL-1 on chondrocyte functions have been be ascribed, at least in part, to the production of nitric oxide (NO) (see Fig. 1).

*Address for correspondence: Tel.: +33 3 83 59 26 22; Fax: +33 3 83 59 26 21; E-mail: jouzeau@facmed.u-nancy.fr.

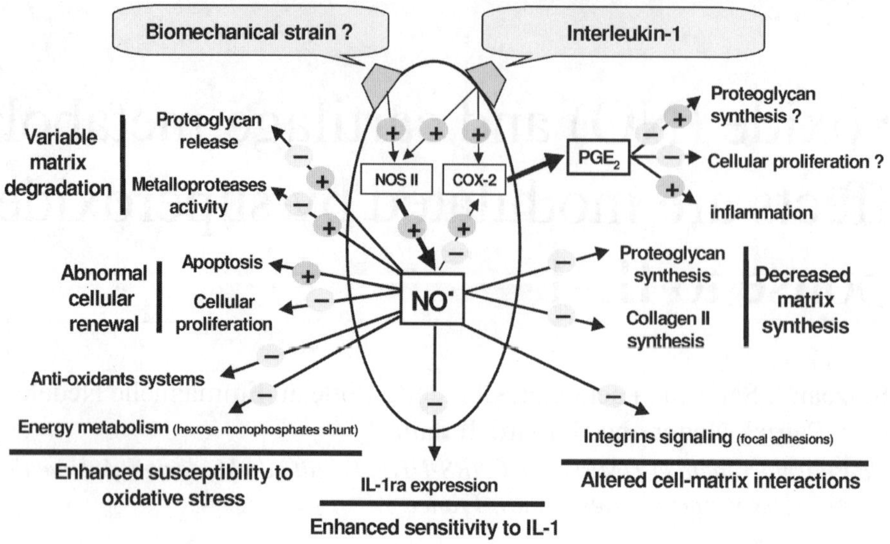

Fig. 1. Overall contribution of NO as a mediator of IL-1 effects on chondrocyte functions.

1.2. How nitric oxide was suspected

NO is a short half-life inorganic free radical, derived from the oxidation of the guanidino group of L-arginine by molecular oxygen. It behaves as an intercellular messenger at a low physiological level but is thought to have pathophysiological relevance when it is produced in huge amounts by the NO synthase (NOS) type 2 [13]. NOS 2 is highly inducible by pro-inflammatory cytokines in articular cells [29]. In humans, high levels of NO end-products [14] and nitrosoproteins have been found in serum and synovial fluid of RA patients and, less intensively, OA patients (see Table 1). NO overproduction is thought to originate from all articular tissues since NOS 2 expression was observed in inflamed synovium [41] and OA cartilage, although a "neuronal-like" isoform was also reported in human OA cartilage [1]. In addition, the production of NO by chondrocytes was stimulated by fluid-induced shear [12] and high peak stresses [22] but it was decreased by dynamic compressive strain [21], showing that it was a mediator of mechanotransduction processes. In the latter case, constitutive rather than inducible NOS activity is likely to be involved by a Ca^{++}-dependent mechanism [21].

1.3. Peculiarities of nitric oxide biochemistry

NO is a highly reactive compound, able to react with radicals and transition metals. Thus, it can interact with many proteins having a pivotal role in the body: (i) hemeproteins, leading to activation of guanylate cyclase, interaction with hemoglobin and modulation of cyclooxygenases; (ii) iron–sulfur enzymes, leading to inhibition of aconitase, ribonucleotide reductase and mitochondrial complexes with subsequent blockade of electron transport chains; (iii) poly(ADP-ribosyl)polymerase (PARP), whose indirect activation leads to a dramatic fall in energy store. Despite its short half life, NO is able to exert prolonged effects through its high affinity for thiol functions and the subsequent formation of nitrosoproteins and nitrosoglutathion [4]. The diversity of NO biology is further supported by its lack of receptor-mediated signalling pathways, ability to diffuse between cells depending on the pericellular environment and sensitivity of various transcription factors to the cellular redox status [6]. Of particular

Table 1

Indirect proofs that NO has pathological relevance in human joint diseases

Biological fluid/Tissue	Parameter	RA	OA
Serum/plasma	nitrite	++	+
	nitrite + nitrate	++	+
	3-nitrotyrosine	+	−
	nitrosoproteins	++	+
Synovial fluid	nitrite ± nitrate	+++	++
	3-nitrotyrosine	++	+ or −
	nitrosoproteins	++	+
Urines	nitrate/nitrite	+	Not determined
Leukocytes	NOS II protein	+	−
	NOS activity	+	Not determined
Synovial membrane	NOS II mRNA	+ or ++	+
	NOS II protein	+ or ++	+
	nitrite ± nitrate	+	+ or −
Cartilage	NOS II protein	+	+
	nitrite	+	+

\+ = presence or increase: moderate (+), important (++), major (+++); − = absence.

importance, NO is able to combine with superoxide anion ($O_2^{-\bullet}$) to produce peroxynitrite ($ONOO^-$) whose formation is rate limited by the enzymes NADPH oxidase, a major source of $O_2^{-\bullet}$, and superoxide dismutase (SOD), a major dismutator of $O_2^{-\bullet}$ to the non radical products O_2 and H_2O_2 [39]. Interestingly, peroxynitrite can either decompose into inactive compounds, or generate hydroxyl radical or nitrate proteins. When peroxynitrite formation occurs, tyrosine are nitrated subsequently, a finding which is widely used as a hallmark of its formation within tissues [16].

1.4. Lessons from animal models

Experimental models have provided compelling evidence that inhibition of NO synthesis reduces the severity of joint lesions in arthritic or osteoarthritic animals [33,47]. Such beneficial effect was reported both in mono-articular, i.e., antigen-induced arthritis or anterior cruciate ligament section OA model, and in polyarticular, i.e., adjuvant-induced, collagen-induced or streptococcal cell wall-induced, arthritides. Interestingly, the severity of arthritis diminished whatever the selectivity of NOS inhibitors for NOS 2 [45], although some of the beneficial effects of L-arginine-methyl ester (L-NAME), a preferential NOS 3 inhibitor, could be explained by a reduced blood flow in synovial microcirculation. However, these data must be interpreted when considering that cartilage protection was associated with a general anti-inflammatory effect, either in OA [33] or in arthritis [47], suggesting that NOS inhibitors provided chondroprotection indirectly. Moreover, NOS inhibitors had a favourable effect when they were administered from the day of sensitization in polyarthritis but were ineffective, or even had a deleterious influence, when they were used in established arthritis [15,48]. One can suggest that NOS inhibitors interfere with the immunological spreading of arthritis. This is further supported by the different kinetics of circulating nitrite levels illustrating the different immunological responses during collagen-induced and adjuvant-induced polyarthritis [8].

A direct protection of cartilage by NOS inhibitors was demonstrated in the IL-1 arthritis model in rat [38]. Thus, a single intra-articular administration or a systemic sustained release of L-NMMA diminished the inhibition of proteoglycan synthesis induced by a local injection of IL-1β in rat. Cartilage protection was more effective when L-NMMA was administered systemically but, as mentioned below, the restoration of proteoglycan synthesis remained partial despite the normalization of nitrite levels in the joint cavity [38]. Alternative experiments in NOS $2^{-/-}$ deficient mice showed that inhibition of proteoglycan synthesis was lacking when animals were injected intra-articularly with IL-1 or zymosan and that patellar cartilage lacked the IGF1-hyporesponsiveness observed classically in arthritis [54]. These data support that NO is a crucial regulator of cartilage metabolism by promoting chondrocyte impairment and reducing repair capability by growth factors. Nonetheless, chondrocyte death and proteoglycan loss remained similar to those observed in wild-type animals, suggesting that, in contrast to what could be expected from *in vitro* experiments (see Fig. 1 adapted from [19]), NO disturbed cartilage anabolism exquisitely rather than the overall functions of chondrocytes.

2. Purpose

We and others demonstrated previously that N^{ω}-monomethyl-L-arginine (L-NMMA) spared the ability of chondrocytes to synthesize proteoglycans in the presence of IL-1 [10,38]. However, if the restoration of proteoglycan synthesis was quite complete in chondrocytes embedded in alginate beads, the inhibitory effect of IL-1 was corrected only in part in patellas and femoral head caps exposed *in vitro* to the cytokine [10]. We obtained similar data after a local injection of IL-1 into the knee joints, i.e., the *ex vivo* synthesis of proteoglycan in patellar cartilage was only partly restored by a sustained release of L-NMMA, despite a quite complete reduction of nitrite levels in synovial fluid [38]. We suggested therefore that NO could be an important but not exclusive mediator of IL-1 effect. In the present work, we investigated the contribution of superoxide anion ($O_2^{-\bullet}$) as an additional effector of IL-1 in chondrocytes with special care to its ability to combine with NO to form peroxynitrite. We studied also its possible implication in the inhibition of proteoglycan synthesis induced by IL-17, a regulatory cytokine shown recently to activate the NOS 2 pathway.

3. Materials and methods

3.1. Materials

All culture medium, cell culture reagents and heat-inactivated fetal calf serum (FCS) were purchased from Life technologies. Pronase, collagenase and nitrate reductase were from Bœhringer-Mannheim. Type II collagen goat antibody and FITC conjugate were from Jackson immunochemicals. Human recombinant interleukin-1β (IL-1β) or -17 (IL-17) and NS-398 were from Tebu. $Na_2^{35}SO_4$ was obtained from Amersham, ultima gold and soluene 350 from Packard. Alginate, Cu(II) (3,5-diisopropylsalicylate)$_2$ (CuDips), N^{ω}-monomethyl-L-arginine (L-NMMA), N^{ω}-monomethyl-D-arginine (D-NMMA), indomethacin, 2,3-diaminonaphtalene, N-naphtylethylenediamine dihydrochloride, cetylpyridinium chloride, formic acid, sulfanilamide and chondroitinase ABC were from Sigma-Aldrich chemicals. Anti-nitrotyrosine antibody was from Upstate, anti-NOS 2 antibody was from Santa Cruz and kit for immuno-staining (Novostain®) was from Novocastra. Immunoenzymatic assay kit for PGE_2 was from Assay Design (Bershire, UK).

3.2. Isolation and culture of chondrocytes

A pool of articular chondrocytes was obtained from hip joints of healthy male Wistar rats (Charles River) killed under dissociative anesthesia. Chondrocytes were isolated from cartilage pieces by sequential digestion with pronase and collagenase. Primary cultures, either in monolayers or in alginate beads, were maintained in DMEM/Ham's F12 medium supplemented with 10% (v/v) heat-inactivated FCS, 2 mM L-glutamine, 50 μg/ml gentamicin and 0.25 μg/ml amphotericin B, in a humidified atmosphere of 5% CO_2 at 37°C. In all experiments, chondrocytes were used at passage 2 or 3 and their phenotype was controlled by the ability to synthesize type II collagen, as assessed by immunofluorescence labelling. Experiments with IL-1, NO synthase inhibitors, non-steroidal anti-inflammatory drugs (NSAIDs) and oxygen-derived radical species scavengers were done in a low FCS (0.5 to 2.5%) containing culture medium with or without phenol red. Alginate beads were obtained by suspending chondrocytes (6×10^6 cells/ml) in sterile filtered low viscosity alginate solution (1.2%, w/v) and expressing slowly the mixture through a 22 gauge needle into a sterile 102 mM $CaCl_2$ solution. Beads were allowed to further their polymerisation for 10 min in $CaCl_2$, then washed three times in sterile saline and sub-cultured for 6 days before use.

3.3. In vitro experiments with NOS inhibitors, radical scavengers and NSAIDs

Proteoglycan synthesis, NO and PGE_2 production were assessed in chondrocytes beads and culture supernatants respectively, after an incubation of 48 h with IL-1β (2.5 ng/ml/day) or IL-17 (30 ng/ml) in the presence or absence of tested substances. L-NMMA, as a non-selective NOS inhibitor, and D-NMMA, as its inactive isomer, were used at a concentration of 1 mM. CuDips, as a superoxide dismutator, was used at a concentration of 10 μM. NSAIDs, i.e., indomethacin as a non-selective COX inhibitor and NS-398 as a selective COX-2 inhibitor, were used at a concentration of 1 μM.

For immuno-histochemical examination, chondrocytes were cultured on coverslips (25×10^4 cells/well) for 12 to 24 h in the presence of IL-1β (2.5 ng/ml) or IL-17 (30 ng/ml) and tested substances (1 mM L-NMMA, 10 μM CuDips).

3.4. In vivo experiment with IL-1

Rats were injected bilaterally with 1 μg IL-1β in their knee joints in a final volume of 50 μl. Ten hours after cytokine administration, rats were killed under dissociative anesthesia and synovial fluid was collected with small pieces of filter paper as described elsewhere in details [24]. Cartilage samples (tibial plateaus and femoral condyles) were collected concomitantly and processed for immuno-staining as described below. Other batches of rats were killed 48 h post-cytokine injection in order to check for proteoglycan synthesis in patellar cartilage.

3.5. Proteoglycan synthesis in chondrocytes beads

Beads were incubated in culture medium supplemented with 10 μCi/ml of $Na_2^{35}SO_4$ for 4 h at 37°C in a 5% CO_2 atmosphere, then washed five times with 0.15 M NaCl and solubilised in 55 mM/20 mM citrate–EDTA buffer. Isolated chondrocytes were further digested overnight at 60°C in papain buffer (0.2 M NaH_2PO_4, 0.01 M EDTA, 0.01 M cysteine and 6.75 U/ml papain). The [^{35}S]-proteoglycan content was measured by liquid scintillation counting (Ultima gold) with a LKB 1214 counter (Wallac).

3.6. Proteoglycan synthesis in patellar cartilage

Patellas from IL-1 injected rats were pulsed for 3 h at 37°C in a humidified atmosphere in RPMI-Hepes 1640 medium supplemented with 2 mM L-glutamine, 100 μg/ml streptomycin, 100 IU/ml penicillin and 0.6 μCi/ml of $Na_2^{35}SO_4$. After 5 washings in saline, patellas were fixed overnight at room temperature in 0.5% cetylpyridinium chloride in 10% (v/v) formalin, then decalcified in 5% (v/v) formic acid for 6 h at room temperature. Biopsy punches, made in the central part of the patellas, were then dissolved overnight in soluene 350 and [^{35}S]-proteoglycan content was measured as described above.

3.7. Immuno-staining for nitrotyrosines in chondrocytes

Cells were fixed in methanol/acetone (1/1, v/v) for 10 min at −20°C, permeabilised with triton X-100 (0.5%) for 45 min at room temperature, then immersed for 1 h in immune goat serum to prevent non-specific binding. Chondrocytes were incubated subsequently with anti-nitrotyrosine rabbit polyclonal antibody (1.5 μg/ml) for 2 h at 37°C, washed 3 times with phosphate buffered saline, then incubated with biotinylated goat anti-rabbit IgG for 45 min at room temperature. After extensive washings, amplification was performed with preformed avidin-biotinylated horseradish peroxidase complexes for 45 min at room temperature and staining was obtained by addition of 3,3′-diaminobenzidine (0.05% w/v in 0.01% w/v hydrogen peroxide). Chondrocytes incubated without anti-nitrotyrosine antibody or with immune rabbit serum (1.5 μg/ml) served as controls. All slides were counterstained with eosin.

3.8. Immuno-staining for nitrotyrosines and NOS 2 in cartilage

Femoral condyles and tibial plateaus were fixed in 4% paraformaldehyde for at least 24 h, decalcified for 45 min in RDO rapid bone decalcifier, postfixed for 24 h in 4% paraformaldehyde, then embedded in paraffin. Paraffin sections, 5 μm thick, were permeabilised first with chondroitinase ABC (0.25 U/ml) for 90 min at 37°C, then with triton X-100 (0.5%) for 45 min at room temperature and further proccessed as described above except that anti-nitrotyrosine rabbit polyclonal antibody was used at 10 μg/ml and anti-NOS 2 rabbit polyclonal antibody at 1.3 μg/ml.

3.9. Assays of mediators

In culture supernatants, NO_2^- levels were measured by the Griess colorimetric assay. Briefly, 100 μl of culture supernatant were mixed with 100 μl of Griess reagent [sulfanilamide (1%, w/v) in 5% (v/v) of H_3PO_4 and N-naphtylethylenediamine dihydrochloride (0.1%) in H_2O] for 5 min. Optical density was read at 550 nm with a MR 5000 (Dynatech) microplate reader and data were expressed as cumulated nitrite level over 48 h from a standard curve of sodium nitrite solutions ranging from 5 to 50 μM. PGE_2 levels were assessed using a commercially available enzyme-linked immunosorbent assay (Assay Design Inc.) according to the manufacturer's instructions. The assay showed no cross-reactivity with other prostanoids and sensitivity was around 30 pg/ml.

In synovial fluid, NO production was measured as $NO_2^- + NO_3^-$ levels using a spectrofluorimetric assay with 2,3-diaminonaphtalene (DAN) as reagent. Briefly, NO_3^- were reduced with nitrate reductase from *Aspergillus niger* (0.28 IU/ml in 20 mM Tris-HCl in the presence of 40 μM NADPH) in a final volume of 500 μl, then samples were incubated for 10 min at room temperature after addition of 100 μl of DAN reagent (0.05 mg/ml in 0.62 N HCl) and the reaction was stopped 10 min later by adding 50 μl of 2.8 N NaOH. Fluorescence was measured at 462 nm in glass tubes with a Hitachi F-2000 spectrofluorimeter, using 363 nm as excitation wavelength. Data were expressed as $NO_2^- + NO_3^-$ levels from a standard curve of sodium nitrate solutions ranging from 0.5 to 50 μM and processed as the samples.

3.10. Statistical analysis

Data are expressed as mean ± SEM of at least 5 samples. Nitrite production is expressed as cumulated (*in vitro*) or instantaneous (*in vivo*) levels vs controls. Proteoglycan synthesis is expressed as percentage of variation of radiolabelled sodium sulphate incorporation *vs* controls. Comparisons between groups were performed by analysis of variance followed by a Fisher test. A level of $P < 0.05$ was considered as significant.

4. Results

4.1. $O_2^{-\cdot}$ contributes to IL-1-mediated inhibition of PGs and combines with NO in vitro

Chondrocytes embedded in alginate beads responded extensively to IL-1 by producing NO, as assessed by NO_2^- levels in supernatants (Table 2). This increase was both time- (obvious from 12 hrs of stimulation) and dose-dependent (proportional increase from 0.25 to 250 UI/ml of IL-1β) and resulted from the expression of the inducible NOS 2 isoform (data not shown). NO_2^- levels were reduced almost completely when chondrocytes were stimulated with IL-1β in the presence of L-NMMA, whereas D-NMMA remained ineffective. In these experimental conditions, CuDips reduced only moderately the NO_2^- levels, whilst a return to basal levels was observed when chondrocytes were treated with a mixture of L-NMMA and CuDips (Table 2). Proteoglycan synthesis, as assessed by radiolabelled sulphate incorporation, was reduced by exposure to IL-1β (-24%) and this effect was corrected by L-NMMA and CuDips, alone or in combination, but not D-NMMA (Table 2). At a time consistent with NO overproduction (12 hrs), chondrocytes stimulated with IL-1β exhibited a marked staining for 3-nitrotyrosines which was prevented by incubation with L-NMMA or CuDips (Fig. 2a–d).

4.2. $O_2^{-\cdot}$ combines with NO to mediate IL-1-induced activation of COX-2 in vitro

Chondrocytes beads responded extensively to IL-1 by producing prostaglandins, as assessed by PGE_2 levels in supernatants (Table 3). As for NO, this increase was time- (obvious from 6 h of stimulation) and dose-dependent and resulted from the expression of an inducible isoform, i.e., COX-2 (data not shown). PGE_2 levels were completely suppressed by NSAIDS, independently of their selectivity for COX-2. Prostaglandin synthesis was also significantly inhibited by L-NMMA and CuDips, alone or in combination, but not by D-NMMA (Table 3). The reduction of PGE_2 levels by L-NMMA and CuDips was similar when the drugs were added 8 hrs after IL-1 instead of from the beginning of the cytokine challenge (data not shown). At concentrations inhibiting prostaglandin synthesis almost completely (1 μM), indomethacin and NS-398 failed to correct the inhibitory effect of IL-1 on proteoglycan synthesis (data not shown).

4.3. $O_2^{-\cdot}$ distinguishes IL-1 and IL-17 for their inhibitory effect on PGs synthesis in vitro

Chondrocytes beads produced NO in response to IL-17, although a concentration twelve-fold higher than for IL-1 was required to obtain comparable NO_2^- levels (Table 4). As for IL-1, L-NMMA reduced

Table 2

Effect of L-NMMA, D-NMMA and CuDips on IL-1 mediated inhibition of proteoglycan synthesis in rat chondrocytes beads

	PGs synthesis (%)	NO_2^- levels (μM)
Controls	100	9.4 ± 0.7
IL-1β 2.5 ng/ml	76	22.0 ± 1.9
IL-1β + L-NMMA 10^{-3} M	103*	6.1 ± 1.3*
IL-1β + D-NMMA 10^{-3} M	78	18.5 ± 0.6
IL-1β + CuDips 10^{-5} M	95*	19.1 ± 0.7
IL-1β + L-NMMA 10^{-3} M + CuDips 10^{-5} M	106*	5.8 ± 0.6*

*$p < 0.05$ *versus* IL-1 treated cells.

a c b d e

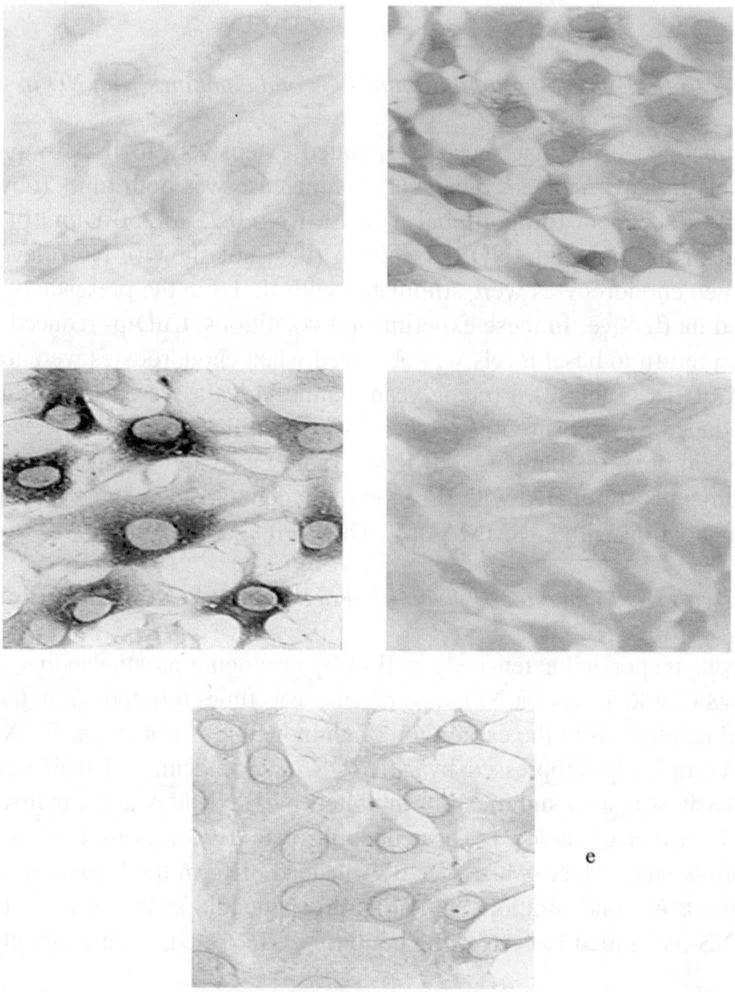

Fig. 2. Detection of nitrotyrosines in rat chondrocytes stimulated with rhIL-1β or IL-17 for 12 hours. (a) Unstimulated controls; (b) IL-1β 2.5 ng/ml; (c) IL-1β + CuDips 10^{-5} M; (d) IL-1β + L-NMMA 10^{-3} M; (e) IL-17 30 ng/ml. Rabbit polyclonal 3-nitrotyrosine antibody (Upstate), biotinylated goat conjugate, avidin-biotinylated horseradish peroxidase, di-amino-benzidine (Novostain^{TM}).

Table 3

Effect of L-NMMA, D-NMMA and CuDips on IL-1 mediated stimulation of COX-2 activity in rat chondrocytes beads

	PGE_2 levels (ng/ml)	NO_2^- levels (μM)
Controls	1.6 ± 0.2	6.8 ± 0.6
IL-1β 2.5 ng/ml	12.4 ± 2.0	20.0 ± 3.0
IL-1β + L-NMMA 10^{-3} M	$2.7 \pm 0.3^*$	$4.3 \pm 1.0^*$
IL-1β + D-NMMA 10^{-3} M	12.9 ± 0.9	18.7 ± 1.3
IL-1β + CuDips 10^{-5} M	$1.7 \pm 0.8^*$	$12.0 \pm 1.0^*$
IL-1β + L-NMMA 10^{-3} M + CuDips 10^{-5} M	$2.9 \pm 0.5^*$	$4.0 \pm 0.4^*$

$^*p < 0.05$ *versus* IL-1 treated cells.

Table 4

Effect of L-NMMA and CuDips on IL-17 mediated inhibition of proteoglycan synthesis in rat chondrocytes beads

	PGs synthesis (%)	NO_2^- levels (μM)
Controls	100	9.4 ± 0.7
IL-17 30 ng/ml	72	18.4 ± 0.5
IL-17 + L-NMMA 10^{-3} M	81^*	$6.8 \pm 0.3^*$
IL-17 + CuDips 10^{-5} M	76	20.0 ± 1.0

$^*p < 0.05$ *versus* IL-17 treated cells.

NO_2^- levels to baseline whereas CuDips lacked any effect. At the concentration used (30 ng/ml), IL-17 inhibited proteoglycan synthesis by 28% and this effect was corrected partly (about 30% of restoration) by L-NMMA (Table 4). In sharp contrast with IL-1, this IL-17-mediated inhibition of proteoglycan synthesis was unaffected by the presence of CuDips. Consistent with this finding, chondrocytes stimulated with IL-17 for 12 hrs did not exhibit any staining for 3-nitrotyrosines (Fig. 2e).

4.4. $O_2^{-\cdot}$ combines with NO in the IL-1 arthritis model

Rats injected with IL-1β showed a progressive increase in $NO_2^- + NO_3^-$ levels in their synovial fluids with a peak (4-fold increase) occurring 10 hrs after cytokine administration (data not shown). At this time, a strong immuno-staining for NOS 2 and 3-nitrotyrosines was observed throughout the femoral cartilage. Fourthy height hours after IL-1, proteoglycan synthesis was inhibited by 30% in the patellar cartilage (data not shown).

5. Discussion

The present work was undertaken to assess the ability of superoxide anion ($O_2^{-\cdot}$) to mediate the effects of IL-1 and/or to modulate those of NO in chondrocytes. This assumption was supported by several lines of evidence suggesting that NO was not sufficient *per se* to mediate inhibition of proteoglycan synthesis. Thus, in most *in vitro* studies, the ability of NO synthase inhibitors to counteract the anti-anabolic effect of IL-1 was partial [17] or even absent [49]. *In vivo*, the complete inhibition of NO synthesis in rat articular joints was accompanied by a partial restoration of proteoglycan synthesis in the patellar cartilage of IL-1 injected rats [38], even if no inhibitory effect of IL-1 was reported in NOS $2^{-/-}$ deficient mice under similar circumstances [54]. In addition, the ability of NO to induce chondrocytes apoptosis was shown to warrant ROS scavenging and was balanced with cell necrosis depending on the formation of peroxynitrite [5].

In our experimental conditions, the NOS inhibitor L-NMMA restored the inhibitory effect of IL-1β on proteoglycan synthesis while reducing almost completely the production of NO, confirming that NO contributed to the anti-anabolic activity of IL-1 in rat cartilage. As non-specific effects, including inhibition of heme proteins [35] and anti-oxidant properties [56], were reported for some NOS inhibitors, one may ask whether such restoration was really NO-dependent. The lack of efficacy of the inactive isomer D-NMMA and the reversion of L-NMMA's effect by an excess of L-arginine supports a NO-dependent mechanism. Moreover, the ability of NOS inhibitors to scavenge ROS was demonstrated mainly for aminoguanidine [56] and L-NAME [37], compounds we did not use in the present study and previously shown to be inconsistently effective in rat cartilage [10]. Interestingly, the low molecular weight SOD mimetic, CuDips, was able to restore proteoglycan synthesis to the same extent as L-NMMA. CuDips is lipid soluble and takes advantage over superoxide dismutase to behave as an intracellular scavenger [7]. As NO_2^- levels were not affected dramatically by this compound, such "cartilage sparing" effect was not attributable to an inhibition of NOS activity. Thus, CuDips demonstrated that the IL-1-mediated inhibition of proteoglycan synthesis depended on the tone of superoxide in addition to the production of NO. This result was not surprising owing to the ability of chondrocytes to produce high levels of $O_2^{-\cdot}$ [40], H_2O_2 [51] and OH· [52] in response to pro-inflammatory cytokines or mitogens. Although its physiological relevance remains uncertain, due to the relatively anaerobic environment of chondrocytes within cartilage, the NADPH complex was shown to be functional and to vary under cytokine stimulation in this cell type [18,26,28,51]. The rate of $O_2^{-\cdot}$ production by this complex was lower for chondrocytes than for phagocytic cells [26] but it remained sufficient to account possibly for the spontaneous production of radicals [40,52] and for their increase with the metabolic activity of the cells [40]. In addition, there is an intriguing parallelism between the production of radicals and NO in response to cytokines, since both systems were shown to have ambivalent properties on cartilage depending on the dose, i.e., a stimulation of chondrocyte functions at low "physiological" levels and their inhibition at high "pathological" fluxes [30,46]. We showed further that nitration of tyrosines occurred *in vitro* and *in vivo* after IL-1 challenge, suggesting that $O_2^{-\cdot}$ and NO were able to combine to form peroxynitrite despite a probable difference in their time course of production. There has been some debate to establish if tyrosine nitration was a biological marker of peroxynitrite formation rather than of "reactive nitrogen species" formation [39]. In addition, there was some controversy concerning the ability of peroxynitrite formed in biological systems to nitrate tyrosine as efficiently as chemically synthesized $ONOO^-$ [36,43]. However, the similar topography of the staining for 3-nitrotyrosine and NOS 2 in the cartilage of femoral condyles from IL-1-injected rats, the prevention of nitrotyrosine labelling by L-NMMA and CuDips together with their correcting effect on the inhibitory potency of IL-1 on proteoglycan synthesis *in vitro*, provide a sufficient amount of evidence that $O_2^{-\cdot}$ is a key modulator of NO effects through the formation of peroxynitrite. These data are consistent with the staining for 3-nitrotyrosine reported in the cartilage of tibial plateaus and femoral condyles of osteoarthritic dogs [34], further supporting a pathophysiological role of peroxynitrite in IL-1 mediated joint diseases.

The formation of peroxynitrite, *in vitro* and *in vivo*, provides a plausible mechanism for IL-1 effects in cartilage and opens new therapeutic perspectives. Indeed, reactive oxygen species produced in huge amounts by enzymatic systems were shown to decrease proteoglycan synthesis [3,30] and to alter the aggregating properties of large proteoglycan monomers [30]. As a consequence, it appears logical that most of the IL-1 mediated inhibition of proteoglycan synthesis was supported by peroxynitrite but it raises the possibility that the partial correcting effects of NOS inhibitors in cartilage reflected the deleterious effect of "non-combined" oxygen radicals. Amongst mechanisms accounting likely for IL-1-mediated inhibition of proteoglycan synthesis, NO was shown to act indirectly by decreasing the production of TGFβ1 [50]. However, NADPH oxidase-mediated oxygen radicals were also shown to regulate negatively TGFβ1 transcription in other cell type [55], raising the possibility that peroxynitrite and superoxide could be key regulators of TGFβ1 in cartilage.

When considering the regulation of COX-2 activity by NO, one must underline that very controversial data were reported depending on the tissue and inducing agent [42]. The present finding demonstrates that both NO and $O_2^{-\cdot}$ contributed to COX-2 activation and that this occurred likely through the formation of peroxynitrite. This data is consistent with the ability of peroxynitrite to act as a hydroperoxide initiator for the catalytic activity of COX enzymes [23], all the more so that CuDips lacked any inhibitory potency on isolated COX isoenzymes [20]. As the synthesis of prostanoids by COX-2 was suppressed when L-NMMA or CuDips were added at a time allowing a complete expression of inducible enzymes, it is likely that the regulation of COX was rather at the level of enzyme activity than at the gene level. A direct consequence is that NOS inhibitors and antioxidants may have additional (prostaglandin-dependent) anti-inflammatory properties, a finding agreeing well with the reduced PGE_2 levels in the synovial fluid of arthritic rabbits [25] or osteoarthritic dogs [33] treated with NOS inhibitors. In addition, one must keep in mind that the bovine Cu,ZnSOD orgotein® showed promising results as a therapeutic for human joint disorders [27], and that it was abandoned due to its inability to enter the cells, instability in aqueous solutions and immunogenicity responsible for side effects. However, spin-trapping nitrones, metalloporphyrins, SOD mimics and peroxynitrite decomposition catalysts have proven efficacious in experimental models of inflammation [11], further supporting the contribution of reactive oxygen species, alone or in combination with NO, in these processes.

Whether IL-17 distinguishes from IL-1 in terms of proteoglycan synthesis despite activating inducible NOS pathway as a common way remains unclear. Our data show that the production of $O_2^{-\cdot}$ and the subsequent formation of peroxynitrite were lacking in chondrocytes stimulated with IL-17, supporting that oxygen radicals were not involved in the inhibitory potency of this cytokine on proteoglycans. However, the moderate efficacy of L-NMMA suggested further that the decrease in proteoglycan synthesis generated by IL-17 depended also on NO-independent mechanisms. This result differs somewhat from experiments with mouse cartilage having shown that both inhibition of NO production by (L-N^5-[1-iminoethyl]ornithine) or NOS2 deficiency prevented completely the suppressive effect of IL-17 [24]. Further studies are required to assess if this discrepancy can be attributed to species differences, as was partly the case for IL-1, but a variable contribution of NO to the inhibitory effects of IL-1 and IL-17 on proteoglycan synthesis could also be explained by the activation of distinct signalling pathways in chondrocytes [44].

To sum up, the present study demonstrates that some of the pro-inflammatory and anti-anabolic effects of IL-1 in chondrocytes are mediated by the formation of peroxynitrite but that this is not the case for IL-17. Our data support the general meaning that a combined therapy with NOS inhibitors and antioxidants or the use of peroxynitrite decomposition catalysts may be promising to protect cartilage against cytokine-mediated injury.

Acknowledgment

This work was supported by grants from the European Community (QLRT-1999-02072).

References

[1] A. Amin, P.E. DiCesare, P. Vyas, M. Attur, E. Tzeng, T.R. Billiar, S.A. Stuchin and S.B. Abramson, The expression and regulation of nitric oxide synthase in human osteoarthritis-affected chondrocytes: evidence for up-regulated neuronal nitric oxide synthase, *J. Exp. Med.* **182** (1995), 2097–2102.

[2] W.P. Arend and J.M. Dayer, Cytokines and cytokine inhibitors or antagonists in rheumatoid arthritis, *Arthritis Rheum.* **33** (1990), 305–315.

[3] E.J. Bates, C.C. Johnson and D.A. Lowther, Inhibition of proteoglycan synthesis by hydrogen peroxide in cultured bovine articular cartilage, *Biochim. Biophys. Acta* **838** (1985), 221–228.

[4] J.S. Beckman and W.H. Koppenol, Nitric oxide, superoxide, and peroxynitrite: the good, the bad, and ugly, *Am. J. Physiol.* **271** (1996), C1424–1437.

[5] F.J. Blanco, R.L. Ochs, H. Schwarz and M. Lotz, Chondrocyte apoptosis induced by nitric oxide, *Am. J. Pathol.* **146** (1995), 75–85.

[6] C. Bogdan, Nitric oxide and the regulation of gene expression, *Trends Cell. Biol.* **11** (2001), 66–75.

[7] R.H. Burdon, D. Alliangana and V. Gill, Hydrogen peroxide and the proliferation of BHK-21 cells, *Free Radic. Res.* **23** (1995), 471–486.

[8] G.W. Cannon, S.J. Openshaw, J.B. Hibbs, J.R. Hoidal, T.P. Huecksteadt and M.M. Griffiths, Nitric oxide production during adjuvant-induced and collagen-induced arthritis, *Arthritis Rheum.* **39** (1996), 1677–1684.

[9] J.-P. Caron, J.C. Fernandes, J. Martel-Pelletier, G. Tardif, F. Mineau, C. Geng and J.-P. Pelletier, Chondroprotective effect of intraarticular injections of interleukin-1 receptor antagonist in experimental osteoarthritis, *Arthritis Rheum.* **39** (1996), 1535–1544.

[10] C. Cipolletta, J.-Y. Jouzeau, P. Gegout-Pottie, N. Presle, K. Bordji, P. Netter and B. Terlain, Modulation of IL-1-induced cartilage injury by NO synthase inhibitors: a comparative study with rat chondrocytes and cartilage entities, *Br. J. Pharmacol.* **124** (1998), 1719–1727.

[11] S. Cuzzocrea, D.P. Riley, A.P. Caputi and D. Salvemini, Antioxidant therapy: a new pharmacological approach in shock, inflammation, and ischemia/reperfusion injury, *Pharmacol. Rev.* **53** (2001), 135–159.

[12] P. Das, D.J. Schurman and R.L. Smith, Nitric oxide and G proteins mediate the response of bovine articular chondrocytes to fluid-induced shear, *J. Orthop. Res.* **15** (1997), 87–93.

[13] C.H. Evans, S.C. Watkins and M. Stefanovic-Racic, Nitric oxide and cartilage metabolism, *Meth. Enzymol.* **269** (1996), 75–88.

[14] A.J. Farrell, D.R. Blake, R.M.J. Palmer and S. Moncada, Increased concentrations of nitrite in synovial fluid and serum samples suggest increased nitric oxide synthesis in rheumatic diseases, *Ann. Rheum. Dis.* **51** (1992), 1219–1222.

[15] D.S. Fletcher, W.R. Widmer, S. Luell, A. Christen, C. Orevillo, S. Shah and D. Visco, Therapeutic administration of a selective inhibitor of nitric oxide synthase does not ameliorate the chronic inflammation and tissue damage associated with adjuvant-induced arthritis in rats, *J. Pharmacol. Exp. Ther.* **284** (1998), 714–721.

[16] B. Halliwell, What nitrates tyrosine? Is nitrotyrosine specific as a biomarker of peroxynitrite formation in vivo?, *FEBS Lett.* **411** (1997), 147–160.

[17] H.J. Häuselmann, L. Oppliger, B.A. Michel, M. Stefanovic-Racic and C.H. Evans, Nitric oxide and proteoglycan biosynthesis by human articular chondrocytes in alginate cultures, *FEBS Lett.* **352** (1994), 361–364.

[18] T.S. Hiran, P.J. Moulton and J.T. Hancock, Detection of superoxide and NADPH oxidase in porcine articular chondrocytes, *Free Radic. Biol. Med.* **23** (1997), 736–743.

[19] D. Jang and G.A.C. Murrell, Nitric oxide in arthritis, *Free Rad. Biol. Med.* **24** (1998), 1511–1519.

[20] L.H. Landino, B.C. Crews, M.D. Timmons, J.D. Morrow and L.J. Marnett, Peroxynitrite, the coupling product of nitric oxide and superoxide activates prostaglandin biosynthesis, *Proc. Natl. Acad. Sci. USA* **93** (1996), 15 069–15 074.

[21] D.A. Lee, T. Noguchi, S.P. Frean, P. Lees and D.L. Bader, The influence of mechanical loading on isolated chondrocytes seeded in agarose constructs, *Biorheology* **37** (2000), 149–161.

[22] A.M. Loening, I.E. James, M.E. Levenston, A.M. Badger, E.H. Frank, I. Kurz, M.E. Nuttall, M.E. Hung, S.M. Blake, A.J. Grodzinsky and M.W. Lark, Injurious mechanical compression of bovine articular cartilage induces chondrocyte apoptosis, *Arch. Biochem. Biophys.* **15** (2000), 205–212.

[23] G. Lu, A.L. Tsai, H.E. Van Wart and R.J. Kulmacz, Comparison of the peroxidase reaction kinetics of prostaglandin H synthase-1 and -2, *J. Biol. Chem.* **274** (1999), 16 162–16 167.

[24] E. Lubberts, L.A.B. Joosten, F.A.J. Van de Loo, L.A. Van den Bersselaar and W.B. Van den Berg, Reduction of interleukin-17-induced inhibition of chondrocyte proteoglycan synthesis in intact murine articular cartilage by interleukin-4, *Arthritis Rheum.* **43** (2000), 1300–1306.

[25] S.B. Mello, G.S. Novaes, I.M. Laurindo, M.N. Muscara, F.M. Maciel and W. Cossermelli, Nitric oxide synthase inhibitor influences prostaglandin and interleukin-1 production in experimental arthritic joints, *Inflamm. Res.* **46** (1997), 72–77.

[26] P.J. Moulton, T.S. Hiran, M.B. Goldring and J.T. Hancock, Detection of protein and mRNA of various components of the NADPH oxidase complex in an immortalized human chondrocyte line, *Br. J. Rheumatol.* **36** (1997), 522–529.

[27] Y. Niwa, K. Somiya, A.M. Michelson and K. Puget, Effect of liposomal-encapsulated superoxide dismutase on active oxygen-related human disorders. A preliminary study, *Free Radic. Res. Commun.* **1** (1985), 137–153.
[28] M. Oh, K. Fukuda, S. Asada, Y. Yasuda and S. Tanaka, Concurrent generation of nitric oxide and superoxide inhibits proteoglycan synthesis in bovine articular cartilage: involvement of peroxynitrite, *J. Rheumatol.* **25** (1998), 2169–2174.
[29] R.M.J. Palmer, M.S. Hickery, I.G. Charles, S. Moncada and M.T. Bayliss, Induction of nitric oxide synthase in human chondrocytes, *Biochem. Biophys. Res. Commun.* **193** (1993), 398–405.
[30] A. Panasyuk, E. Frati, D. Ribault and D. Mitrovic, Effect of reactive oxygen species on the biosynthesis and structure of newly synthesized proteoglycans, *Free Radic. Biol. Med.* **16** (1994), 157–167.
[31] J.-P. Pelletier, J.A. DiBattista, P. Roughley, R. McCollum and J. Martel-Pelletier, Cytokines and inflammation in cartilage degradation, *Rheum. Dis. Clin. North America* **19** (1993), 545–568.
[32] J.-P. Pelletier, J.-P. Caron, C. Evans, P.D. Robbins, H.I. Georgescu, D. Jovanovic, J.C. Fernandes and J. Martel-Pelletier, In vivo suppression of early experimental osteoarthritis by interleukin-1 receptor antagonist using gene therapy, *Arthritis Rheum.* **40** (1997), 1012–1019.
[33] J.-P. Pelletier, D. Jovanovic, J.C. Fernandes, P. Manning, J.R. Connor, M.G. Currie, J.A. Di Battista and J. Martel-Pelletier, Reduced progression of experimental osteoarthritis in vivo by selective inhibition of nitric oxide synthase, *Arthritis Rheum.* **41** (1998), 1275–1286.
[34] J.-P. Pelletier, V. Lascau-Coman, D. Jovanovic, J.C. Fernandes, P. Manning, J.R. Connor, M.G. Currie and J. Martel-Pelletier, Selective inhibition of inducible nitric oxide synthase in experimental osteoarthritis is associated with reduction in tissue levels of catabolic factors, *J. Rheumatol.* **26** (1999), 2002–2014.
[35] D.A. Peterson, D.C. Peterson, S. Archer and E.K. Weir, The non specificity of specific nitric oxide synthase inhibitors, *Biochem. Biophys. Res. Commun.* **187** (1992), 797–801.
[36] S. Pfeiffer, K. Schmidt and B. Mayer, Dityrosine formation outcompetes tyrosine nitration at low steady-state concentrations of peroxynitrite, *J. Biol. Chem.* **275** (2000), 6346–6352.
[37] S. Pou, L. Keaton, W. Surichamorm and G.M. Rosen, Mechanism of superoxide generation by neuronal nitric-oxide synthase, *J. Biol. Chem.* **274** (1999), 9573–9580.
[38] N. Presle, C. Cipolletta, J.-Y. Jouzeau, A. Abid, P. Netter and B. Terlain, Cartilage protection by nitric oxide synthase inhibitors after intraarticular injection of interleukin-1β in rat, *Arthritis Rheum.* **42** (1999), 2094–2102.
[39] R. Radi, G. Peluffo, M.N. Alvarez, M. Naviliat and A. Cayota, Unraveling peroxynitrite formation in biological systems, *Free Rad. Biol. Med.* **30** (2001), 463–488.
[40] C. Rathakrishnan and M.L. Tiku, Lucigenin-dependent chemiluminescence in articular chondrocytes, *Free Radic. Biol. Med.* **15** (1993), 143–149.
[41] H. Sakurai, H. Kohsaka, M.F. Liu, H. Higashiyama, Y. Hirata, K. Kanno, I. Saito and N. Miyasaka, Nitric oxide production and inducible nitric oxide synthase expression in inflammatory arthritides, *J. Clin. Invest.* **96** (1995), 2357–2363.
[42] D. Salvemini and J. Masferrer, Interactions of nitric oxide with cylooxygenases: in vitro, ex vivo, and in vivo studies, *Meth. Enzymol.* **269** (1996), 1225–1265.
[43] T. Sawa, T. Akaike and H. Maeda, Tyrosine nitration by peroxynitrite formed from nitric oxide and superoxide generated by xanthine oxydase, *J. Biol. Chem.* **275** (2000), 32 467–32 474.
[44] T. Shalom-Barak and J. Quach, M. Lotz, Interleukin-17-induced gene expression in articular chondrocytes is associated with activation of mitogen-activated protein kinases and NF-κB, *J. Biol. Chem.* **273** (1998), 27 467–27 473.
[45] G.J. Southan and C. Szabo, Selective pharmacological inhibition of distinct nitric oxide synthase isoforms, *Biochem. Pharmacol.* **51** (1996), 383–394.
[46] J. Stadler, M. Stefanovic-Racic, T.R. Billiar, R.D. Curran, L.A. McIntyre, H.I. Georgescu, R.L. Simmons and C.H. Evans, Articular chondrocytes synthesize nitric oxide in response to cytokines and lipopolysaccharide, *J. Immunol.* **147** (1991), 3915–3920.
[47] M. Stefanovic-Racic, K. Meyers, C. Meschter, J.W. Coffrey, R.A. Hoffman and C.H. Evans, N-monomethyl arginine, an inhibitor of nitric oxide synthase, suppresses the development of adjuvant arthritis in rats, *Arthritis Rheum.* **37** (1994), 1062–1069.
[48] M. Stefanovic-Racic, K. Meyers, C. Meschter, J.W. Coffey, R.A. Hoffman and C.H. Evans, Comparison of the nitric oxide synthase inhibitors methylarginine and aminoguanidine as prophylactic and therapeutic agents in rat adjuvant arthritis, *J. Rheumatol.* **22** (1995), 1922–1928.
[49] M. Stefanovic-Racic, T.I. Morales, D. Taskiran, L.A. McIntyre and C.H. Evans, The role of nitric oxide in proteoglycan turnover by bovine articular cartilage organ cultures, *J. Immunol.* **156** (1996), 1213–1220.
[50] R.K. Studer, H.I. Georgescu, L.A. Miller and C.H. Evans, Inhibition of transforming growth factor β production by nitric oxide-treated chondrocytes. Implications for matrix synthesis, *Arthritis Rheum.* **42** (1999), 248–257.
[51] M.L. Tiku, J.B. Liesch and F.M. Robertson, Production of hydrogen peroxide by rabbit articular chondrocytes. Enhancement by cytokines, *J. Immunol.* **145** (1990), 690–696.
[52] M.L. Tiku, Y.P. Yan and K.Y. Chen, Hydroxyl radical formation in chondrocytes and cartilage as detected by electron paramagnetic resonance spectroscopy using spin trapping reagents, *Free Radic. Res.* **29** (1998), 177–187.

[53] F.A.J. Van de Loo, O.J. Arntz, I.G. Otterness and W.B. Van den Berg, Protection against cartilage proteoglycan synthesis inhibition by antiinterleukin 1 antibodies in experimental arthritis, *J. Rheumatol.* **19** (1992), 348–356.
[54] F.A.J. Van de Loo, O.J. Arntz, F.H.J. Van Enckevort, P.L.E.M. Van Lent and W.B. Van den Berg, Reduced cartilage proteoglycan loss during zymosan-induced gonarthritis in NOS2-deficient mice and in anti-interleukin-1-treated wild-type mice with unabated joint inflammation, *Arthritis Rheum.* **41** (1998), 634–646.
[55] G. Wolf, T. Hannken, R. Schroeder, G. Zahner, F.N. Ziyadeh and R.A. Stahl, Antioxidant treatment induces transcription and expression of transforming growth factor β in cultured renal proximal tubular cells, *FEBS Lett.* **488** (2001), 154–159.
[56] G. Yildiz, A.T. Demiryürek, I. Sahin-Erdemli and I. Kanzik, Comparison of antioxidant activities of aminoguanidine, methylguanidine and guanidine by luminol-enhanced chemiluminescence, *Br. J. Pharmacol.* **124** (1998), 905–910.

Biorheology 39 (2002) 215–220
IOS Press

Growth factors in cartilage tissue engineering

Gerjo J.V.M. van Osch [a,b,*], Erik W. Mandl [a], Willem J.C.M. Marijnissen [a], Simone W. van der Veen [a,b], Henriette L. Verwoerd-Verhoef [b] and Jan A.N. Verhaar [a]

[a] *Erasmus University Medical Centre Rotterdam, Department of Orthopaedics, Rotterdam, The Netherlands*

[b] *Erasmus University Medical Centre Rotterdam, Department of Otorhinolaryngology, Rotterdam, The Netherlands*

Abstract. Tissue engineering of cartilage consists of two steps. Firstly, the cells from a small biopsy of patient's own tissue have to be multiplied. During this multiplication process they lose their cartilage phenotype. In the second step, these cells have to be stimulated to re-express their cartilage phenotype and produce cartilage matrix. Growth factors can be used to improve cell multiplication, redifferentiation and production of matrix. The choice of growth factors should be made for each phase of the tissue engineering process separately, taking into account cell phenotype and the presence of extracellular matrix. This paper demonstrates some examples of the use of growth factors to increase the amount, the quality and the assembly of the matrix components produced for cartilage tissue engineering. In addition it shows that the "culture history" (e.g., addition of growth factors during cell multiplication or preculture period in a 3-dimensional environment) of the cells influences the effect of growth factor addition. The data demonstrate the potency as well as the limitations of the use of growth factors in cartilage tissue engineering.

1. Introduction

The availability of extra cartilage to reconstruct defects in articular joints or cartilaginous structures in the head and neck area would be of great benefit. Therefore much effort is put into the generation of autologous (patient's own) cartilage using tissue engineering techniques. Generally, tissue engineering involves a two step procedure (Fig. 1). First cells are isolated from a small biopsy of patient's own cartilage and expanded *in vitro* to obtain the cell number required. This process of cell expansion is routinely carried out in monolayer culture. During this culture the cells gradually lose their cartilage phenotype in a process called dedifferentiation [14]. The second step in the tissue engineering process involves seeding of the expanded cells in a 3-dimensional carrier and implantation in the defect. An *in vitro* culture period can be performed before implantation *in vivo* to induce re-expression of the cartilaginous phenotype and to stimulate formation of functional extracellular matrix. The formation of functional extracellular matrix is of utmost importance for the success of the graft. The matrix provides the graft with adequate mechanical properties in order to withstand the forces it is subjected to *in vivo*. Formation of functional matrix can be stimulated by the use of growth factors during tissue engineering process.

In vivo, growth factors are synthesised within a variety of tissues, where they act on their cell of origin (autocrine) or on adjacent cells (paracrine). Certain growth factors are also present in systemic circulation, and the same factor may act both as a systemic and as a local regulator of tissue metabolism. Growth factors are important for the metabolism of chondrocytes. They are present in synovial fluid and reach the

*Address for correspondence: Gerjo J.V.M. van Osch, Dept. Orthopaedics and Otorhinolaryngology, Erasmus University Rotterdam, Room Ee1659, PO Box 1738, 3000 DR Rotterdam, The Netherlands. Tel.: +31 10 4087661; Fax: +31 10 4089441; E-mail: vanosch@kno.fgg.eur.nl.

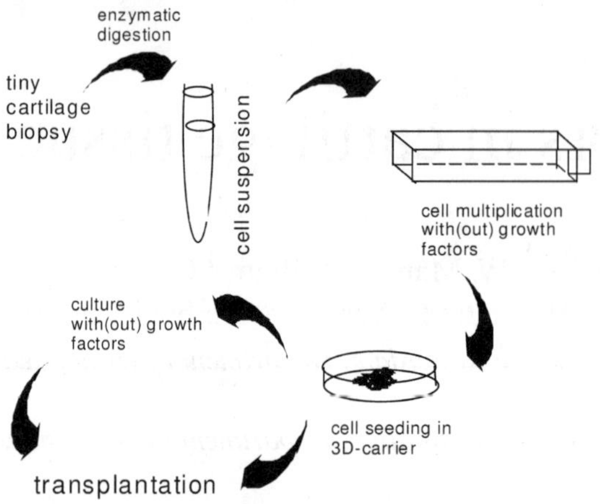

Fig. 1. Schematic overview of a cartilage tissue engineering process. First, cells have to be isolated from a small biopsy of cartilage using enzymatic digestion of the matrix. The cells are expanded in monolayer culture and when sufficient cells are obtained they are seeded in a 3-dimensional carrier (biomaterial). The construct can then be transplanted either directly or after *in vitro* culture to induce redifferentiation and matrix production.

Small biopsy of cartilage

↓

Cell suspension
Differentiated chondrocytes
No extracellular matrix

↓

Cell Expansion in monolayer
Chondrocytes dedifferentiate = lose their phenotype

↓

3-D culture in alginate
Redifferentiation of chondrocytes
Production of extracellular matrix

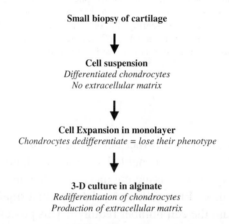

Fig. 2. The cellular phenotype and extracellular matrix during different stages of the cartilage tissue engineering process.

chondrocytes by diffusion. Furthermore, growth factors are produced by the chondrocytes themselves. The extracellular matrix of cartilage acts as a store of growth factors that can be released when needed. Various growth factors have been demonstrated to be able to affect chondrocyte multiplication, phenotype and extracellular matrix production and thus can be used during both steps of the tissue engineering process (Fig. 1). One has to realise however, that growth factors can influence extra-cellular matrix formation and cellular phenotype but that a certain growth factor can have different effects depending on the chondrocyte phenotype and the presence of extracellular matrix. During the cartilage tissue engineering process, both cellular phenotype and the extra-cellular matrix change (Fig. 2). Thus the choice of growth factors must be made for each stage of the tissue engineering process while taking into account the desired effect as well as the phenotype and presence of extracellular matrix of the cells.

We here present a compilation of experiments demonstrating that growth factors can improve the quantity and quality of extracellular matrix components produced in a tissue engineering process. The data show that the effect of a growth factor depends on possible treatments in previous steps of the procedure. The results may help to recognise the potency as well as the limitations of the use of growth factors for cartilage tissue engineering.

2. Materials and methods

In this paper a compilation of experiments using different types of cartilage is presented. Adult bovine articular cartilage was obtained from metacarpophalangeal joint; Rabbit ear cartilage was prepared from 3 month old New Zealand White rabbits; Adult human articular cartilage was obtained post-mortem from the femoral condyle of road victims without any previous history of joint diseases. Young human ear cartilage was obtained from children undergoing surgery for correction of protruding ears.

In all experiments, cells were isolated from full-thickness cartilage using pronase and collagenase digestion. For multiplication they were cultured in monolayer for 4 passages. To achieve matrix production the chondrocytes were seeded in 1.2% alginate (LV) and cultured for 3 weeks. The amount of glycosaminoglycan was determined using dimethylmethylene blue (Farndale assay). The amount of DNA was determined using Hoechst 33258 in a fluorimetric assay. Immunohistochemistry was performed on cytospin preparations with antibodies against collagen type II (II-II6B3, Developmental Studies Hybridoma Bank), procollagen type I (m38; Developmental Studies Hybridoma Bank) and a membrane marker for fibroblasts (11-fibrau, Imgen distributed by ITK diagnostics, Uithoorn, The Netherlands).

For some of the monolayer cultures rhFGF2 (Instruchemie, Delfzijl, The Netherlands) was used. For cultures in alginate ITS+ (1 : 100; Becton Dickinson Labware, Bedford, MA), rhIGF1 (Roche Diagnostics, Mannheim, Germany) and rhTGFβ2 (R&D systems, Abingdon, UK) were used.

3. Results and discussion

3.1. Matrix production

For production of cartilage matrix a three-dimensional environment is required. Culture in alginate gel is a generally accepted method. To further increase the production of extracellular matrix, growth factors can be added to medium with 10% serum. Insulin-like growth factor 1 (IGF1) is described to be the major anabolic factor for cartilage. Transforming Growth Factor β (TGFβ) has been found to be able to have both stimulating and inhibitory effects on cartilage. Addition of IGF1 or TGFβ (in a dose of 10 or 25 ng/ml) to freshly isolated, differentiated chondrocytes in alginate increases matrix production with $\sim$50% [8,9]. However, the data are of limited use for tissue engineering procedures because normally we use multiplied chondrocytes that have lost their phenotype in stead of freshly isolated, differentiated chondrocytes. When added to dedifferentiated chondrocytes, TGFβ stimulates glycosaminoglycan production even better than in differentiated cells (2.2$\times$ vs 1.6$\times$) [9]. However, it does not stimulate re-expression of cartilage specific collagen type II.

3.2. Serum-free culture

In the serum used for previous studies, many growth factors are present in variable and unknown concentrations. In addition, many inhibitors of growth factors are also present. It has been demonstrated that the presence of serum can influence the effect of growth factors [13]. Therefore, the use of serum in experiments with growth factors creates a very badly defined system with results that are difficult to predict and reproduce. This is why we prefer a serum-free culture system for tissue engineering procedures, especially when growth factors are supplemented. When serum in our alginate culture experiments is replaced by 10 ng/ml IGF1 and 10 ng/ml TGFβ, production of collagen type II in dedifferentiated cells can be stimulated. This stimulation of redifferentiation in serum-free medium with IGF1 and TGFβ2 is shown for various types of chondrocytes [references for rabbit ear: 10; rabbit nasal septum: 12; bovine articular: 5; human ear: 10; human nasal septum: 12; human articular: 15], although the amount of stimulation was variable.

3.3. *Matrix assembly*

Besides the amount of matrix produced, the assembly of matrix components is also important for the functionality of the matrix. The extra-cellular matrix of cartilage is structured in a pericellular compartment ($\sim$1–2 μm surrounding the cell), a territorial matrix ($\sim$5 μm) and an interterritorial matrix. The alginate culture system offers the unique possibility to quantify the accumulation of matrix components in cell-associated compartment (which is believed to correspond to the pericellular and the territorial matrix) and further-removed compartment (believed to correspond to the interterritorial matrix) [2]. The cell-associated matrix is in close contact with the cell and therefore largely influences cell behaviour. The interterritorial matrix is supposed to be most important for the gross mechanical properties of the cartilage. Addition of growth factors to medium with serum has been demonstrated to change the assembly of glycosaminoglycans. After addition of IGF1 to differentiated bovine articular chondrocytes in alginate relatively more of the produced GAG is laid down in the cell-associated matrix than in the further-removed matrix compartment. On the contrary, addition of TGFβ shifts the deposition of GAG from the cell-associated matrix more towards the further-removed matrix compartment [8]. Furthermore, dedifferentiated chondrocytes cultured in alginate in medium with 10% serum divide the newly synthesised glycosaminoglycans differently over the two compartments than differentiated chondrocytes do; relatively more GAG is laid down in the cell-associated matrix of dedifferentiated cells [5]. This might be a way for the cell to keep its direct environment as optimal as possible. When serum is replaced by IGF1 and TGFβ, the division of glycosaminoglycans over the two compartments by dedifferentiated cells is comparable with differentiated cells [5]. This demonstrates that growth factors not only affect the production of matrix components, but also the assembly of these components in the matrix.

3.4. *"Culture history"*

The effect of a growth factor not only depends on cellular phenotype but also in a more general way the "culture history" of the cells before addition of a growth factor is important for the effect of this growth factor. This not only involves the difference between freshly isolated, differentiated cells and culture expanded, dedifferentiated cells, but also more subtle differences. For example, when cells are surrounded by extracellular matrix, the response of the cell to growth factor addition can be altered. This might be explained, for example, by an interaction between integrins and growth factor receptors [3,7]. In the experiments described above, growth factors were added starting from the moment the cells were cultured in alginate. Before the cells were seeded into alginate they were either freshly isolated from the cartilage tissue or multiplied in monolayer culture in medium with 10% serum. In another set of experiments we demonstrated that TGFβ2 stimulates matrix production when supplemented to alginate cultures immediately, but has no effect on matrix production when addition starts of for the first time after 3 weeks culture in alginate [9]. This appears to be due to the presence of extracellular matrix surrounding the cell, produced during the 3 weeks pre-culture in alginate. Removing this previously synthesised matrix, makes the cells again respond to TGFβ by increased matrix production.

Beside the 3D culture history, the history in 2D monolayer can also influence the response to growth factors during alginate culture. Addition of FGF2 in monolayer has been demonstrated to affect the capacity of the cells to re-express their phenotype [4,6]. We added 5 ng/ml FGF2 to serum-containing medium for expansion of human articular chondrocytes. In the subsequent alginate culture, these cells have increased production of GAG and collagen II upon stimulation with IGF1 and TGFβ, compared to cells expanded in medium with serum alone. Finally, the effect of IGF and TGFβ on GAG and collagen II production in alginate culture is identical for cells expanded in medium in which serum had been replaced by ITS+ and 100 ng/ml FGF2. This shows that ITS+ and FGF2 can imitate the effect of serum. However, next to evaluating chondrocyte phenotype using the most common determinants (GAG and collagen II) we have previously demonstrated the usefulness of additional markers, f.e. the anti-fibroblast marker 11-fibrau, that can be thought of as a dedifferentiation marker [11]. The 11-fibrau antibody does not bind to differentiated chondrocytes and binding is increased with time in monolayer culture when cells dedifferentiate. 11-fibrau still binds to chondrocytes that have been expanded in medium with serum and subsequently "redifferentiated" in alginate in serum-free medium with IGF and TGFβ. This indicates that these cells, although expressing collagen II, are not (yet) fully redifferentiated. However, when using a serum-free medium with ITS+ and 100 ng/ml FGF2 to expand the chondrocytes, expression of fibroblast marker 11-fibrau is lost after alginate culture. This implicates that addition of FGF2 during expansion culture influences the response of cells to IGF/TGFβ during alginate culture. A relation between the group of fibroblast growth factors, Sox-9 and chondrocyte matrix production is suggested [1]. Sox-9 is a transcription factor, which is responsible for the differentiation of mesenchymal stem cells to chondrocytes. Also, sox-9 promotes upregulation of several chondrocyte-specific marker genes including the genes for collagen type II and aggrecan. Fibroblast growth factors, including FGF2, can enhance the expression of sox-9 and can alter the response of cells to other growth factors.

3.5. Conclusion and future perspectives

In summary, growth factors can be used in different phases of the tissue engineering process to stimulate the formation of a functional tissue to obtain better results for cartilage grafting procedures. The aim is to improve mechanical properties of the graft and of the interface between graft and adjacent cartilage. This is achieved by stimulating both the formation and proper assembly of matrix components. The optimal choice of growth factors will vary with time during the tissue engineering process and depends on the changing phenotype and presence of extracellular matrix components. This is a highly complex process.

Until now all *in vitro* experiments to evaluate the effect of growth factors are performed in free-floating (unloaded) cultures. This provides valuable information for the use of growth factors during such an *in vitro* procedure. However, this does not necessarily mean that these growth factors will be valuable for *in vivo* use in the joint, for example, coupled on a biomaterial or in a gene delivery system. In the joint, the construct will be mechanically loaded. Although nothing is known about the effects of growth factors during simultaneous loading it is very likely that there will be some effects. This will be an important topic of future studies in our laboratory.

Acknowledgements

This work was partly funded by the Dutch Technology Foundation (STW), applied science division of NWO and the technology programme of the Ministry of Economic Affairs and partly by the Dutch Ministry of Economic Affairs in a grant of Senter (BTS 00021).

References

[1] B. de Crombrugghe, V. Lefebvre, R.R. Behringer, W. Bi, S. Murakami and W. Huang, Transcriptional mechanisms of chondrocyte differentiation, *Matrix Biol.* **19** (2000), 389–394.

[2] H.J. Hauselmann, M.B. Aydelotte, B.L. Schumacher, K.E. Kuettner, S.H. Gitelis and E.J. Thonar, Synthesis and turnover of proteoglycans by human and bovine adult articular chondrocytes cultured in alginate beads, *Matrix* **12** (1992), 116–129.

[3] T.M. Hering, Regulation of chondrocyte gene expression, *Front Biosci.* **4** (1999), D743–761.

[4] M. Jakob, O. Demarteau, D. Schafer, B. Hintermann, W. Dick, M. Heberer and I. Martin, Specific growth factors during the expansion and redifferentiation of adult human articular chondrocytes enhance chondrogenesis and cartilaginous tissue formation in vitro, *J. Cell. Biochem.* **81** (2001), 368–377.

[5] W.J.C.M. Marijnissen, G.J.V.M. van Osch and J.A.N. Verhaar, Effect of culture conditions on glycosaminoglycan assembly in cell-associated and further removed matrix of differentiated and dedifferentiated chondrocytes, *Trans. ORS* **24** (1999), 701.

[6] I. Martin, G. Vunjak-Novakovic, J. Yang, R. Langer and L.E. Freed, Mammalian chondrocytes expanded in the presence of fibroblast growth factor 2 maintain the ability to differentiate and regenerate three-dimensional cartilaginous tissue, *Exp. Cell. Res.* **253** (1999), 681–688.

[7] M. Shakibaei, T. John, P. De Souza, R. Rahmanzadeh and H.J. Merker, Signal transduction by beta1 integrin receptors in human chondrocytes in vitro: collaboration with the insulin-like growth factor-I receptor, *Biochem. J.* **342** (1999), 615–623.

[8] G.J.V.M. van Osch, W.B. van den Berg, E.B. Hunziker and H.J. Häuselmann, Differential effects of IGF1 and TGFβ2 on proteoglycan assembly of pericellular and territorial proteoglycan assembly by cultured bovine articular chondrocytes, *Osteoarthritis Cartilage* **6** (1998), 187–195.

[9] G.J.V.M. van Osch, S.W. van der Veen, P. Buma and H.L. Verwoerd-Verhoef, Effect of transforming growth factor β on proteoglycan synthesis by chondrocytes in relation to differentiation stage and the presence of pericellular matrix, *Matrix Biology* **17** (1998), 413–424.

[10] G.J.V.M. van Osch, S.W. van der Veen and H.L. Verwoerd-Verhoef, In-vitro redifferentiation of culture expanded rabbit and human auricular chondrocytes for cartilage reconstruction, *Plastic Reconstr. Surg.* **107** (2001), 433–440.

[11] G.J.V.M. van Osch, S.W. van der Veen, W.J.C.M. Marijnissen and J.A.N. Verhaar, Monoclonal antibody 11-Fibrau: a useful marker to characterize chondrocyte differentiation stage, *Biochem. Biophys. Res. Comm.* **280** (2001), 806–812.

[12] G.J.V.M. van Osch, S.W. van der Veen and H.L. Verwoerd-Verhoef, The potency of culture expanded nasal septum chondrocytes for tissue engineering of cartilage, *Am. J. Rhinoloy* **15** (2001), 187–192.

[13] D. Vivien, P. Galera, G. Loyau and J.P. Pujol, Differential response of cultured rabbit articular chondrocytes (RAC) to transforming growth factor beta (TGF-beta)-evidence for a role of serum factors, *Eur. J. Cell. Biol.* **54** (1991), 217–223.

[14] K. Von der Mark, V. Gauss, H. Von der Mark and P. Müller, Relationship between cell shape and type of collagen synthesized as chondrocytes lose their cartilage phenotype in culture, *Nature* **267** (1977), 531–532.

[15] P.C. Yaeger, T.L. Masi, J.L. Buck de Ortiz, F. Binette, R. Tubo and J.M. McPherson, Synergistic action of transforming growth factor-β and insulin-like growth factor-I induces expression of type II collagen and aggrecan genes in adult human articular chondrocytes, *Exp. Cell. Res.* **237** (1997), 318–325.

Biorheology 39 (2002) 221–235
IOS Press

Animal models of osteoarthritis

Kenneth D. Brandt
Rheumatology Division, Indiana University School of Medicine, and Indiana University Multipurpose Arthritis and Musculoskeletal Diseases Center, 1110 West Michigan Street, Room LO 545, Indianapolis, IN 46202, USA
Tel.: +1 317 274 4225; Fax: +1 317 274 7792; E-mail: kbrandt@iupui.edu

Abstract. Animal models have proved to be of considerable importance in elucidating mechanisms underlying joint damage in osteoarthritis (OA) and providing proof of concept in the development of pharmacologic and biologic agents that may modify structural damage in the OA joint. The utility of animal models in predicting the response to an intervention with a drug or biologic agent in humans, however, can be established only after evidence is obtained of a positive effect of the agent in humans. To date, no agent has been shown unequivocally to have such an effect, although diacerhein and glucosamine have recently been reported to lower the rate of loss of articular cartilage in patients with hip OA and knee OA, respectively, based on measurements of the rate of joint space narrowing in plain radiographs. Furthermore, the predominant manifestation of OA – and the feature that leads people with radiographic changes of the disease to decide to seek medical attention and contributes to the enormous medicoeconomic and socioeconomic burden imposed by the disease – is joint pain. Notably, none of the animal models of OA is a good indicator of the analgesic effects of pharmacologic agents. Indeed, it should not be assumed *a priori* that reduction in the rate of progression of joint damage in OA will be associated with a reduction in joint pain.

1. Introduction

Animal models of osteoarthritis (OA) are useful in studying the evolution of structural changes in joint tissues, determining how various risk factors may initiate or promote these changes, and evaluating therapeutic interventions. Because it is difficult to obtain joint tissues from humans with OA until pathological changes are far advanced, and to obtain them sequentially, use of animal models provides the only practical way to examine the processes involved in initiation of OA, and how, e.g., the balance of matrix cartilage synthesis and degradation influences OA progression. However, OA is not a disease only of articular cartilage – all tissues of the joint are affected, i.e., the subchondral bone, synovium, menisci, ligaments, periarticular muscles and the afferent nerves whose terminals lie both within and external to the joint capsule. To date, studies of pharmacologic agents in animal models have focused chiefly on their effects on articular cartilage. The effects, e.g., of quadriceps weakness, a major cause of disability in humans with knee OA and a risk factor for knee OA in humans, have received essentially no attention in animal models of knee OA. Indeed, if such studies were performed in quadripeds, their relevance to OA in humans would be questionable. Furthermore, because joint pain is the chief clinical feature leading the human subject with OA to seek medical attention, the effects of a drug on joint pain in humans cannot be studied adequately in any animal model.

Great interest currently exists in therapeutic agents which may slow the progression of OA, i.e., disease-modifying OA drugs (DMOADs) [50]. Several drugs have been shown to slow progression, or prevent development, of OA in animals. Clinical trials of diacerhein and of doxycycline and other matrix metalloproteinase (MMP) inhibitors and of an antiresorptive agent (targeting bone rather than articular cartilage) have been undertaken in patients with OA, based, in large part on the demonstration of DMOAD effects in animal models. Diacerhein has exhibited structure-modifying effects in a spontaneous OA model in the guinea pig [7,19], a canine cruciate-deficiency model of OA [92], a lapine post-contusion model [55], and an ovine model in which OA is induced by menisectomy [42]. Doxycycline treatment has also led to positive results in canine cruciate-deficiency models [105] and the above guinea pig model of OA. In general, the greater the similarity of OA changes in the model to those in humans, the more desirable the model. However, practical issues, such as the availability of the animal species, the rate at which changes of OA develop, the ease of handling and maintenance, and cost all are important considerations. A comprehensive, although not exhaustive, list of animal models of OA is provided in Table 1, modified from the excellent recent review of the subject by Doherty et al. [2,3,5,8, 10,14,18,21–25,27,30–33,37,38,41,46–49,52,54,56,62,66,71,72,81,85,93–100,102,103].

Whether surgical models of OA and models that involve intra-articular injection of a degradative enzyme to destabilize the joint by damaging articular cartilage and other structures are more relevant to post-traumatic OA than to primary (idiopathic) OA in humans, at least with regard to their suitability for evaluating therapeutic agents, remains to be seen. The spontaneous development of knee OA in a variety of animal species provides alternatives to surgically or chemically induced models (Table 1). However, because no animal model can mirror fully all of the variables which may modify the disease in humans, use of models for discovery of DMOADs represents "... an expensive gamble that what can be achieved in animals will translate to the OA diseases of man" [11]. This paper draws extensively on two recent excellent comprehensive reviews: one [11], emphasizes the advantages attached to the use of animal models of OA; the other [28] deals with limitations in the use of models.

Models based on surgically induced laxity of the knee joint, altering the mechanical stresses, have most frequently utilized rabbits and dogs. The most extensively studied surgical model is the cruciate-deficient dog [54,81], which has been characterized in detail in our unit [14]. In the rabbit, a variety of surgical procedures on ligaments and menisci have been described [61], including cruciate ligament transection, with and without medial and collateral ligament section; total and partial meniscectomy; and creation of bucket-handle tears of the meniscus. In the guinea pig, partial meniscectomy and cruciate and collateral ligament transection have been studied [9,59]. Partial meniscectomy leads rapidly to osteophyte formation and extensive cartilage degeneration, greatly accelerating the underlying spontaneous disease in this species.

Because the natural history of OA in the canine cruciate-deficiency model has been well elucidated and the pathology, not only in the articular cartilage but also in other joint tissues, has been characterized to a greater extent than that in other animal models of OA [14], the remainder of this review will focus on that model.

2. The natural history of OA as seen in the cruciate-deficient dog

Until recently, because of an apparent absence of progressive changes and lack of full-thickness cartilage ulceration (as seen in human OA), the cruciate-deficient dog was widely viewed with skepticism as a model of OA. Studies by Marshall and Olsson [54] had suggested that the articular cartilage changes 2 years after ligament transection were no more severe than those seen only a few months after surgery. It was presumed that capsular fibrosis and buttressing osteophytes stabilized the cruciate-deficient knee and prevented progressive cartilage breakdown. The apparent lack of progressive cartilage changes led a number of investigators to contend that this was merely a model of cartilage injury and repair, rather than of OA.

Table 1
Examples of animal models of OA*

Underlying mechanism	Species/strain	Inducing agent	Reference
Spontaneous OA	Guinea pig, Hartley	Age/obesity	[8,10]
	Mouse, STR/ORT, STR/INS	Unidentified genetic predisposition	[25,30,31]
	Mouse, C57 black	Unidentified genetic predisposition	[72,93,94,97]
	Mouse	Type II collagen mutation	[37]
	Mouse	Type IX collagen mutation	[66]
	Dog	Hip dysplasia	[3]
	Primate	Unidentified genetic predisposition	[2,18,21]
Chemically-induced	Chicken	Intra-articular iodoacetate	[46]
	Guinea pig	Intra-articular papain	[23,24,52]
	Dog	Intra-articular chymopapain	[49]
	Guinea pig	Intra-articular papain	[95]
	Mouse	Intra-articular papain	[98]
	Mouse	Intra-articular collagenase	[98]
	Mouse	Intra-articular TGF-β	[96]
	Rabbit	Intra-articular hypertonic saline	[99]
Physically-induced	Dog	Anterior cruciate ligament transection (unilateral)	[14,81]
	Dog	Anterior cruciate ligament transection (bilateral)	[54]
	Rabbit	Anterior cruciate ligament transection	[22,102]
	Sheep	Meniscectomy	[38]
	Rabbit	Meniscectomy	[32,33,62]
	Guinea pig	Meniscectomy	[10]
	Guinea pig	Myectomy	[5,27,48]
	Rabbit	Patellar contusion	[56,71]
	Rabbit	Immobilization	[47,100]
	Dog	Immobilization	[41,85]
	Dog	Denervation followed by anterior cruciate ligament transection	[103]

*Modified from N. Doherty et al. [28].

However, our studies of unstable knee over a lengthier period of observation than had been employed previously clearly validate the cruciate-deficient dog as a model of OA [12]. Furthermore, they emphasize the striking capacity of the chondrocyte to repair cartilage damage in the earlier stages of OA [1], as described below:

McDevitt and Muir had noted that an increase in proteoglycan (PG) synthesis by chondrocytes in articular cartilage of the unstable knee occurs within only days after ligament transection [57,58]. We found that cartilage from the OA knee, while showing typical biochemical, metabolic and histological changes of OA, rapidly became thicker than normal and remained thicker 3 years or more after the onset of knee instability [1], and showed that this thickening was associated with a sustained increase in PG synthesis and increases in both the content and concentration of PGs in the OA cartilage (Figs 1–3). By magnetic resonance imaging (MRI) [15], we showed that this hypertrophic cartilage was maintained for as long as 3 years after cruciate transection, but that progressive loss of articular cartilage then occurred, so that 45 months after cruciate ligament transection extensive areas of the joint surface were devoid

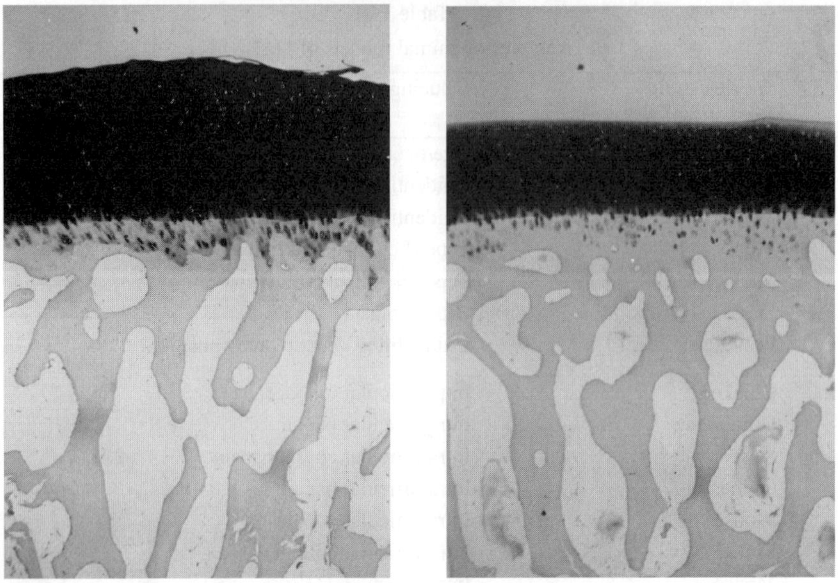

Fig. 1. Full thickness sections of articular cartilage from a dog whose anterior cruciate ligament was transected 12 weeks earlier. The thickness of the articular cartilage and intensity of staining with safranin-O are greater in the OA knee (left side of figure) than in the contralateral knee (right side of figure). Original magnification ×40.

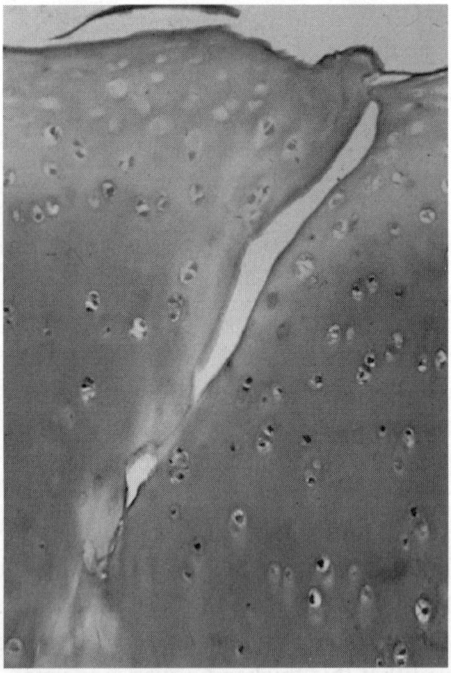

Fig. 2. Defect in articular cartilage of the unstable knee of a dog 12 weeks after transection of the anterior cruciate ligament. Integrity of the surface has been disrupted and a vertical fissure extending into the depths of the cartilage is apparent (fibrillation). A decrease in the intensity of staining of the extracellular matrix with safranin-O is apparent along the margins of the fissure, reflecting a decrease in proteoglycan concentration in this area. Note also the empty lacunae in the upper zone of the cartilage, representing loss of chondrocytes and the early cloning of chondrocytes, especially near the area of fibrillation.

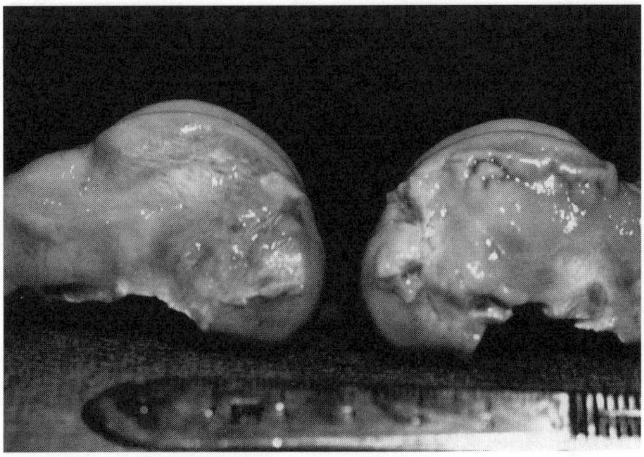

Fig. 3. Prominent osteophytosis developing along the trochlear ridge in the unstable knee of a dog 12 weeks after transection of the anterior cruciate ligament (right hand side of figure). Note the absence of osteophytes in the stable contralateral knee (left hand side of figure).

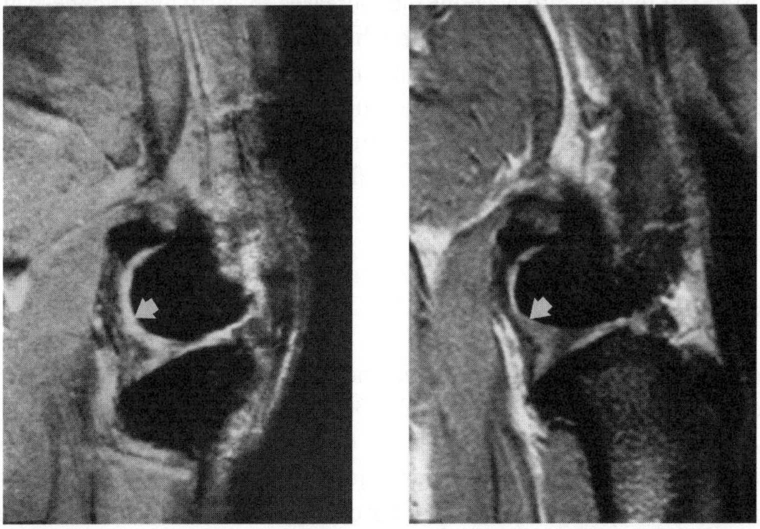

Fig. 4. Magnetic resonance imaging (MRI) of a dog 36 months after anterior cruciate ligament transection (left hand panel). The arrow indicates the persistence of cartilage hypertrophy. The right hand panel depicts an MRI of the same knee in the left hand panel, performed 9 months later. The arrow in this photo points to extensive full-thickness loss of articular cartilage on the posterior aspect of the medial femoral condyle.

of articular cartilage (Fig. 4). Studies of dogs sacrificed 54 months after cruciate ligament transection provided pathological confirmation of our MRI observations (Fig. 5). In summary, given sufficient time, the changes of OA in this model are progressive; this is indeed a model of OA.

Although the phenomenon of hypertrophic repair of articular cartilage in OA is illustrated particularly well by the canine cruciate deficiency model, it is by no means unique to that model. It was recognized first in human OA cartilage by Bywaters [17] and subsequently by Johnson [45], and evidence of hypertrophic repair may be seen in OA cartilage from rabbits subjected to partial meniscectomy [101] and from Rhesus macaques developing OA spontaneously [20,46].

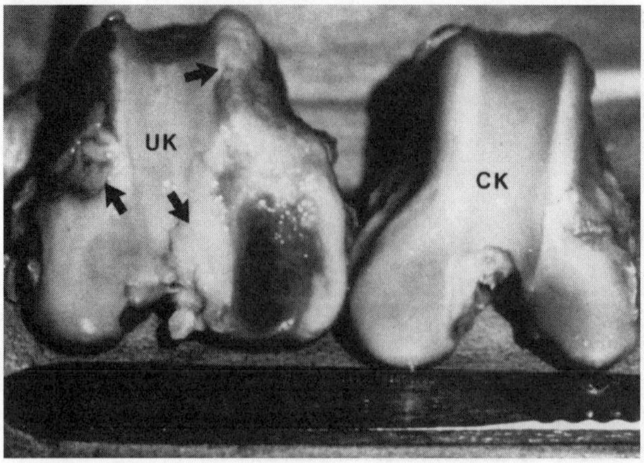

Fig. 5. Full thickness loss of articular cartilage on the femoral condyle of the unstable knee (UK) 4 years after transection of the anterior cruciate ligament. No pathologic changes are apparent in the stable contralateral knee (CK) of the same animal.

Although contemporary descriptions of the pathology of OA emphasize the progressive loss of articular cartilage, they often fail to take into account the increases in cartilage thickness and PG synthesis in the earlier stages of OA, which are consistent with a homeostatic phase of compensated, stabilized OA. Repair may keep pace with cartilage breakdown and maintain the joint in a reasonably functional state for years. The repair tissue, however, does not hold up as well to mechanical stress as normal hyaline cartilage and, eventually, at least in some cases, the rate of PG synthesis falls, and the cells can no longer successfully maintain the matrix and end-stage OA develops, with full-thickness loss of the articular cartilage [12].

3. Neurogenic acceleration of OA

As described above, in the canine cruciate-deficiency model of OA cartilage degeneration develops gradually. Full thickness loss of cartilage may take 3–5 years. However, if cruciate ligament transection is preceded by dorsal root ganglionectomy [103] or articular nerve neurectomy [69], full-thickness ulceration of the cartilage occurs within only weeks [68] (Fig. 6). This observation is directly relevant to Charcot (neuropathic) arthropathy, a joint disorder characterized by severe joint destruction, with osteochondral fractures, loose bodies, effusions, ligament instability and formation of new bone and new cartilage within the joint. The general concept of the pathogenesis of Charcot arthropathy is that interruption of sensory input from the extremity by a neurologic disease (e.g., tabes dorsalis, syringomyelia) deprives the central nervous system of information from nociceptive or proprioceptive nerve endings, leading to recurrent episodes of joint trauma and, ultimately, to joint breakdown.

A problem exists with this proposed mechanism, however: it has not been possible to consistently produce Charcot arthropathy experimentally by any neurosurgical procedure which interrupts the sensory nerve supply from an extremity. For example, we were unable to induce knee joint pathology in normal dogs subjected to unilateral L4-S1 dorsal root ganglionectomy to extensively deafferent the ipsilateral hind limb [68]. On the other hand, if we transected the ipsilateral anterior cruciate ligament in dogs which had previously undergone dorsal root ganglionectomy, cartilage in the ipsilateral knee broke down extensively within only a few weeks. These results stand in sharp contrast to the very slowly progressive joint changes seen in the neurologically intact dog with a cruciate-deficient knee (see above).

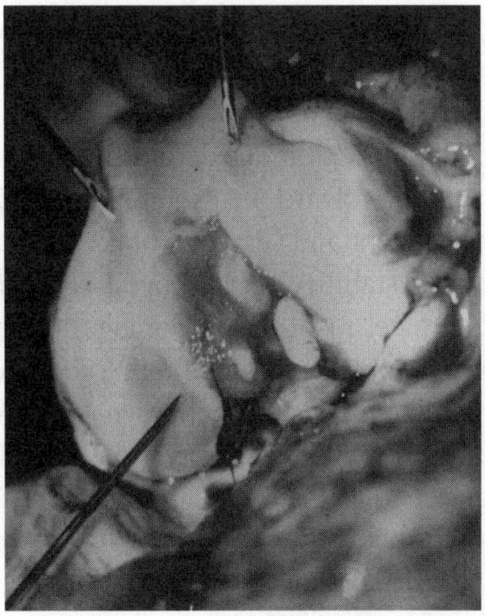

Fig. 6. Marked thinning of articular cartilage on the femoral condyle and trochlea of a dog only 3 weeks after the ipsilateral anterior cruciate ligament had been transected in a dog which had previously undergone a neurosurgical procedure to extensively interrupt sensory input from the limb.

The accelerated canine model of OA has proved very useful for evaluation of pharmacologic agents, whose effect on progressive joint damage can be observed within only a few weeks [105] (Fig. 2). However, the pathogenetic processes underlying the initiation of OA in the accelerated model, i.e., the homeostatic phase of hypertrophic repair (see above), may be different from those associated with the stage of progressive joint damage [13,26,92]. A drug may show efficacy in the slowly progressing (neurologically intact) cruciate-deficiency model but not in the accelerated model, as was the case with diacerhein [13, 92]; the converse may also be expected.

4. The importance of periarticular muscle

Several years ago we showed that when the hind limb of a normal dog was immobilized in an orthopedic cast, the articular cartilage rapidly developed striking atrophic changes, with a loss of thickness, decrease in proteoglycan concentration, reduction in net proteoglycan synthesis and defect in proteoglycan aggregation [75]. Notably, if an unstable knee is immobilized immediately after cruciate ligament transection, OA does not develop, but changes of cartilage atrophy are seen [76], identical to those noted after immobilization of the normal stable knee. The results emphasize the importance of altered joint mechanics in the genesis of OA in this model.

Studies aimed at elucidating the pathogenesis of the cartilage changes seen with immobilization indicated that these were due to reduction in the contraction of the periarticular muscles that span the joint (e.g., hamstrings, quadriceps), and stabilize the limb during stance [77]. In patients with knee OA, quadriceps weakness is common and is generally believed to be due to atrophy of the muscle, secondary to unloading of the painful extremity. However, epidemiologic data suggest an additional explanation for the quadriceps weakness in subjects with knee OA. Among community cohort of elderly individuals we found that quadriceps weakness was common in women with radiographic changes of knee OA – even in the absence of a history of knee pain. Furthermore, hypertrophy (rather than atrophy) of the quadriceps muscle was seen, due to the obesity of these subjects [89]. Furthermore, longitudinal analysis strongly suggested that baseline quadriceps weakness was a risk factor for incident radiographic changes of knee OA (i.e., structural damage) [90].

The proposed mechanism by which quadriceps weakness in humans may predispose to knee joint damage is as follows: Although hamstring contraction decelerates the forward swing of the leg during gait, at the end of the swing phase the leg is pulled toward the ground (toward heel-strike) by gravity. Quadriceps action at this point retards the rate of descent and serves to brake the fall of the leg. Furthermore, because the quadriceps provides anteroposterior stability to the knee [44], quadriceps weakness may alter sites of mechanical loading of the joint surface, resulting in damage to the articular cartilage. Experimental paralysis of the quadriceps in a normal subject provided a marked heelstrike transient and an estimated 5-fold increase in the forces acting at the knee at touchdown [43]. Among subjects with normal knee radiographs, those with knee pain exhibited a significantly greater heelstrike transient than those without knee pain [84]. The authors suggested that the gait pattern in the former group, the consequence of which was repeated impact loading of the lower extremity, led to a "pre-osteoarthritic" state, although no evidence was provided that those with knee pain developed OA or that the increased heelstrike transient was the cause, rather than the result, of knee pain.

Experimentally, some support exists for the above hypothesis: repeated acute (50 msec) impact loading of the knee in rabbits resulted in damage to both the articular cartilage and subchondral bone [82] where impulsive loads of comparable or even greater magnitude, it applied more gradually (500 msec), were innocuous (Radin, E., personal communication). Very rapid application of load does not allow sufficient time for the periarticular muscles, the major shock absorbers protecting the joint, to absorb the load through eccentric contraction [39,83]. Why some normal subjects generate a large heelstrike transient when walking, while others do not, is unknown. It may reflect individual differences in central program generators, i.e., neurologic mechanisms based in the central nervous system which coordinate complex limb movements during gait [67].

Finally, the importance of muscle spindles (sensory nerve endings within muscle) in modulating muscle tone in response to changes in the position of a limb, thereby protecting the joint from injury, has received attention recently [43]. Joint position sense, measured both as the ability to reproduce passive positioning (proprioception) and to detect the onset of movement (kinesthesia), was significantly impaired in normal subjects after completion of a strenuous exercise protocol [88]. Hence, maintenance of quadriceps strength appears to be important also in preserving the integrity of muscular reflexes that protect the knee form damage.

Our data suggest that among women with incident OA, the greater their body weight, the poorer their quadriceps function. Regardless of the cause, our findings raise the possibility that the well-recognized association of obesity with knee OA in women is mediated through quadriceps weakness. Clearly, quadriped models of OA are unlikely to prove suitable for study of the relationship between periarticular muscle strength and development of OA; whether primate models will prove suitable in this respect remains to be seen.

5. Subchondral bone

Some investigators have suggested that stiffening of the subchondral bone may be of primary importance in the etiopathogenesis of OA in man [82]. Radiographic studies of dogs that had undergone cruciate ligament transection revealed typical bony changes of OA, including subchondral sclerosis, by

24 months after surgery [104]. However, direct examination by computerized tomographic microdensitometry of the subchondral bone of dogs maintained as long as 72 weeks after cruciate ligament transection showed no increase in thickness of the subchondral plate and thinning of the underlying cancellous bone [26]. Only later was a trend noted for thickening of the subchondral plate. The osteopenia seen in the subchondral trabeculae is presumably due to the decrease in loading of the unstable limb after cruciate ligament transection [70] and to synovitis [63].

Our observations make it clear that typical articular cartilage changes of OA can occur in the presence of osteopenia; stiffening of subchondral bone is not a requisite for *initiation* of the early cartilage changes in this OA model. Indeed, loss of subchondral bone could theoretically increase mechanical strain in the overlying articular cartilage, leading to degeneration [16]. On the other hand, thickening of the subchondral plate, a relatively late phenomenon in the canine model, could contribute to failure of intrinsic repair mechanisms and to *progression* of cartilage breakdown. We have shown that administration of a bisphosphonate can effectively inhibit resorption and formation of subchondral bone in the canine cruciate-deficiency model of OA [65], although it had no effect on osteophyte formation. Formation of new subchondral bone in the OA knee is presumably coupled to osteoclastic resorption, which can be inhibited by bisphonates [51,86]. However, osteophyte formation involves bony metaplasia of new fibrous connective tissue and bears no relationship to bone resorption. Therefore, absence of a pharmacologic effect on osteophytosis, while formation of subchondral bone was blocked by the bisphosphonate, is not surprising. A multi-center clinical trial to evaluate the effectiveness of a drug employed as a DMOAD in patients with knee OA is now in progress.

6. Demonstration of structure-modifying effects of drugs in animal models of OA

Despite the costs associated with the canine cruciate-deficiency model, several drug studies have been undertaken in it. The NSAID, tiaprofenic acid, was reported to reduce both the severity of pathologic changes of OA and protease levels within the cartilage [78], although evidence for a DMOAD effect in humans with this or any other NSAID is lacking. Tenidap, a nonsteroidal anti-inflammatory drug with cytokine-inhibitory properties, was reported to prevent cartilage damage in this model [34], but the extent of cartilage damage in that study was unusually severe only 8 weeks after cruciate ligament transection and confirmation of the results is needed. Corticosteroid treatment in relatively high doses has been shown to prevent osteophyte formation and to reduce levels of stromelysin in extracts of articular cartilage [78–80] in this model, but when prednisone was given orally in a lower dose (the equivalent of 5 mg/d for a 70 kg human), no effect on OA pathology was noted [64]. This model [4], and the lapine partial medial meniscectomy model [40], have been used to demonstrate the efficacy of glycosaminoglycan polysulfate acid ester. Co-administration of pentosan sulfate and IGF-1 protected against cartilage damage and resulted in a reduction in the levels of MMP and increase in the levels of TIMP within the cartilage [87]. Finally, we have shown that oral doxycycline administration strikingly inhibited joint breakdown in the accelerated canine cruciate-deficiency model (i.e., in which sensory input from the ipsilateral limb was interrupted by L4-5 dorsal root ganglionectomy prior to ligament transection, see above) [105]. Doxycycline was effective also in the standard canine cruciate-deficiency model (i.e., without deafferentation). This treatment prevented the extensive loss of cartilage seen in this model and strikingly reduced the levels of MMP within the cartilage. Based to a large extent on these results in the animal models of OA, a multicenter placebo-controlled trial of doxycycline has been initiated in humans with OA. As indicated above, diacerhein exhibited protective effects in the canine cruciate-deficiency model of OA [92], although such effects were mitigated when the rate of development of joint damage and severity of joint pathology were greatly accelerated by interruption of sensory input from the ipsilateral hind limb prior to transection of the cruciate ligament [13]. This suggests that a DMOAD that is effective in the earlier stages of OA (i.e., during the homeostatic phase of hypertrophic repair) may not be effective later in the disease (i.e., during the phase of decompensation) and *vice versa*.

The utility of animal models as predictors of the effects of a putative DMOAD in humans can, obviously, be established only after demonstration of an effect of that agent in humans. Because, as described above, diacerhein exhibited protective effects in several animal models of OA, it is particularly notable that it slowed the rate of articular cartilage loss, as reflected by joint space narrowing in the plain radiograph [29], in a recent placebo-controlled trial in humans with hip OA.

7. Pitfalls associated with use of animal models in the discovery of DMOADs

Because the etiology of OA is poorly understood, it is impossible to measure how well *any* animal model of OA mimics the disease in humans. It is also impossible to predict how closely a therapeutic effect in a model will mirror therapeutic activity in man. In particular, the importance of inflammation in OA is uncertain. Therefore, use of models in which joint damage is mediated by an inflammatory response may give misleading indications of efficacy. Similar uncertainty concerning the importance of other etiologic factors, such as a systemic increase in bone density, ligamentous laxity, inherited defects in the structure of cartilage collagen and periarticular muscle weakness.

Major differences exist among animal species with respect to the relative contribution of various mediators, receptors or enzymes to the pathologic changes of OA. Extrapolation to humans of therapeutic activity seen in animal models, therefore, may be hazardous [28]. For example, results in rodent models of inflammatory arthritis, in which NSAID treatment profoundly inhibits the disease [73], led to *over*estimation of the efficacy of NSAIDs in treatment of arthritis in humans, in whom prostaglandins do not play the same fundamental role in pathogenesis as they do in rodents and in whom clinical benefits of NSAIDs are, therefore, limited to treatment of symptoms rather than modification of joint damage.

On the other hand, *under*estimation of the therapeutic potential of novel agents on the basis of animal data can lead to abandonment of potentially useful agents. For example, gold, penicillamine, chloroquine and sulfasalazine all have some efficacy in the treatment of rheumatoid arthritis, but are ineffective in the animal models routinely employed to screen antirheumatic drugs [106]. Discovery of their efficacy in treatment of patients with rheumatoid arthritis was serendipitous and their mechanisms of action in this disease remain uncertain. Excessive reliance on animal models of disease in drug screening could lead researchers to discard novel mechanism-based agents with potential utility in the treatment of arthritis in humans.

A species difference particularly relevant to the development of DMOADs relates to cartilage collagenase, an enzyme widely assumed to contribute importantly to cartilage damage in OA. Although inhibitors of interstitial collagenase (collagenase-1, MMP-1) are frequently evaluated in rodent models, a rodent homologue of collagenase-1 has not been identified and probably does not exist. Specific inhibitors of human collagenase-1 would, therefore, not exhibit therapeutic activity in a rodent model of arthritis. Fortunately, most of the MMP inhibitors developed to date do not have great selectivity [6] and will inhibit also the collagenase now known to be involved in cartilage damage in rodents (collagenase-3, MMP-13). Furthermore, human collagenase-3 [35], which is expressed in human osteoarthritic cartilage, may play an important role in human OA [60].

Finally, even assuming that a given mediator, receptor or enzyme plays a role in the pathogenesis of OA in an animal model similar to that it plays in human OA, the ability of a drug to interfere with a molecular target may differ among species [28]. For example, although the chemotactic potency of leukotriene B_4 for human, murine and lapine neutrophils is comparable, leukotriene B_4 antagonists have shown differences in potency as great as 1,000-fold in some of these species. In the evaluation of a mechanism-based agent, such species-specific differences in potency can be detected and methods developed for *in vivo* pharmacodynamic studies, for example, by evaluating the effects of compounds which inhibit an exogenously administered human enzyme or mediator. This technique has been used to evaluate MMP inhibitors by determining the ability of drugs to inhibit release of proteoglycans from articular cartilage after injection of human stromelysin into the lapine knee [36].

Despite the fact that results obtained from animal models of OA may be misleading in the evaluation of potential DMOADs, models play an important role in basic research. By examining the pathogenetic mechanisms underlying joint damage in animal models, the potential roles of enzymes, mediators or receptors in human OA may be identified. On the other hand, evidence of the importance a molecular mechanism of joint damage in an animal model provides no assurance it is important in man. The strongest evidence of the pathogenetic importance of a postulated disease mechanism is the demonstration of a therapeutic effect after pharmacologic interference with that mechanism. Therefore, development of mechanism-based drugs and demonstration of their efficacy in animal models of OA may be integral to our understanding of the pathogenesis of the disease; however, validation of a molecular target in human disease can be obtained only after positive results are obtained in Phase III clinical trials *in humans*.

8. Conclusion

Animal models have proved extremely useful in elucidating pathogenetic mechanisms underlying damage to joint structures. Knowledge of the pathology of OA may be useful in testing mechanism-based drugs designed to prevent, retard, or reverse pathologic changes in the OA joint. However, differences among species exist that may influence the results and evidence of a positive therapeutic effect in an animal model of OA provides no assurance that this effect will be seen also in humans with OA. On the other hand, care must be taken that a negative result in an animal model will not the result in abandonment of pharmacologic agents which could have utility in treatment of patients. Notably, the structure-modifying effects of diacerhein that have been demonstrated in several animal models are supported by the results of a long-term placebo-controlled clinical trial in humans with hip OA.

Finally, basic scientists involved with the development of DMOADs need to be aware that the real problem is not OA, but *painful* OA, (indeed, all of us will develop OA). We know very little about the risk factors for pain and disability in OA or why an individual with obvious radiographically structural changes of OA decides one day to become a patient and thereby, contributes to the large medicoeconomic and socioeconomic impact of this disease. There is no *a priori* reason that a DMOAD will have a beneficial effect on joint pain. Indeed, there is a danger that the interest in development of DMOADs will jeopardize the search for, and development of, better agents for OA pain than currently exist.

Acknowledgements

Supported in part by NIH grant AR 20582-23. Kathie Lane provided invaluable assistance in the production of this review.

References

[1] M.E. Adams and K.D. Brandt, Hypertrophic repair of canine articular cartilage in osteoarthritis after anterior cruciate ligament transection, *J. Rheumatol.* **18** (1991), 428–435.

[2] C.J. Alexander, Utilisation of joint movement range in arboreal primates compared with human subjects: an evolutionary frame for primary osteoarthritis, *Ann. Rheum. Dis.* **53** (1994), 720–725.

[3] J.W. Alexander, The pathogenesis of canine hip dysplasia, *Vet. Clin. North America: Small Animal Practice* **22** (1992), 503–511.

[4] R.D. Altman, D.D. Dean, O.E. Muniz and D.S. Howell, Therapeutic treatment of canine osteoarthritis with glycosaminoglycan polysulfuric acid ester, *Arthritis Rheum.* **32** (1989), 1300–1307.

[5] C.L. Arsever and G.G. Bole, Experimental osteoarthritis induced by selective myectomy and tendotomy, *Arthritis Rheum.* **29** (1986), 251–261.

[6] N.R.A. Beeley, P.R.J. Ansell and A.J.P. Docherty, Inhibitors of matrix metalloproteinases (MMPs), *Curr. Opin. Ther. Patents* **4** (1994), 7–16.

[7] A.M. Bendele, R.A. Bendele, J.F. Hulman and B.P. Swann, Effets bénéfiques d'un traitement par la diacerhéine chez des cobayes atteints d'arthrose, *Rev. Praticien* **46**(Suppl. to n° 6) (1996), S35–39.

[8] A.M. Bendele, J.F. Hulman and J.S. Bean, Spontaneous osteoarthritis in Hartley albino guinea pigs: effects of dietary and surgical manipulations, *Arthritis Rheum.* **32** (1989), S106.

[9] A.M. Bendele and S.L. White, Early histopathologic and ultrastructural alterations in femorotibial joints of partial medial meniscectomised guinea pigs, *Vet. Pathol.* **24** (1987), 436–443.

[10] A.M. Bendele, Progressive chronic osteoarthritis in femorotibial joints of partial medial meniscectomized guinea pigs, *Vet. Pathol.* **24** (1987), 444–448.

[11] M.E.J. Billingham, Advantages afforded by the use of animal models for evaluation of potential disease-modifying osteoarthritis drugs (DMOADs), in: *Osteoarthritis*, K.D. Brandt, M. Doherty and L.S. Lohmander, eds, Oxford University Press, Oxford, 1998, pp. 429–438.

[12] K.D. Brandt, E.M. Braunstein, D.M. Visco, B. O'Connor, D. Heck and M. Albrecht, Anterior (cranial) cruciate ligament transection in the dog: a bona fide model of canine osteoarthritis, not merely of cartilage injury and repair, *J. Rheumatol.* **18** (1991), 436–446.

[13] K.D. Brandt, G. Smith, S.Y. Kang, S. Myers, B. O'Connor and M. Albrecht, Effects of diacerhein in an accelerated canine model of osteoarthritis, *Osteoarthritis Cart.* **5** (1997), 438–449.

[14] K.D. Brandt, Insights into the natural history of osteoarthritis provided by the cruciate-deficient dog: An animal model of osteoarthritis, *Ann. N. Y. Acad. Sci.* **732** (1994), 199–205.

[15] E.M. Braunstein, K.D. Brandt and M. Albrecht, MRI demonstration of hypertrophic articular cartilage repair in osteoarthritis, *Skeletal Radiol.* **19** (1990), 335–339.

[16] T.D. Brown, E.L. Radin, R.B. Martin and D.B. Burr, Finite element studies of some justaarticular stress changes due to localized subchondral stiffening, *J. Biomech.* **17** (1984), 11–24.

[17] E.G.L. Bywaters, Metabolism of joint tissues, *J. Pathol. Bacteriol.* **44** (1937), 247–268.

[18] C.S. Carlson, R.F. Loeser, M.J. Jayo, D.S. Weaver, M.R. Adams and C.P. Jerome, Osteoarthritis in Cynomolgus macaques: a primate model of naturally occurring disease, *J. Orthop. Res.* **12** (1994), 331–339.

[19] S.L. Carney, C.A. Hicks, B. Tree and R.J. Broadmore, An *in vivo* investigation of the effect of anthroquinones on the turnover of aggrecans in spontaneous osteoarthritis in the guinea pig, *Inflamm. Res.* **44** (1995), 182–186.

[20] J. Châteauvert, K.P. Pritzker, M.J. Kessler and M.D. Grynpas, Spontaneous osteoarthritis in rhesus macaques: I. Chemical and biochemical studies, *J. Rheumatol.* **16** (1989), 1098–1104.

[21] J.M. Châteauvert, M.D. Grynpas, M.J. Kessler and K.P. Pritzker, Spontaneous osteoarthritis in rhesus macaques. II. Characterization of disease and morphometric studies, *J. Rheumatol.* **17** (1990), 73–83.

[22] S.B. Christensen, Localization of bone-seeking agents in developing experimentally induced osteoarthritis in the knee joint of the rabbit, *Scand. J. Rheumatol.* **12** (1983), 343–349.

[23] Y. Coulais, G. Marcelon, J. Cros and R. Guiraud, An experimental model of osteoarthritis. II Biochemical study of collagen and proteoglycans, *Pathol. Biol.* **32** (1984), 23–28.

[24] Y. Coulais, G. Marcelon, J. Cros and R. Guiraud, Studies on an experimental model for osteoarthritis. I. Induction and ultrastructural investigation, *Pathol. Biol.* **31** (1983), 577–582.

[25] E.P. Das-Gupta, T.J. Lyons, J.A. Hoyland, D.M. Lawton and A.J. Freemont, New histological observations in spontaneously developing osteoarthritis in the STR/ORT mouse questioning its acceptability as a model of human osteoarthritis, *Int. J. Exp. Pathol.* **74** (1993), 627–634.

[26] D.K. Dedrick, S.A. Goldstein, K.D. Brandt, B.L. O'Connor, R.W. Goulet and M. Albrecht, A longitudinal study of subchondral plate and trabecular bone in cruciate-deficient dogs with osteoarthritis followed up for 54 months, *Arthritis Rheum.* **36** (1993), 1460–1467.

[27] D.K. Dedrick, R. Goulet, L. Huston, S.A. Goldstein and G.G. Bole, Early bone changes in experimental osteoarthritis using microscopic computed tomography, *J. Rheumatol.* **27S** (1991), 44–45.
[28] N.S. Doherty, R.J. Griffiths and E.R. Pettipher, The role of animal models in the discovery of novel disease-modifying osteoarthritis drugs (DMOADs), in: *Osteoarthritis*, K.D. Brandt, M. Doherty and L.S. Lohmander, eds, Oxford University Press, Oxford, 1998, pp. 439–449.
[29] M. Dougados, M. Nguyen, L. Berdah, B. Mazieres, E. Vignon, M. Lequesne, for the ECHODIAH Investigators study group, Evaluation of the structure-modifying effects of diacerein in hip osteoarthritis: ECHODIAH, a 3-year placebo-controlled trial, *Arthritis Rheum.* (Nov. 2001), in press.
[30] J. Dunham, M.G. Chambers, M.K. Jasani, J. Bitensky and J. Chayen, Quantitative criteria for evaluating the early development of osteoarthritis and the effect of diclofenac sodium, *Agents Actions* **28** (1989), 1–2.
[31] J. Dunham, M.G. Chambers, M.K. Jasani, L. Bitensky and J. Chayen, Changes in the orientation of proteoglycans during the early development of natural murine osteoarthritis, *J. Orthop. Res.* **8** (1990), 101–104.
[32] M.G. Ehrlich, H.J. Mankin, H. Jones, A. Grossman, C. Crispen and D. Ancona, Biochemical confirmation of an experimental osteoarthritis model, *J. Bone Joint Surg. [Am.]* **57** (1975), 392–396.
[33] A.G. Fam, I. Morava-Protzner, C. Purcell, B.D. Young, P.S. Bunting and A.J. Lewis, Acceleration of experimental lapine osteoarthritis by calcium pyrophosphate microcrystalline synovitis, *Arthritis Rheum.* **38** (1995), 201–210.
[34] J.C. Fernandes, J. Martel-Pelletier, L.G. Otterness, A. Lopez-Anaya, F. Mineau, G. Tardiff et al., Effects of tenidap on canine experimental osteoarthritis. I. Morphologic and metalloprotease analysis, *Arthritis Rheum.* **38** (1995), 1290–1303.
[35] J.M.P. Freije, I. Diez-Itza, M. Balbib, L.M. Snachez, R. Blasco, J. Tolivia et al., Molecular cloning and expression of collagenase-3, a novel human matrix metalloproteinase produced by breast carcinomas, *J. Biol. Chem.* **269** (1994), 16 766–16 773.
[36] V. Ganu, D. Parker, L. MacPherson, S.-I. Hu, R. Goldberg, A. Raychaudhuri et al., Biochemical and pharmacological profile of a non-peptidic orally active inhibitor of matrix metalloproteinases, *Osteoarthritis Cart.* **2**(S1) (1994), 34.
[37] S. Garofalo, E. Vuorio, M. Metsaranta, R. Rosati, D. Toman, J. Vaughan et al., Reduced amounts of cartilage collagen fibrils and growth plate anomalies in transgenic mice harboring a glycine-to-cysteine mutation in the mouse type II procollagen α_1-chain gene, *Proc. Natl. Acad. Sci. (USA)* **88** (1991), 9648–9652.
[38] P. Ghosh, S. Armstrong, R. Read, Y. Numata, S. Smith, P. McNair et al., Animal models of early osteoarthritis: their use for the evaluation of potential chondroprotective agents, *Agents Actions* **39S** (1993), 195–206.
[39] A.V. Hill, Production and absorption of work by muscle, *Science* **131** (1960), 897.
[40] D.S. Howell, M.R. Carreno, J.P. Pelletier and O.E. Muniz, Articular cartilage breakdown in a lapine model of osteoarthritis: action of glycosaminoglycan polysulfate ester (GAGPS) on proteoglycan degrading enzyme activity, hexuronate and cell counts, *Clin. Orthop.* **21** (1986), 69–76.
[41] D.S. Howell, F. Muller and D.H. Manicourt, A mini review: proteoglycan aggregate profiles in the Pond-Nuki dog model of osteoarthritis and in canine disuse atrophy, *Br. J. Rheumatol.* **31** (1992), 7–11.
[42] S.-Y. Hwa, D. Burkhardt, C. Little and P. Ghosh, The effects of orally administered diacerein on cartilage and subchondral bone in an ovine model of osteoarthritis, *J. Rheumatol.* **28** (2001), 825–834.
[43] R.J. Jefferson, J.J. Collins, M.W. Whittle, E.L. Radin and J.J. O'Connor, The role of the quadriceps in controlling impulsive forces around heel strike, *Proc. Inst. Mech. Eng. [H]* **204** (1990), 21–28.
[44] H. Johansson, P. Sjölander and P. Sojka, A sensory role for the cruciate ligaments, *Clin. Orthop.* **268** (1991), 161–178.
[45] L.C. Johnson, Kinetics of osteoarthritis, *Lab. Invest.* **8** (1959), 1223–1241.
[46] D.A. Kalbhen, Chemical model of osteoarthritis – a pharmacological evaluation, *J. Rheumatol.* **130** (1987), 130–131.
[47] A. Langenskiold, J.E. Michelsson and T. Videman, Osteoarthritis of the knee in the rabbit produced by immobilization. Attempts to achieve a reproducible model for studies on pathogenesis and therapy, *Acta Orthop. Scand.* **50** (1979), 1–14.
[48] M.W. Layton, C. Arsever and G.G. Bole, Use of the guinea pig myectomy osteoarthritis model in the examination of cartilage–synovium interactions, *J. Rheumatol.* **125** (1987), 125–126.
[49] H.R. Leipold, R.L. Goldberg and G. Lust, Canine serum keratan sulfate and hyaluronate concentrations. Relationship to age and osteoarthritis, *Arthritis Rheum.* **32** (1989), 312–321.
[50] M. Lequesne, K. Brandt, N. Bellamy, R. Moskowitz, C.J. Menkes, J.P. Pelletier and R.D. Altman, Guidelines for testing slow-acting and disease-modifying drugs in osteoarthritis, *J. Rheumatol.* **21**(S41) (1994), 65–71.
[51] J.H. Lin, Bisphosphonates: a review of their pharmacokinetic properties, *Bone* **18** (1996), 75–85.
[52] G. Marcelon, J. Cros and R. Guiraud, Activity of anti-inflammatory drugs on an experimental model of osteoarthritis, *Agents Actions* **6** (1976), 191–194.
[53] J.L. Marshall and S.-E. Olsson, Instability of the knee: a long-term experimental study in dogs, *J. Bone Joint Surg.* **53A** (1971), 1561–1570.
[54] K.W. Marshall and A.D. Chan, Bilateral canine model of osteoarthritis, *J. Rheumatol.* **23** (1996), 344–350.
[55] B. Mazieres, Diacerhein in a post-contusive model of osteoarthritis. Structural results with "prophylactic" and "curative" regimens, *Osteoarthritis Cartilage* **5**(Suppl. A) (1997).

[56] B. Mazieres, E. Maheu, M. Thiechart and G. Vallieres, Effects of N-acetyl-hydroxyproline (oxaceprol (R)) on an experimental post-contusive model of osteoarthritis. A pathological study, *J. Drug Dev.* **3** (1990), 135–142.

[57] C. McDevitt, E. Gilbertson and H. Muir, An experimental model of osteoarthritis: early morphological and biochemical changes, *J. Bone Joint Surg.* **59B** (1977), 24–35.

[58] C.A. McDevitt, H. Muir and M.J. Pond, Canine articular cartilage in natural and experimentally induced osteoarthrosis, *Biochem. Soc. Trans.* **1** (1973), 287–289.

[59] S.C.R. Meacock, J.L. Bodmer and M.E.J. Billingham, Experimental osteoarthritis in guinea pigs, *J. Exp. Path.* **71** (1990), 279–293.

[60] P.G. Mitchell, H.A. Magna, L.M. Reeves, L.L. Lopresti-Morrow, S.A. Yocum, P.J. Rosner et al., Cloning of matrix metalloproteinase-13 (MMP-1, collagenase-3) from human chondrocytes, expression of MMP-13 by osteoarthritic cartilage and activity of the enzyme on type II collagen, *J. Clin. Invest.* **97** (1996), 761–768.

[61] R.W. Moskowitz, Experimental models of osteoarthritis, in: *Osteoarthritis: Diagnosis and Medical/Surgical Management*, R.W. Moskowitz, D.S. Howell and V.M. Goldberg, eds, W.B. Saunders, Philadelphia, PA, 1992, pp. 213–232.

[62] R.W. Moskowitz and V.M. Goldberg, Studies of osteophyte pathogenesis in experimentally induced osteoarthritis, *J. Rheumatol.* **14** (1987), 311–320.

[63] S.L. Myers, K.D. Brandt, B.L. O'Connor, D.M. Visco and M.E. Albrecht, Synovitis and osteoarthritic changes in canine articular cartilage after cruciate ligament transection. Effect of surgical hemostasis, *Arthritis Rheum.* **33** (1990), 1406–1415.

[64] S.L. Myers, K.D. Brandt and B.L. O'Connor, "Low-dose" prednisone treatment does not reduce the severity of osteoarthritis in dogs after anterior cruciate ligament transection, *J. Rheumatol.* **18** (1991), 1856–1862.

[65] S.L. Myers, K.D. Brandt, D.B. Burr, B.L. O'Connor and M. Albrecht, Effects of a bisphosphonate on bone histomorphometry and dynamics in the canine cruciate-deficiency model of osteoarthritis, *J. Rheumatol.* **26** (1999), 2645–2653.

[66] K. Nakata, K. Ono, J.-I. Miyazaki, B. Olsen, Y. Muragaki, E. Adachi et al., Osteoarthritis associated with mild chondrodysplasia in transgenic mice expressing α_1(IX) collagen chains with a central deletion, *Proc. Natl. Acad. Sci. (USA)* **90** (1993), 2870–2874.

[67] B. O'Connor and K.D. Brandt, Neurogenic factors in the etiopathogenesis of osteoarthritis, in: *Rheumatic Disease Clinics of North America*, R.W. Moskowitz, ed., W.B. Saunders Co., Philadelphia, 1993, pp. 450–468.

[68] B.L. O'Connor, M.J. Palmoski and K.D. Brandt, Neurogenic acceleration of degenerative joint lesions, *J. Bone Joint Surg.* **67A** (1985), 562–572.

[69] B.L. O'Connor, D.M. Visco, K.D. Brandt, S.L. Myers and L. Kalasinski, Neurogenic acceleration of osteoarthritis: The effects of prior articular nerve neurectomy on the development of osteoarthritis after anterior cruciate ligament transection in the dog, *J. Bone Jt. Surg.* **74A** (1992), 367–376.

[70] B.L. O'Connor, D.M. Visco, D.A. Heck, S.L. Myers and K.D. Brandt, Gait alterations in dogs after transection of the anterior cruciate ligament, *Arthritis Rheum.* **32** (1989), 1142–1147.

[71] T.R.J. Oegema, J.L.J. Lewis and R.C.J. Thompson, Role of acute trauma in development of osteoarthritis, *Agents Actions* **40** (1993), 3–4.

[72] T. Okabe, Experimental studies on the spontaneous osteoarthritis in C57 black mice, *J. Tokyo Med. Coll.* **47** (1989), 546–557.

[73] I.G. Otterness, D. Larson and J.G. Lombardino, An analysis of piroxicam in rodent models of arthritis, *Agents Actions* **12** (1982), 308–312.

[74] M. Palmoski and K. Brandt, Running inhibits the reversal of atrophic changes in canine knee cartilage after removal of a leg cast, *Arthritis Rheum.* **24** (1981), 1329–1337.

[75] M. Palmoski, E. Perricone and K.D. Brandt, Development and reversal of a proteoglycan aggregation defect in normal canine knee cartilage after immobilization, *Arthritis Rheum.* **22** (1979), 508–517.

[76] M.J. Palmoski and K.D. Brandt, Immobilization of the knee prevents osteoarthritis after anterior cruciate ligament transection, *Arthritis Rheum.* **25** (1982), 1201–1208.

[77] M.J. Palmoski, R.A. Colyer and K.D. Brandt, Joint motion in the absence of normal loading does not maintain normal articular cartilage, *Arthritis Rheum.* **23** (1980), 325–334.

[78] J.P. Pelletier and J. Martel-Pelletier, *In vivo* protective effects of prophylactic treatment with tiaprofenic acid or intraarticular corticosteroids on osteoarthritic lesions in the experimental dog model, *J. Rheumatol.* **27S** (1991), 127–130.

[79] J.P. Pelletier and J. Martel-Pelletier, Protective effects of corticosteroids on cartilage lesions and osteophyte formation in the Pond-Nuki dog model of osteoarthritis, *Arthritis Rheum.* **32** (1989), 181–193.

[80] J.P. Pelletier, F. Mineau and J.P. Raynauld, Intraarticular injections with methyl-prednisolone acetate reduce osteoarthritic lesions in parallel with chondrocyte stromelysin synthesis in experimental osteoarthritis, *Arthritis Rheum.* **37** (1994), 414–423.

[81] M.J. Pond and G. Nuki, Experimentally-induced osteoarthritis in the dog, *Ann. Rheum. Dis.* **32** (1973), 387–388.

[82] E.L. Radin, R.D. Boyd, R.B. Martin, D.B. Burr, B. Caterson and C. Goodwin, Mechanical factors influencing cartilage damage, in: *Osteoarthritis: Current Clinical and Fundamental Problems*, J.G. Peyron, ed., Geigy, Paris, 1985, pp. 90–99.

[83] E.L. Radin and I.L. Paul, Does cartilage compliance reduce skeletal impact loads? The relative force attenuating properties of articular cartilage, synovial fluid, periarticular soft-tissue and bone, *Arthritis Rheum.* **13** (1970), 139.

[84] E.L. Radin, K.H. Yang, C. Riegger, V.L. Kish and J.J. O'Connor, Relationship between lower limb dynamics and knee joint pain, *J. Orthop. Res.* **9** (1991), 398–405.

[85] A. Ratcliffe, P.J. Beauvais and F. Saed-Nejad, Differential levels of synovial fluid aggrecan aggregate components in experimental osteoarthritis and joint disuse, *J. Orthop. Res.* **12** (1994), 464–473.

[86] G.A. Rodan and H.A. Fleisch, Bisphosphonates: mechanisms of action, *J. Clin. Invest.* **97** (1996), 2692–2696.

[87] R.A. Rogachevsky, D.D. Dean, D.S. Howell and R.D. Altman, Treatment of canine osteoarthritis with insulin-like growth factor (IGF1) and sodium pentosan polysulfate, *Ann. N. Y. Acad. Sci.* **732** (1994), 392–394.

[88] H.B. Skinner, M.P. Wyatt, J.A. Hodgdon, D.W. Conard and R.L. Barrack, Effect of fatigue on joint position sense of the knee, *J. Orthop. Res.* **4** (1986), 112–118.

[89] C. Slemenda, K.D. Brandt, D.K. Heilman, S. Mazzuca, E.M. Braunstein, B.P. Katz and F.D. Wilinsky, Quadriceps weakness and osteoarthritis of the knee, *Ann. Int. Med.* **127** (1997), 97–104.

[90] C. Slemenda, D.K. Heilman, K.D. Brandt, B.P. Katz, S. Mazzuca, E.M. Braunstein and D. Byrd, Reduced quadriceps strength relative to body weight. A risk factor for knee osteoarthritis in women?, *Arthritis Rheum.* **41** (1998), 1951–1959.

[91] G. Smale, A.M. Bendele and W.E. Horton, Comparison of age-associated degeneration of articular cartilage in Wistar and Fischer 344 rats, *Lab. Anim. Sci.* **45** (1995), 191–194.

[92] G.N. Smith, Jr., S.L. Myers, K.D. Brandt, E.A. Mickler and M.E. Albrecht, Diacerhein treatment reduces the severity of osteoarthritis in the canine cruciate-deficiency model of osteoarthritis, *Arthritis Rheum.* **42** (1999), 545–554.

[93] R. Stanescu, A. Knyszynski, M.P. Muriel and V. Stanescu, Early lesions of the articular surface in a strain of mice with very high incidence of spontaneous osteoarthritis-like lesions, *J. Rheumatol.* **20** (1993), 102–110.

[94] A. Takahama, Histological study on spontaneous osteoarthritis of the knee in C57 black mouse, *J. Jpn. Orthop. Assoc.* **64** (1990), 271–281.

[95] H. Tanaka, Y. Kitoh, T. Katsuramaki, M. Tanaka, N. Kitabayashi, S. Fujimori et al., Effects of SL-1010 (sodium hyaluronate with high molecular weight) on experimental osteoarthritis induced by intra-articularly applied papain in guinea pigs, *Folia Pharmacol. Jpn.* **100** (1992), 77–86.

[96] W.B. van den Berg, Growth factors in experimental osteoarthritis: transforming growth factor beta pathogenic?, *J. Rheumatol.* **43S** (1995), 143–145.

[97] P.M. van der Kraan, E.L. Vitters, H.M. van Beuningen, L.B. van de Putte and W.B. van den Berg, Degenerative knee joint lesions in mice after a single intra-articular collagenase injection. A new model of osteoarthritis, *J. Exp. Pathol.* **71** (1990), 19–31.

[98] P.M. van der Kraan, E.L. Vitters and W.B. van den Berg, Development of osteoarthritis models in mice by "mechanical" and "metabolical" alterations in the knee joints, *Arthritis Rheum.* **32** (1989), S107.

[99] V. Vasilev, H.J. Merker and N. Vidinov, Ultrastructural changes in the synovial membrane in experimentally-induced osteoarthritis of rabbit knee joint, *Histol. Histopathol.* **7** (1992), 119–127.

[100] T. Videman, Experimental osteoarthritis in the rabbit: comparison of different periods of repeated immobilization, *Acta Orthop. Scand.* **53** (1982), 339–347.

[101] E. Vignon, M. Arlot, D. Hartmann, B. Moyen and G. Ville, Hypertrophic repair of articular cartilage in experimental osteoarthrosis, *Ann. Rheum. Dis.* **42** (1983), 82–88.

[102] E. Vignon, P. Mathieu, J. Bejui, J. Descotes, D. Hartmann, L.M. Patricot et al., Study of an inhibitor of plasminogen activator (tranexamic acid) in the treatment of experimental osteoarthritis, *J. Rheumatol.* **27S** (1991), 131–133.

[103] J.A. Vilensky, B.L. O'Connor, K.D. Brandt, E.A. Dunn and P.I. Rogers, Serial kinematic analysis of the canine knee after L4–S1 dorsal root ganglionectomy: implications for the cruciate deficiency model of osteoarthritis, *J. Rheumatol.* **21** (1994), 2113–2117.

[104] D.M. Visco, M.A. Hill, W.R. Widmer, B. Johnstone and S.L. Myers, Experimental osteoarthritis in dogs: a comparison of the Pond-Nuki and medial arthrotomy methods, *Osteoarthritis Cart.* **4** (1996), 9–22.

[105] L.P. Yu, G.N. Smith, Jr., K.D. Brandt, S.L. Myers, B.L. O'Connor and D.A. Brandt, Reduction of the severity of canine osteoarthritis by prophylactic treatment with oral doxycycline, *Arthritis Rheum.* **35** (1992), 1150–1159.

[106] J. Zhang, B.M. Weichman and A.J. Lewis, Role of animal models in the study of rheumatoid arthritis: An overview, in: *Mechanisms and Models in Rheumatoid Arthritis*, B. Henderson, J.C.W. Edwards and E.R.P. Pettipher, eds, Academic Press, New York, 1995, pp. 363–372.

Biorheology 39 (2002) 237–246
IOS Press

The role of cytokines in osteoarthritis pathophysiology

Julio C. Fernandes *, Johanne Martel-Pelletier and Jean-Pierre Pelletier
Osteoarthritis Research Unit, Centre hospitalier de l'Université de Montréal, Hôpital Notre-Dame, Montréal, Québec, Canada

Abstract. Morphological changes observed in OA include cartilage erosion as well as a variable degree of synovial inflammation. Current research attributes these changes to a complex network of biochemical factors, including proteolytic enzymes, that lead to a breakdown of the cartilage macromolecules. Cytokines such as IL-1 and TNF-alpha produced by activated synoviocytes, mononuclear cells or by articular cartilage itself significantly up-regulate metalloproteinases (MMP) gene expression. Cytokines also blunt chondrocyte compensatory synthesis pathways required to restore the integrity of the degraded extrecellular matrix (ECM). Moreover, in OA synovium, a relative deficit in the production of natural antagonists of the IL-1 receptor (IL-1Ra) has been demonstrated, and could possibly be related to an excess production of nitric oxide in OA tissues. This, coupled with an upregulation in the receptor level, has been shown to be an additional enhancer of the catabolic effect of IL-1 in this disease.

IL-1 and TNF-α significantly up-regulate MMP-3 steady-state mRNA derived from human synovium and chondrocytes. The neutralization of IL-1 and/or TNF-α up-regulation of MMP gene expression appears to be a logical development in the potential medical therapy of OA. Indeed, recombinant IL-1receptor antagonists (ILRa) and soluble IL-1 receptor proteins have been tested in both animal models of OA for modification of OA progression. Soluble IL-1Ra suppressed MMP-3 transcription in the rabbit synovial cell line HIG-82. Experimental evidence showing that neutralizing TNF-α suppressed cartilage degradation in arthritis also support such strategy. The important role of TNF-α in OA may emerge from the fact that human articular chondrocytes from OA cartilage expressed a significantly higher number of the p55 TNF-alpha receptor which could make OA cartilage particularly susceptible to TNF-alpha degradative stimuli. In addition, OA cartilage produces more TNF-α and TNF$\angle\alpha$ convertase enzyme (TACE) mRNA than normal cartilage. By analogy, an inhibitor to the p55 TNF-α receptor may also provide a mechanism for abolishing TNF-α-induced degradation of cartilage ECM by MMPs. Since TACE is the regulator of TNF-α activity, limiting the activity of TACE might also prove efficacious in OA. IL-1 and TNF-α inhibition of chondrocyte compensatory biosynthesis pathways which further compromise cartilage repair must also be dealt with, perhaps by employing stimulatory agents such as transforming growth factor-beta or insulin-like growth factor-I.

Certain cytokines have antiinflammatory properties. Three such cytokines – IL-4, IL-10, and IL-13 – have been identified as able to modulate various inflammatory processes. Their antiinflammatory potential, however, appears to depend greatly on the target cell. Interleukin-4 (IL-4) has been tested *in vitro* in OA tissue and has been shown to suppress the synthesis of both TNF-α and IL-1β in the same manner as low-dose dexamethasone. Naturally occurring antiinflammatory cytokines such as IL-10 inhibit the synthesis of IL-1 and TNF-α and can be potential targets for therapy in OA. Augmenting inhibitor production *in situ* by gene therapy or supplementing it by injecting the recombinant protein is an attractive therapeutic target, although an *in vivo* assay in OA is not available, and its applicability has yet to be proven. Similarly, IL-13 significantly inhibits lipopolysaccharide (LPS)-induced TNF-α production by mononuclear cells from peripheral blood, but not in cells from inflamed synovial fluid. IL-13 has important biological activities: inhibition of the production of a wide range of proinflammatory cytokines in monocytes/macrophages, B cells, natural killer cells and endothelial cells, while increasing IL-1Ra production. In OA synovial membranes treated with LPS, IL-13 inhibited the synthesis of IL-1β, TNF-α and stromelysin, while increasing IL-1Ra production.

In summary, modulation of cytokines that control MMP gene up-regulation would appear to be fertile targets for drug development in the treatment of OA. Several studies illustrate the potential importance of modulating IL-1 activity as a means to reduce the progression of the structural changes in OA. In the experimental dog and rabbit models of OA, we have demonstrated that *in vivo* intraarticular injections of the IL-Ra gene can prevent the progression of structural changes in OA. Future directions in the research and treatment of osteoarthritis (OA) will be based on the emerging picture of pathophysiological events that modulate the initiation and progression of OA.

Keywords: OA, proinflammatory cytokines, antiinflammatory cytokines, cytokine antagonists

*Address for correspondence: Julio C. Fernandes, M.D., M.Sc., Associate Professor of Medicine, Hôpital Notre-Dame du Centre hospitalier de l'Université de Montréal, 1560 east rue Sherbrooke, Montréal, Québec, Canada H2L 4M1. Tel.: +1 514 890 8000, ext. 25114; E-mail: julio.fernandes@sympatico.ca.

1. Introduction

Osteoarthritis (OA) is believed to be a consequence of mechanical and biological events that destabilize the normal coupling of degradation and synthesis within articular joint tissues. In primary OA, no trauma or other predisposing factor is identified, and intrinsic alterations of the articular tissue, or response to normal cumulative stresses, are presumed responsible [43]. The disease process affects not only the articular cartilage, but also the entire joint structure including the subchondral bone, ligaments, capsule, synovial membrane and periarticular muscles. Morphological changes observed in OA include cartilage erosion as well as a variable degree of synovial inflammation. Current research attributes these changes to a complex network of biochemical factors, including proteolytic enzymes, which lead to a breakdown of the cartilage macromolecules. This disease process involves a disturbance in the normal balance of degradation and repair in articular cartilage, synovial membrane and subchondral bone [33, 37,53,62].

2. Cytokines and osteoarthritis

It is believed that cytokines and growth factors play an important role in the pathophysiology of OA. They are closely associated with functional alterations in synovium, cartilage and subchondral bone, and are produced both spontaneously and following stimulation by the joint tissue cells. Cytokines such as IL-1 and TNF-alpha produced by activated synoviocytes, mononuclear cells or by articular cartilage itself significantly up-regulate metalloproteinases (MMP) gene expression. Cytokines also blunt chondrocyte compensatory synthesis pathways required to restore the integrity of the degraded extracellular matrix (ECM). Moreover, in OA synovium, a relative deficit in the production of natural antagonists of the IL-1 receptor (IL-1Ra) has been demonstrated, and could possibly be related to an excess production of nitric oxide in OA tissues. This, coupled with an upregulation in the receptor level, has been shown to be an additional enhancer of the catabolic effect of IL-1 in this disease. In OA synovial membrane, the synovial lining cells are key inflammatory effectors. Once cartilage degradation has begun, the synovial membrane phagocytoses the breakdown products released into the synovial fluid. Consequently, the membrane becomes hypertrophic and hyperplasic. Several studies have reported inflammatory changes in the synovial membrane of patients with OA that, on occasion, were almost indistinguishable from those in patients with an inflammatory arthritis such as rheumatoid arthritis (RA) [33,37,53,62].

2.1. *Proinflammatory cytokines*

Proinflammatory cytokines are believed to play a pivotal role in the initiation and development of this disease process [19,68,87]. IL-1 and TNF-alpha can induce joint articular cells, such as chondrocytes and synovial cells, to produce other cytokines such as IL-8, IL-6, as well as stimulate proteases and prostaglandin E_2 (PGE_2) production. IL-1beta and TNF-alpha have also been shown to increase osteoclastic bone resorption *in vitro* [12]. Blocking IL-1 activity with the IL-1 receptor antagonist (IL-1Ra) in synovial fibroblasts *in vitro* reduced IL-6 and IL-8 production, but not TNF-alpha [34]. Moreover, adding anti-TNF-alpha antibodies to synovial cells greatly reduced the production of other proinflammatory cytokines such as IL-1, GM-CSF, IL-6 [15].

IL-1beta is primarily synthesized as a 31 kilodalton (kD) precursor and released in the active form of 17.5 kD [61,79]. In synovial membrane, synovial fluid and cartilage, IL-1beta has been found in the active form. *Ex vivo* OA synovial membrane secretes this cytokine [67]. Several serine proteases can process the pro-IL-1beta to bioactive form [13]. In mammals only one protease, belonging to the cysteine-dependent protease family and named IL-1beta converting enzyme (ICE or Caspase-1), can specifically generate the mature 17.5 kD cytokine [13,51]. ICE is a pro-enzyme polypeptide of 45 kD [p45] located in the plasma membrane [91]. The biological activation of cells by IL-1 is mediated through association with two specific cell-surface receptors (IL-1R), named type I and type II [80]. Type I receptor binds IL-1beta more than IL-1alpha and appears responsible for signal transduction [8,59,71]. Type II IL-1R has a greater affinity for IL-1alpha than IL-1beta. The number of type I IL-1R is significantly increased in OA chondrocytes and synovial fibroblasts [59,71]. This appears to be responsible for the higher sensitivity of these cells to stimulation by IL-1 [59].

Both types of IL-1R can also be shed from the cell surface, and exist extracellularly in truncated forms; they are named IL-1 soluble receptors (IL-1sR). The shed receptor may function as a receptor antagonist because the ligand-binding region is preserved, thus enabling it to compete with the membrane-associated receptors of the target cells. Similarly, the shedding of surface receptors may decrease the responsiveness of target cells to the ligand. It is suggested that type II IL-1R serves as the main precursor for shed soluble receptors. The binding affinity of IL-1sR to both IL-1 isoforms and IL-1Ra differs. Type II IL-1sR binds IL-1beta more readily than IL-1Ra; in contrast, type I IL-1sR binds IL-1Ra with high affinity [8,29,82].

TNF-alpha appears to be an important mediator of matrix degradation and a pivotal cytokine in synovial membrane inflammation in OA, although being detected in OA articular tissue low levels. TNF-alpha is synthesized as a precursor protein comprising 76 amino acids. Proteolytic cleavage takes place at the cellular surface via a TNF-alpha converting enzyme named TACE [14]. This enzyme is also required for shedding the TNF receptors. An upregulation of TACE mRNA in human OA cartilage has recently been reported [7]. Human TNF-alpha is converted to the 157-residue (17 kD) secreted protein that oligomerizes to form trimers [3].

TNF-alpha binds to two specific receptors on the cell membrane [28,73,76]. These two TNF-R have molecular masses of 55 to 60 kD and 75 to 80 kD [54,72], and are named according to their molecular weight; TNF-R55 and TNF-R75. Their extracellular domains share 28% identity [17,45]. There is a complete absence of homology between the intracellular domains of the two TNF-R and any other known protein receptor [52,54,72,81]. In articular tissue cells, TNF-R55 seems to be the dominant receptor responsible for mediating TNF-alpha activity. In OA chondrocytes and synovial fibroblasts, enhanced expression of TNF-R55 has been reported [4,90]. Both receptor types appear to be actively involved in signal transduction [4,42,63,77,85]. Each receptor type has been shown to induce a specific subset of TNF-alpha activities [44,84]. TNF-R55 and TNF-R75 are linked to distinct intracellular second-messengers. TNF-R75 may regulate the rate of TNF-alpha association to TNF-R55 [83]. TNF-R75/TNF-alpha complex may exhibit enhanced and/or specific intracellular function. Heterogeneity in the TNF-alpha response may also be caused by different postreceptor signal transduction pathways [75]. It is not clear, however, whether TNF-alpha receptor trimerization is necessary for activation, or whether receptor dimerization is sufficient, or if receptor trimerization triggers other and/or additional intracellular pathways.

Proteolytic cleavage of the extracellular domain of each TNF-R produces two soluble receptors, TNF-sR55 and TNF-sR75. OA synovial fibroblasts and chondrocytes release a significantly elevated level of TNF-sR75 [4,16]. A higher ratio of TNF-sR75/TNF-sR55 is noted in the more severe cases of arthritis [21,22,70]. At low concentrations, TNF-sR seems to stabilize the trimeric structure of TNF-alpha, thereby increasing the half-life of bioactive TNF-alpha [2], while at high concentrations, TNF-sR reduce the bioactivity of TNF-alpha by competing for TNF binding with cell-associated receptors [40]. However, as the affinity of both TNF-sR is similar to that of the plasma membrane receptor, large amounts of these inhibitors are required to decrease TNF-alpha activity.

The balance between cytokine-driven anabolic and catabolic processes determines the integrity of articular joint tissue. Other proinflammatory cytokines such as IL-6, IL-8, LIF, IL-11, and IL-17 have been shown to be expressed in OA tissue, and have therefore been considered potential contributing factors in the pathogenesis of this disease.

IL-6 has been proposed as a contributor to the OA pathological process by: (1) increasing the number of inflammatory cells in synovial tissue [35]; (2) stimulating the proliferation of chondrocytes; and (3) inducing an amplification of the IL-1 effects on the increased synthesis of metalloproteases (MMP) and inhibiting proteoglycan production [64]. However, as IL-6 can induce the production of TIMP [55], and not MMP, it is believed that this cytokine is involved in the feedback mechanism that limits proteolytic damage.

Interleukin-8 is a potent chemotactic cytokine for polymorphonuclear neutrophils (PMN), stimulating their chemotaxis and generating reactive oxygen metabolites [93]. This chemokine is synthesized by a variety of cells including monocytes/macrophages, chondrocytes and fibroblasts [41,49,50,86]. IL-8 plays an important role in the acute inflammatory reaction. In synovial culture, TNF-alpha stimulates the production of IL-8 in a time- and dose-dependent manner [41]. In OA patients, IL-beta, IL-6, TNF-alpha and IL-8 coexist in the synovial fluid. IL-8 enhances the release of IL-1beta, IL-6 and TNF-α [93]. The presence of IL-8 in the lining cell layers could explain the high amount of IL-8 in the synovial fluid [27]. IL-8 is also present in the chondrocytes, and has been shown to enhance the production of oxidative and 5-lipoxygenase products [74]. Stimulated human articular chondrocytes express the IL-8 gene and secrete bioactive IL-8 [57].

Leukemia inhibitory factor (LIF) is a single-chain glycoprotein that has diverse effects, including induction of acute-phase protein synthesis and the inhibition of lipoprotein lipase activity. LIF level has been detected in synovial fluid of OA patients [25]. LIF has been shown to enhance IL-1beta and IL-8 expression in chondrocytes, and IL-1beta and TNF-alpha in synovial fibroblasts [89]. IL-1beta and TNF-alpha upregulate LIF production [18,36,56]. LIF regulates the metabolism of connective tissue such as cartilage and bone [1,69], induces expression of collagenase and stromelysin but not tissue inhibitor of metalloproteases, TIMP [56]. This cytokine stimulates cartilage proteoglycan resorption [20] as well as nitric oxide (NO) production.

The IL-11 receptor shares the gp 130 domain with the LIF and IL-6 receptors, suggesting that they may have similar actions. This cytokine was originally identified as a stromal cell-derived lymphoietic and hematopoietic factor, but can also be induced in articular chondrocyte and synovial fibroblast cultures [58,65]. IL-11 has been found to decrease the release of PGE_2 from OA synovial fibroblasts [6], suggesting that IL-11 can prevent the excessive extracellular matrix degeneration induced by synovial inflammation.

IL-17 is a newly discovered cytokine of 20–30 kD present as a homodimer with variable glycosylated polypeptides [92]. The tissue distribution of IL-17R appears ubiquitous, and it is not yet known whether all cells expressing IL-17R respond to its ligand. IL-17 upregulates a number of gene products involved in cell activation, including the proinflammatory cytokines IL-1beta, TNF-alpha and IL-6, as well as MMP in target cells such as human macrophages [46]. IL-17 also increases the production of NO in chondrocyte cultures [9,60].

2.2. *Antiinflammatory cytokines*

Three antiinflammatory cytokines (IL-4, IL-10, and IL-13) are spontaneously elaborated by synovial membrane and cartilage, and are found in increased levels in the synovial fluid of OA patients. The antiinflammatory properties of these cytokines include decreased production of IL-1beta, TNF-alpha and MMP, upregulation of IL-1Ra and TIMP-1, and inhibition of PGE_2 release [5,30,31,38,39,47,78,88]. It was found that IL-10 modulated TNF-alpha production by increasing the release of the TNF-sR from monocytes in culture, while downregulating the receptor surface expression [38]. In human OA synovial fibroblasts, IL-10 also downregulated the TNF-R density, while increasing the release of TNF-alpha-induced TNF-Rs75. In these cells, however, IL-4 upregulated TNF-R, and enhanced TNF-alpha-induced TNF-sR75 [5]. In mononuclear cells from RA synovial fluid, both TNF-R55 and TNF-R75 are upregulated by IL-4 [23], and contrasts with data from monocytes, where this antiinflammatory cytokine downregulated both the membrane and soluble TNF-R [48].

IL-13 has been shown to have important biological activities such as inhibiting the production of a wide range of proinflammatory cytokines, while increasing IL-1Ra production [24,26]. In human synovial membrane specimens from OA patients treated with LPS, *in vitro* IL-13 inhibited the synthesis of IL-1beta, TNF-alpha and stromelysin, and increased production of IL-1Ra [47], but not in cells recovered from the synovial fluid of OA and RA patients [24]. The TNF receptor system does not appear to be a target for IL-13 in OA synovial fibroblasts [5].

IL-1Ra is a competitive inhibitor of IL-1R. This molecule does not bind to IL-1, is not a binding protein, nor does it stimulate target cells. IL-1Ra can block many of the effects observed during the pathological process of OA, including PGE_2 synthesis in synovial cells, collagenase production by chondrocytes, and cartilage matrix degradation. Three forms of IL-1Ra were found, one extracellular and termed soluble IL-1Ra (IL-1sRa), and two intracellular, icIL-1RaI and icIL-1RaII [8]. Both the soluble and icIL-1Ra can bind to IL-1R, but with about 5-fold less affinity for the latter. Although intensive research is underway, the biological actions of icIL-1Ra remain elusive. *In vitro* experiments have revealed that an excess of 10–100 times the amount of IL-1Ra is necessary to inhibit IL-1beta activity whereas, *in vivo*, 100–2000 times more IL-1Ra is needed [8,67]. This may likely explain why, even though a higher level of IL-1Ra is found in OA articular tissue, there is a relative deficit of IL-1Ra to IL-1beta in this tissue. This in turn may cause the increased level of IL-1beta activity.

3. Potential therapeutic applications of cytokine modulation in OA

The neutralization of IL-1 and/or TNF-α up-regulation of MMP gene expression appears to be a logical development in the potential medical therapy of OA. Indeed, recombinant IL-1 receptor antagonists (ILRa) and soluble IL-1 receptor proteins have been tested in both animal models of OA for modification of OA progression. Soluble IL-1Ra suppressed MMP-3 transcription in the rabbit synovial cell line HIG-82. Experimental evidence showing that neutralizing TNF-α suppressed cartilage degradation in arthritis also supports such strategy. The important role of TNF-α in OA may emerge from the fact that human articular chondrocytes from OA cartilage expressed a significantly higher number of the p55 TNF-alpha receptor that could make OA cartilage particularly susceptible to TNF-alpha degradative stimuli. In addition, OA cartilage produces more TNF-α and TNF-α convertase enzyme (TACE) mRNA than normal cartilage. By analogy, an inhibitor to the p55 TNF-α receptor may also provide a mechanism for abolishing TNF-α-induced degradation of cartilage ECM by MMPs. Since TACE is the regulator of TNF-α activity, limiting the activity of TACE might also prove efficacious in OA. IL-1 and TNF-α inhibition of chondrocyte compensatory biosynthesis pathways which further compromise cartilage repair must also be dealt with, perhaps by employing stimulatory agents such as transforming growth factor-beta or insulin-like growth factor-I.

The capacity of IL-1Ra to reduce *in vitro* and *in vivo* cartilage degradation, MMP production and the progression of OA lesions [19,66] has elicited much attention concerning the use of this molecule in OA therapy, and more particularly in regard to gene therapy. Using the MFG retrovirus, the IL-1Ra gene has been successfully transferred into animal and human synovial cells using an *ex vivo* technique [10,32]. One such study using the experimental dog model of OA showed *in vivo* that the progression of structural changes of OA was significantly reduced [66]. It has also been demonstrated *in vitro* that the human IL-1Ra gene can be successfully transferred into chondrocytes using the Ad.RSV adenovirus, and that the resulting increase in production of IL-1Ra can protect the OA cartilage explants from degradation induced by IL-1 [10].

A novel and interesting approach to controlling proinflammatory cytokine production and/or activity is the use of biological molecules possessing antiinflammatory properties. Augmenting inhibitor production *in situ* by gene therapy or supplementing it by injecting the recombinant protein is an attractive therapeutic target, although an *in vivo* assay in OA is not available, and its applicability has yet to be proven. As such, recombinant human IL-4 (rhIL-4) has been tested *in vitro* on OA synovial tissue, and has been shown to suppress the synthesis of both IL-1beta and TNF-alpha in the same manner as low-dose dexamethasone [11]. To date, of the antiinflammatory cytokines, only IL-10 is employed in clinical trials for the treatment of RA in humans. Results from IL-13 experimentation on human synovial membrane from OA patients [47] indicate it is potentially useful in the treatment of this disease.

4. Conclusion

Although the primary etiology of OA remains undetermined, it is now believed that cartilage integrity is maintained by a balance obtained from cytokine-driven anabolic and catabolic processes. In OA, the specific causative for the pathological process has not been identified, but synovial inflammation at the clinical stage is now a well-documented phenomenon. An excess of proinflammatory cytokines is thought to be responsible for many clinical manifestations of OA. Other cytokines having anti-inflammatory properties could modulate this pathological process; therefore, these cytokines may prevent inflammation in OA.

In summary, modulation of cytokines that control MMP gene up-regulation would appear to be fertile targets for drug development in the treatment of OA. Several studies illustrate the potential importance of modulating IL-1 activity as a means to reduce the progression of the structural changes in OA. In the experimental dog and rabbit models of OA, we have demonstrated that *in vivo* intraarticular injections of the IL-Ra gene can prevent the progression of structural changes in OA. Future directions in the research and treatment of osteoarthritis (OA) will be based on the emerging picture of pathophysiological events that modulate the initiation and progression of OA.

References

[1] E. Abe, H. Tanaka, Y. Ishimi, C. Miyaura, T. Hayashi, H. Nagasawa, M. Tomida, Y. Yamaguchi, M. Hozumi and T. Suda, Differentiation-inducing factor purified from conditioned medium of mitogen-treated spleen cell cultures stimulates bone resorption, *Proc. Natl. Acad. Sci. USA* **83** (1986), 5958–5962.

[2] D. Aderka, H. Engelmann, Y. Maor, C. Brakebusch and D. Wallach, Stabilization of the bioactivity of tumor necrosis factor by its soluble receptors, *J. Exp. Med.* **175** (1992), 323–329.

[3] B.B. Aggarwal, W.J. Kohr, P.E. Hass, B. Moffat, S.A. Spencer, W.J. Henzel, T.S. Bringman, G.E. Nedwin, D.V. Goeddel and R.N. Harkins, Human tumor necrosis factor. Production, purification, and characterization, *J. Biol. Chem.* **260** (1985), 2345–2354.

[4] N. Alaaeddine, J.A. Di Battista, J.P. Pelletier, J.M. Cloutier, K. Kiansa, M. Dupuis and J. Martel-Pelletier, Osteoarthritic synovial fibroblasts possess an increased level of tumor necrosis factor-receptor 55 (TNF-R55) that mediates biological activation by TNF-alpha, *J. Rheumatol.* **24** (1997), 1985–1994.

[5] N. Alaaeddine, J.A. Di Battista, J.P. Pelletier, K. Kiansa, J.M. Cloutier and J. Martel-Pelletier, Inhibition of tumor necrosis factor alpha-induced prostaglandin E2 production by the antiinflammatory cytokines interleukin-4, interleukin-10, and interleukin-13 in osteoarthritic synovial fibroblasts: distinct targeting in the signaling pathways, *Arthritis Rheum.* **42** (1999), 710–718.

[6] N. Alaaeddine, J.A. Di Battista, J.P. Pelletier, K. Kiansa, J.M. Cloutier and J. Martel-Pelletier, Differential effects of IL-8, LIF (pro-inflammatory) and IL-11 (anti-inflammatory) on TNF-alpha-induced PGE_2 release and on signaling pathways in human OA synovial fibroblasts, *Cytokine* (1999) in press.

[7] A.R. Amin, Regulation of tumor necrosis factor-alpha and tumor necrosis factor converting enzyme in human osteoarthritis, *Osteoarthritis Cart.* **7** (1999), 392–394.

[8] W.P. Arend, Interleukin-1 receptor antagonist. [Review], *Adv. Immunol.* **54** (1993), 167–227.

[9] M.G. Attur, R.N. Patel, S.B. Abramson and A.R. Amin, Interleukin-17 up-regulation of nitric oxide production in human osteoarthritis cartilage, *Arthritis Rheum.* **40** (1997), 1050–1053.

[10] V.M. Baragi, R.R. Renkiewicz, H. Jordan, J. Bonadio, J.W. Harman and B.J. Roessler, Transplantation of transduced chondrocytes protects articular cartilage from interleukin 1-induced extracellular matrix degradation, *J. Clin. Invest.* **96** (1995), 2454–2460.

[11] A. Bendrups, A. Hilton, A. Meager and J.A. Hamilton, Reduction of tumor necrosis factor alpha and interleukin-1 beta levels in human synovial tissue by interleukin-4 and glucocorticoid, *Rheumatol. Int.* **12** (1993), 217–220.

[12] D.R. Bertolini, G.E. Nedwin, T.S. Bringman, D.D. Smith and G.R. Mundy, Stimulation of bone resorption and inhibition of bone formation *in vitro* by human tumour necrosis factors, *Nature* **319** (1986), 516–518.

[13] R.A. Black, S.R. Kronheim, M. Cantrell, M.C. Deeley, C.J. March, K.S. Prickett, J. Wignall, P.J. Conlon, D. Cosman, T.P. Hopp and D.Y. Mochizuki, Generation of biologically active interleukin-1 beta by proteolytic cleavage of the inactive precursor, *J. Biol. Chem.* **263** (1988), 9437–9442.

[14] R.A. Black, C.T. Rauch, C.J. Kozlosky, J.J. Peschon, J.L. Slack, M.F. Wolfson, B.J. Castner, K.L. Stocking, P. Reddy, S. Srinivasan, N. Nelson, N. Boiani, K.A. Schooley, M. Gerhart, R. Davis, J.N. Fitzner, R.S. Johnson, R.J. Paxton, C.J. March and D.P. Cerretti, A metalloproteinase disintegrin that releases tumour-necrosis factor-alpha from cells, *Nature* **385** (1997), 729–733.

[15] F.M. Brennan, D. Chantry, A. Jackson, R.N. Maini and M. Feldmann, Inhibitory effect of TNF alpha antibodies on synovial cell interleukin-1 production in rheumatoid arthritis, *Lancet* **2** (1989), 244–247.

[16] F.M. Brennan, D.L. Gibbons, A.P. Cope, P. Katsikis, R.N. Maini and M. Feldmann, TNF inhibitors are produced spontaneously by rheumatoid and osteoarthritic synovial joint cell cultures: evidence of feedback control of TNF action, *Scand. J. Immunol.* **42** (1995), 158–165.

[17] D. Camerini, G. Walz, W.A. Loenen, J. Borst and B. Seed, The T cell activation antigen CD27 is a member of the nerve growth factor/tumor necrosis factor receptor gene family, *J. Immunol.* **147** (1991), 3165–3169.

[18] I.K. Campbell, P. Waring, U. Novak and J.A. Hamilton, Production of leukemia inhibitory factor by human articular chondrocytes and cartilage in response to interleukin-1 and tumor necrosis factor alpha, *Arthritis Rheum.* **36** (1993), 790–794.

[19] J.P. Caron, J.C. Fernandes, J. Martel-Pelletier, G. Tardif, F. Mineau, C. Geng and J.P. Pelletier, Chondroprotective effect of intraarticular injections of interleukin-1 receptor antagonist in experimental osteoarthritis: suppression of collagenase-1 expression, *Arthritis Rheum.* **39** (1996), 1535–1544.

[20] G.J. Carroll and M.C. Bell, Leukaemia inhibitory factor stimulates proteoglycan resorption in porcine articular cartilage, *Rheumatol. Int.* **13** (1993), 5–8.

[21] I.C. Chikanza, P. Roux-Lombard, J.M. Dayer and G.S. Panayi, Tumour necrosis factor soluble receptors behave as acute phase reactants following surgery in patients with rheumatoid arthritis, chronic osteomyelitis and osteoarthritis, *Clin. Exp. Immunol.* **92** (1993), 19–22.

[22] A.P. Cope, D. Aderka, M. Doherty, H. Engelmann, D. Gibbons, A.C. Jones, F.M. Brennan, R.N. Maini, D. Wallach and M. Feldmann, Increased levels of soluble tumor necrosis factor receptors in the sera and synovial fluid of patients with rheumatic diseases, *Arthritis Rheum.* **35** (1992), 1160–1169.

[23] A.P. Cope, D.L. Gibbons, D. Aderka, B.M. Foxwell, D. Wallach, R.N. Maini, M. Feldmann and F.M. Brennan, Differential regulation of tumour necrosis factor receptors (TNF-R) by IL-4; upregulation of P55 and P75 TNF-R on synovial joint mononuclear cells, *Cytokine* **5** (1993), 205–212.

[24] R. de Waal Malefyt, C.G. Figdor, R. Huijbens, S. Mohan-Peterson, B. Bennett, J.A. Culpepper, W. Dang, G. Zurawski and J.E. de Vries, Effects of IL-13 on phenotype, cytokine production, and cytotoxic function of human monocytes. Comparison with IL-4 and modulation by IFN-gamma or IL-10, *J. Immunol.* **151** (1993), 6370–6381.

[25] J. Dechanet, J.L. Taupin, P. Chomarat, M.C. Rissoan, J.F. Moreau, J. Banchereau and P. Miossec, Interleukin-4 but not interleukin-10 inhibits the production of leukemia inhibitory factor by rheumatoid synovium and synoviocytes, *Eur. J. Immunol.* **24** (1994), 3222–3228.

[26] T. Defrance, P. Carayon, G. Billian, J.-C. Guillemot, A. Minty, D. Caput and P. Ferrara, Interleukin 13 is a B cell stimulating factor, *J. Exp. Med.* **179** (1994), 135–143.

[27] B. Deleuran, P. Lemche, M.S. Kristensen, C.Q. Chu, M. Field, J. Jensen, K. Matsushima and K. Stengaard-Pedersen, Localisation of interleukin 8 in the synovial membrane, cartilage-pannus junction and chondrocytes in rheumatoid arthritis, *Scand. J. Rheumatol.* **23** (1994), 2–7.

[28] W. Digel, W. Schoniger, M. Stefanic, H. Janssen, C. Buck, M. Schmid, A. Raghavachar and F. Porzsolt, Receptors for tumor necrosis factor on neoplastic B cells from chronic lymphocytic leukemia are expressed *in vitro* but not *in vivo*, *Blood* **76** (1990), 1607–1613.

[29] C.A. Dinarello, Biologic basis for interleukin-1 in disease, *Blood* **87** (1996), 2095–2147.

[30] R.P. Donnelly, M.J. Fenton, D.S. Finbloom and T.L. Gerrard, Differential regulation of IL-1 production in human monocytes by IFN-gamma and IL-4, *J. Immunol.* **145** (1990), 569–575.

[31] R. Essner, K. Rhoades, W.H. McBride, D.L. Morton and J.S. Economou, IL-4 down-regulates IL-1 and TNF gene expression in human monocytes, *J. Immunol.* **142** (1989), 3857–3861.

[32] C.H. Evans and P.D. Robbins, Gene therapy for arthritis, in: *Gene Therapeutics: Methods and Applications of Direct Gene Transfer*, J.A. Wolff, ed., Birkhauser, Boston, 1994, pp. 320–343.

[33] M.N. Farahat, G. Yanni, R. Poston and G.S. Panayi, Cytokine expression in synovial membranes of patients with rheumatoid arthritis and osteoarthritis, *Ann. Rheum. Dis.* **52** (1993), 870–875.

[34] M. Feldmann, F.M. Brennan and R.N. Maini, Role of cytokines in rheumatoid arthritis, *Annu. Rev. Immunol.* **14** (1996), 397–440.

[35] P.A. Guerne, B.L. Zuraw, J.H. Vaughan, D.A. Carson and M. Lotz, Synovium as a source as interleukin-6 *in vitro*: Contribution to local and systemic manifestations of arthritis, *J. Clin. Invest.* **83** (1989), 585–592.

[36] J.A. Hamilton, P.M. Waring and E.L. Filonzi, Induction of leukemia inhibitory factor in human synovial fibroblasts by IL-1 and tumor necrosis factor-alpha, *J. Immunol.* **150** (1993), 1496–1502.

[37] B. Haraoui, J.P. Pelletier, J.M. Cloutier, M.P. Faure and J. Martel-Pelletier, Synovial membrane histology and immunopathology in rheumatoid arthritis and osteoarthritis, *In vivo* effects of anti-rheumatic drugs, *Arthritis Rheum.* **34** (1991), 153–163.

[38] P.H. Hart, M.J. Ahern, M.D. Smith and J.J. Finlay-Jones, Comparison of the suppressive effects of interleukin-10 and interleukin-4 on synovial fluid macrophages and blood monocytes from patients with inflammatory arthritis, *Immunology* **84** (1995), 536–542.

[39] P.H. Hart, G.F. Vitti, D.R. Burgess, G.A. Whitty, D.S. Piccoli and J.A. Hamilton, Potential antiinflammatory effects of interleukin 4: suppression of human monocyte tumor necrosis factor alpha, interleukin 1, and prostaglandin E_2, *Proc. Natl. Acad. Sci. USA* **86** (1989), 3803–3807.

[40] M. Higuchi and B.B. Aggarwal, Inhibition of ligand binding and antiproliferative effects of tumor necrosis factor and lymphotoxin by soluble forms of recombinant P60 and P80 receptors, *Biochem. Biophys. Res. Commun.* **182** (1992), 638–643.

[41] K. Hirota, T. Akahoshi, H. Endo, H. Kondo and S. Kashiwazaki, Production of interleukin 8 by cultured synovial cells in response to interleukin 1 and tumor necrosis factor, *Rheumatol. Int.* **12** (1992), 13–16.

[42] H.P. Hohmann, M. Brockhaus, P.A. Baeuerle, R. Remy, R. Kolbeck and A.P. van Loon, Expression of the types A and B tumor necrosis factor (TNF) receptors is independently regulated, and both receptors mediate activation of the transcription factor NF-kappa B. TNF alpha is not needed for induction of a biological effect via TNF receptors, *J. Biol. Chem.* **265** (1990), 22 409–22 417.

[43] A.J. Hough, Pathology of osteoarthritis, in: *Arthritis and Allied Conditions*, W.J. Koopman, ed., Williams and Wilkins, Baltimore, 1997, pp. 1945–1968.

[44] O.M. Howard, K.A. Clouse, C. Smith, R.G. Goodwin and W.L. Farrar, Soluble tumor necrosis factor receptor: inhibition of human immunodeficiency virus activation, *Proc. Natl. Acad. Sci. USA* **90** (1993), 2335–2339.

[45] N. Itoh, S. Yonehara, A. Ishii, M. Yonehara, S. Mizushima, M. Sameshima, A. Hase, Y. Seto and S. Nagata, The polypeptide encoded by the cDNA for human cell surface antigen Fas can mediate apoptosis, *Cell* **66** (1991), 233–243.

[46] D. Jovanovic, J.A. Di Battista, J. Martel-Pelletier, F.C. Jolicoeur, Y. He, M. Zhang, F. Mineau and J.P. Pelletier, Interleukin-17 (IL-17) stimulates the production and expression of proinflammatory cytokines, IL-beta and TNF-alpha, by human macrophages, *J. Immunol.* **160** (1998), 3513–3521.

[47] D. Jovanovic, J.P. Pelletier, N. Alaaeddine, F. Mineau, C. Geng, P. Ranger and J. Martel-Pelletier, Effect of IL-13 on cytokines, cytokine receptors and inhibitors on human osteoarthritic synovium and synovial fibroblasts, *Osteoarthritis*

Cart. **6** (1998), 40–49.

[48] D.A. Joyce, D.P. Gibbons, P. Green, J.H. Steer, M. Feldmann and F.M. Brennan, Two inhibitors of pro-inflammatory cytokine release, interleukin-10 and interleukin-4, have contrasting effects on release of soluble p75 tumor necrosis factor receptor by cultured monocytes, *Eur. J. Immunol.* **24** (1994), 2699–2705.

[49] A.E. Koch, S.L. Kunkel, J.C. Burrows, H.L. Evanoff, G.K. Haines, R.M. Pope and R.M. Strieter, Synovial tissue macrophage as a source of the chemotactic cytokine IL-8, *J. Immunol.* **147** (1991), 2187–2195.

[50] M.S. Kristensen, K. Paludan, C.G. Larsen, C.O. Zachariae, B.W. Deleuran, P.K. Jensen, P. Jorgensen and K. Thestrup-Pedersen, Quantitative determination of IL-1 alpha-induced IL-8 mRNA levels in cultured human keratinocytes, dermal fibroblasts, endothelial cells, and monocytes, *J. Invest. Dermatol.* **97** (1991), 506–510.

[51] S.R. Kronheim, A. Mumma, T. Greenstreet, P.J. Glackin, K. Van Ness, C.J. March and R.A. Black, Purification of interleukin-1 beta converting enzyme, the protease that cleaves the interleukin-1 beta precursor, *Arch. Biochem. Biophys.* **296** (1992), 698–703.

[52] M. Lewis, L.A. Tartaglia, A. Lee, G.L. Bennett, G.C. Rice, G.H. Wong, E.Y. Chen and D.V. Goeddel, Cloning and expression of cDNAs for two distinct murine tumor necrosis factor receptors demonstrate one receptor is species specific, *Proc. Natl. Acad. Sci. USA* **88** (1991), 2830–2834.

[53] S. Lindblad and E. Hedfors, Arthroscopic and immunohistologic characterization of knee joint synovitis in osteoarthritis, *Arthritis Rheum.* **30** (1987), 1081–1088.

[54] H. Loetscher, Y.C.E. Pan, H.W. Lahm, R. Gentz, M. Brockhaus, H. Tabuchi and W. Lesslauer, Molecular cloning and expression of the human 55 kd tumor necrosis factor receptor, *Cell* **61** (1990), 351–359.

[55] M. Lotz and P.A. Guerne, Interleukin-6 induces the synthesis of tissue inhibitor of metalloproteinases-1/erythroid potentiating activity, *J. Biol. Chem.* **266** (1991), 2017–2020.

[56] M. Lotz, T. Moats and P.M. Villiger, Leukemia inhibitory factor is expressed in cartilage and synovium and can contribute to the pathogenesis of arthritis, *J. Clin. Invest.* **90** (1992), 888-896.

[57] M. Lotz, R. Terkeltaub and P.M. Villiger, Cartilage and joint inflammation. Regulation of IL-8 expression by human articular chondrocytes, *J. Immunol.* **148** (1992), 466–473.

[58] R. Maier, V. Ganu and M. Lotz, Interleukin-11, an inducible cytokine in human articular chondrocytes and synoviocytes, stimulates the production of the tissue inhibitor of metalloproteinases, *J. Biol. Chem.* **268** (1993), 21 527–21 532.

[59] J. Martel-Pelletier, R. McCollum, J.A. Di Battista, M.P. Faure, J.A. Chin, S. Fournier, M. Sarfati and J.P. Pelletier, The interleukin-1 receptor in normal and osteoarthritic human articular chondrocytes. Identification as the type I receptor and analysis of binding kinetics and biologic function, *Arthritis Rheum.* **35** (1992), 530–540.

[60] J. Martel-Pelletier, F. Mineau, D. Jovanovic, J.A. Di Battista and J.P. Pelletier, MAPK and NF-B together regulate the IL-17-induced nitric oxide production in human OA chondrocytes: possible role of transactivating factor MAPKAP-K, *Arthritis Rheum.* (1999) (in press).

[61] B. Mosley, D.L. Urdal, K.S. Prickett, A. Larsen, D. Cosman, P.J. Conlon, S. Gillis and S.K. Dower, The interleukin-1 receptor binds the human interleukin-1 alpha precursor but not the interleukin-1 beta precursor, *J. Biol. Chem.* **262** (1987), 2941–2944.

[62] S.L. Myers, K.D. Brandt, J.W. Ehlich, E.M. Braunstein, K.D. Shelbourne, D.A. Heck and L.A. Kalasinski, Synovial inflammation in patients with early osteoarthritis of the knee, *J. Rheumatol.* **17** (1990), 1662–1669.

[63] B. Naume, R. Shalaby, W. Lesslauer and T. Espevik, Involvement of the 55- and 75-kDa tumor necrosis factor receptors in the generation of lymphokine-activated killer cell activity and proliferation of natural killer cells, *J. Immunol.* **146** (1991), 3045–3048.

[64] J.J. Nietfeld, B. Wilbrink, M. Helle, J.L. van Roy, W. den Otter, A.J. Swaak and O. Huber-Bruning, Interleukin-1-induced interleukin-6 is required for the inhibition of proteoglycan synthesis by interleukin-1 in human articular cartilage, *Arthritis Rheum.* **33** (1990), 1695–1701.

[65] S.R. Paul, F. Bennett, J.A. Calvetti, K. Kelleher, C.R. Wood, R.M.Jr. O'Hara, A.C. Leary, B. Sibley, S.C. Clark and D.A. Williams, Molecular cloning of a cDNA encoding interleukin 11, a stromal cell-derived lymphopoietic and hematopoietic cytokine, *Proc. Natl. Acad. Sci. USA* **87** (1990), 7512–7516.

[66] J.P. Pelletier, J.P. Caron, C.H. Evans, P.D. Robbins, H.I. Georgescu, D. Jovanovic, J.C. Fernandes and J. Martel-Pelletier, *In vivo* suppression of early experimental osteoarthritis by IL-Ra using gene therapy, *Arthritis Rheum.* **40** (1997), 1012–1019.

[67] J.P. Pelletier, R. McCollum, J.M. Cloutier and J. Martel-Pelletier, Synthesis of metalloproteases and interleukin 6 (IL-6) in human osteoarthritic synovial membrane is an IL-1 mediated process, *J. Rheumatol.* **22** (1995), 109–114.

[68] D. Plows, L. Probert, S. Georgopoulos, L. Alexopoulou and G. Kollias, The role of tumour necrosis factor (TNF) in arthritis: studies in transgenic mice, *Rheumatol. Eur.* (Suppl. 2) (1995), 51–54.

[69] L.R. Reid, C. Lowe, J. Cornish, S.J. Skinner, D.J. Hilton, T.A. Willson, D.P. Gearing and T.J. Martin, Leukemia inhibitory factor: a novel bone-active cytokine, *Endocrinology* **126** (1990), 1416–1420.

[70] P. Roux-Lombard, L. Punzi, F. Hasler, S. Bas, S. Todesco, H. Gallati, P.A. Guerne and J.M. Dayer, Soluble tumor necrosis factor receptors in human inflammatory synovial fluids, *Arthritis Rheum.* **36** (1993), 485–489.

[71] M. Sadouk, J.P. Pelletier, G. Tardif, K. Kiansa, J.M. Cloutier and J. Martel-Pelletier, Human synovial fibroblasts coexpress interleukin-1 receptor type I and type II mRNA: The increased level of the interleukin-1 receptor in osteoarthritic cells is related to an increased level of the type I receptor, *Lab. Invest.* **73** (1995), 347–355.

[72] T.J. Schall, M. Lewis, K.J. Koller, A. Lee, G.C. Rice, G.H. Wong, T. Gatanaga, G.A. Granger, R. Lentz and H. Raab, Molecular cloning and expression of a receptor for human tumor necrosis factor, *Cell* **61** (1990), 361–370.

[73] P. Scheurich, B. Thoma, U. Ucer and K. Pfizenmaier, Immunoregulatory activity of recombinant human tumor necrosis factor (TNF)-alpha: induction of TNF receptors on human T cells and TNF-alpha-mediated enhancement of T cell responses, *J. Immunol.* **138** (1987), 1786–1790.

[74] J.M. Schroder, The monocyte-derived neutrophil activating peptide (NAP/interleukin 8) stimulates human neutrophil arachidonate-5-lipoxygenase, but not the release of cellular arachidonate, *J. Exp. Med.* **170** (1989), 847–863.

[75] S. Schutze, K. Potthoff, T. Machleidt, D. Berkovic, K. Wiegmann and M. Kronke, TNF activates NF-kappa B by phosphatidylcholine-specific phospholipase C-induced "acidic" sphingomyelin breakdown, *Cell* **71** (1992), 765–776.

[76] M.R. Shalaby, M.A.Jr. Palladino, S.E. Hirabayashi, T.E. Eessalu, G.D. Lewis, H.M. Shepard and B.B. Aggarwal, Receptor binding and activation of polymorphonuclear neutrophils by tumor necrosis factor-alpha, *J. Leukoc. Biol.* **41** (1987), 196–204.

[77] M.R. Shalaby, A. Sundan, H. Loetscher, M. Brockhaus, W. Lesslauer and T. Espevik, Binding and regulation of cellular functions by monoclonal antibodies against human tumor necrosis factor receptors, *J. Exp. Med.* **172** (1990), 1517–1520.

[78] M. Shingu, S. Miyauchi, Y. Nagai, C. Yasutake and K. Horie, The role of IL-4 and IL-6 in IL-1-dependent cartilage matrix degradation, *Br. J. Rheumatol.* **34** (1995), 101–106.

[79] W.M. Siders, J.C. Klimovitz and S.B. Mizel, Characterization of the structural requirements and cell type specificity of IL-1 and IL-1 secretion, *J. Biol. Chem.* **268** (1993), 22 170–22 174.

[80] J. Slack, C.J. McMahan, S. Waugh, K. Schooley, M.K. Spriggs, J.E. Sims and S.K. Dower, Independent binding of interleukin-1 alpha and interleukin-1 beta to type I and type II interleukin-1 receptors, *J. Biol. Chem.* **268** (1993), 2513–2524.

[81] C.A. Smith, T. Davis, D. Anderson, L. Solam, M.P. Beckmann, R. Jerzy, S.K. Dower, D. Cosman and R.G. Goodwin, A receptor for tumor necrosis factor defines an unusual familx of cellular and viral proteins, *Science* **248** (1990), 1019–1023.

[82] M. Svenson, M.B. Hansen, P. Heegaard, K. Abell and K. Bendtzen, Specific binding of interleukin-1 (IL-1)- and IL-1 receptor antagonist (IL-1ra) to human serum. High-affinity binding of IL-1ra to soluble IL-1 receptor type I, *Cytokine* **5** (1993), 427–435.

[83] L.A. Tartaglia, D. Pennica and D.V. Goeddel, Ligand passing: the 75-kDa tumor necrosis factor (TNF) receptor recruits TNF for signaling by the 55-kDa TNF receptor, *J. Biol. Chem.* **268** (1993), 18 542–18 548.

[84] L.A. Tartaglia, R.F. Weber, I.S. Figari, C. Reynolds, M.A.Jr. Palladino and D.V. Goeddel, The two different receptors for tumor necrosis factor mediate distinct cellular responses, *Proc. Natl. Acad. Sci. USA* **88** (1991), 9292–9296.

[85] B. Thoma, M. Grell, K. Pfizenmaier and P. Scheurich, Identification of a 60-kD tumor necrosis factor (TNF) receptor as the major signal transducing component in TNF responses, *J. Exp. Med.* **172** (1990), 1019–1023.

[86] J. Van Damme, R.A. Bunning, R. Conings, R. Graham, G. Russell and G. Opdenakker, Characterization of granulocyte chemotactic activity from human cytokine-stimulated chondrocytes as interleukin 8, *Cytokine* **2** (1990), 106–111.

[87] F.A.J. Van de Loo, L.A. Joosten, P.L. van Lent, O.J. Arntz and W.B. van den Berg, Role of interleukin-1, tumor necrosis factor alpha, and interleukin-6 in cartilage proteoglycan metabolism and destruction. Effect of *in situ* blocking in murine antigen- and zymosan-induced arthritis, *Arthritis Rheum.* **38** (1995), 164–172.

[88] E. Vannier, L.C. Miller and C.A. Dinarello, Coordinated antiinflammatory effects of interleukin 4: interleukin 4 suppresses interleukin 1 production but up-regulates gene expression and synthesis of interleukin 1 receptor antagonist, *Proc. Natl. Acad. Sci. USA* **89** (1992), 4076–4080.

[89] P.M. Villiger, Y. Geng and M. Lotz, Induction of cytokine expression by leukemia inhibitory factor, *J. Clin. Invest.* **91** (1993), 1575–1581.

[90] C.I. Westacott, R.M. Atkins, P.A. Dieppe and C.J. Elson, Tumour necrosis factor-alpha receptor expression on chondrocytes isolated from human articular cartilage, *J. Rheumatol.* **21** (1994), 1710–1715.

[91] K.P. Wilson, J.A. Black, J.A. Thomson, E.E. Kim, J.P. Griffith, M.A. Navia, M.A. Murcko, S.P. Chambers, R.A. Aldape and S.A. Raybuck, Structure and mechanism of interleukin-1 beta converting enzyme, *Nature* **370** (1994), 270–275.

[92] Z. Yao, S.L. Painter, W.C. Fanslow, D. Ulrich, B.M. Macduff, M.K. Spriggs and R.J. Armitage, Human IL-17: a novel cytokine derived from T cells, *J. Immunol.* **155** (1995), 5483–5486.

[93] C.L. Yu, K.H. Sun, S.C. Shei, C.Y. Tsai, S.T. Tsai, J.C. Wang, T.S. Liao, W.M. Lin, H.L. Chen, H.S. Yu and S.H. Han, Interleukin 8 modulates interleukin-1 beta, interleukin-6 and tumor necrosis factor-alpha release from normal human mononuclear cells, *Immunopharmacology* **27** (1994), 207–214.

Biorheology 39 (2002) 247–258
IOS Press

Differential gene expression analysis in a rabbit model of osteoarthritis induced by anterior cruciate ligament (ACL) section

G. Bluteau [a], J. Gouttenoire [a], T. Conrozier [b], P. Mathieu [b], E. Vignon [b], M. Richard [c], D. Herbage [a] and F. Mallein-Gerin [a,*]

[a] *Institut de Biologie et Chimie des Protéines, UMR 5086, CNRS-UCB Lyon I, 69367 Lyon Cedex 7, France*

[b] *Service Rhumatologie, Centre Hospitalier Lyon-Sud, 69495 Pierre Bénite Cedex, France*

[c] *U189 INSERM, BP12, 69921 Oullins Cedex, France*

Abstract. Osteoarthritis (OA) is the most common of all joint diseases to affect mankind and is characterized by the degradation of articular cartilage. The low availability of normal and pathologic human cartilage and the inability to study the early stages of the disease in humans has led to the development of numerous animal models of OA. The aim of our study was to establish gene expression profiles during the progression of a rabbit model of OA induced by anterior cruciate ligament (ACL) section. Semiquantitative RT–PCR was used to follow expression of several relevant molecules (type II and X collagens, aggrecan, osteonectin, βig-h3, BiP, TIMP-1, MMP-1, -3, -13, aggrecanase-1, -2) during development of OA in articular cartilage. In parallel, we monitored the activities of collagenase, caseinase, phospholipase A2 and glycosyltransferases (xylosyl-, galactosyl-, glucuronyl- and N-acetyl-galactosaminyl-transferase). Novel cDNA clones for rabbit type X collagen, aggrecanase-1 and -2, osteonectin and BiP were constructed to obtain species-specific primers. Ours result show that MMP-13 (collagenase-3) gene expression increased dramatically early after ACL surgery and remained high thereafter. An increase in MMP-1 (collagenase-1) and MMP-3 expression was also noted with an absence of variation for TIMP-1 expression. In addition, the global MMPs activities paralleled the MMP gene expression. These data together characterize at the molecular level the evolution of OA in this rabbit model. Furthermore, we have undertaken a search for identifying differentially expressed genes in normal and OA cartilage in this model, by differential display RT–PCR. We present here preliminary results with the determination of the best technical conditions to obtain reproducible electrophoresis patterns of differential display RT–PCR.

1. Introduction

The function of articular cartilage is to distribute stresses and to provide for a low friction-bearing surface for joint motion [16,32]. Cartilage cells, chondrocytes, are dispersed throughout an abundant extracellular matrix composed of a dense fibrillar collagen network (collagens of type II, IX and XI), a filamentous network of type VI collagen and of several proteoglycans and glycoproteins associated or not to the collagen molecules. Some of these glycoproteins (thrombospondin-1 and -2, tenascin-C, osteopontin, SPARC...) were recently considered as "matricellular" proteins as they play a role as adaptors and modulators of cell–matrix interactions [7]. Their multiple functions derive from their ability to interact with molecules playing a structural role in the matrix (collagen, elastin, fibronectin, decorin...) and with different cell-surface receptors, cytokines, growth factors and proteases [49]. Chondrocyte metabolism

*Address for correspondence: Dr. F. Mallein-Gerin, Institut de Biologie et Chimie des Protéines, UMR CNRS 5086, Université Claude Bernard, 7 passage du Vercors, 69367 Lyon Cedex 07, France. Tel.: +33 4 72 72 26 00; Fax: +33 4 72 72 26 02; E-mail: f.mallein-gerin@ibcp.fr.

is under the control of genetic and environmental factors such as extracellular matrix composition, soluble mediators and mechanical factors [17,37,40,48]. Normal joint loading is required for maintenance of articular cartilage structure and function *in vivo*. However, several clinical observations have shown that abnormal mechanical forces predispose cartilage to degenerative changes as observed in osteoarthritis (OA).

The low availability of normal and pathologic human cartilage and the inability to study the early stages of the disease in human has led to the development of numerous animal models of OA [2]. Among these different models, surgical induction of joint laxity leads to degenerative changes in articular cartilage resembling those occurring in man. For example, section of the anterior cruciate ligament (ACL) of the knee in dog [8,13,34,38,44] or rabbit [19–21,42,43], total or partial meniscectomy in rabbit or guinea pig [31,33] cause progressive ultrastructural and biochemical changes with many of the features observed in human OA including the enzymatic cleavage of matrix molecules.

In our laboratory, we routinely use the rabbit model of OA induced by section of the anterior cruciate ligament (ACL) in the knee. Histological changes including cartilage hypertrophy, reduced cell density, matrix alteration, cystic lesions and enlarged perichondrocytic lacunae were found to precede cartilage fibrillation [42]. 12 weeks after surgery, the activity of acid phosphatase, several glycosidases and neutral proteases were significantly elevated in the operated joint cartilage (femoro-patellar surface) [43]. Thus, this well-defined model permits to study the molecular events occurring at different (early and late) stages of the degenerative disease. We have started a systematic study by histological, biochemical and molecular biology techniques of the cellular phenotype of the articular cartilage in the knee of rabbits sacrificed between 1 and 12 weeks after ACL surgery [5,6]. In this paper, we will summarize our analysis of gene expression for several relevant molecules using semiquantitative reverse transcription (RT)–polymerase chain reaction (PCR) on RNA isolated from operated and control articular cartilage. Several rabbit partial cDNA sequences that had not been previously reported (for type X collagen, aggrecanase 1 and 2, osteonectin and immunoglobulin heavy chain binding protein (BIP, a potential type II collagen chaperone) have been cloned and will be presented here. The gene expression of cartilage degradative enzymes (MMPs and aggrecanases) will be compared to different global enzymatic activities (collagenase, caseinase, PLA2 and different glycosyl transferases) measured after extraction from articular cartilage. Furthermore, we will present preliminary results with the determination of the best technical conditions to obtain reproducible differential display RT–PCR patterns, with the aim of identifying differentially expressed genes in normal and OA cartilage samples.

2. Materials and methods

2.1. Animals and surgical procedures

9–10 Month-old white rabbits, weighing 4.5–5 kg, were anesthetized and the ACL of the two knees was cut through a medial parapatellar incision as described previously [5]. In each series, three animals were sacrificed 1, 3, 9 and 12 weeks after surgery for the enzyme activity analyses and 2, 4 and 9 weeks after surgery for the gene expression analyses. We focused our studies on the articular cartilage of the femoral condyles.

2.2. RNA preparation

Total RNA was isolated from cartilage slices by using a guanidinium isothiocyanate (GIT) procedure [1], modified as reported previously [6]. Briefly, the frozen samples were powdered with a mortar and pestle previously cooled in liquid nitrogen, then homogenized in a solution containing 4 M GIT, 25 mM sodium citrate pH 7, 0.5% N-lauroylsarcosine, 0.1% β-mercaptoethanol for 3 hr at 4°C. After phenol–chloroform extraction and precipitation, pellets were treated with 200 μg/ml proteinase K in 1 M GIT, 25 mM sodium citrate, 2 mM Tris/HCl at 40°C until complete dissolution. The GIT concentration was then adjusted to 4 M and samples were layered on a cushion of cesium trifluoroacetate (CsTFA) with a density of 1.6 g/ml and ultracentrifuged in a Beckman SW60Ti rotor at 33 000 rpm for 18 h at 18°C. In order to eliminate possible traces of genomic DNA, RNA pellets were re-suspended in a total volume of 200 μl for digestion with 6 U of RNAse-free DNAse (Promega) in the presence of 80 U of RNAsin (Perkin Elmer), for 30 min at 37°C. Total RNA was finally recovered after a further phenol–chloroform extraction and sodium acetate precipitation. The pellets were resuspended in 11 μl of diethylpyrocarbonate-treated water and 1 μl was assayed for concentration and purity of RNA by measuring A_{260}/A_{280}.

2.3. Semiquantitative analysis by reverse transcription-polymerase chain reaction (RT–PCR)

The conditions for the semiquantitative RT–PCR analysis and the sequences of the primers used for each marker have been recently reported [5]. For type II collagen, aggrecan, βig-h3, MMP-1, -3, -13, TIMP-1 or GAPDH, specific primers were designed from rabbit DNA sequences available in the databanks. The rabbit DNA sequences for osteonectin, BiP, type X collagen, aggrecanase-1 or aggrecanase-2 were not available in the databanks and were cloned in our laboratory. Semi-quantitative PCR was performed by removing aliquots from the PCR reaction after increasing numbers of cycles. The PCR products were electrophorized on agarose gel, stained with ethidium bromide and in the case of type X collagen, aggrecanase-1 and -2, transferred to a membrane for Southern-blotting with internal specific probes [5]. For semi-quantitative analysis, the cycle number at which the product was first detected on the agarose gel or by autoradiography was taken as a measure of the relative amount of specific mRNA present in the originally isolated RNA [41]. GAPDH was used to verify that equal amounts of RNA were added to the reaction. In order to simplify the figures, photographs of agarose gels are shown only for GAPDH and type II collagen and one photograph of autoradiography is shown for type X collagen (Fig. 1). Table 1 illustrates the variations in expression of the different genes.

2.4. Analysis by Differential Display RT–PCR (DDRT–PCR)

We used the method originally described by Liang and Pardee [24]. For RT, a 40 μl reaction contained 200 ng total RNA, 1 μM oligo(dT$_{12}$VN) primers (V represents A, G or C and N represents A, T, G or C), 100 μM each dNTP, 2 μg BSA, 10 mM DTT, 0.5 U RNAsin (Perkin Elmer). Samples were incubated at 65°C for 5 min then at 42°C for 5 min prior to addition of 200 U SuperScript II RNAse H$^-$ (GIBCO-BRL). Reactions were carried out at 42°C for 55 min, followed by treatment with 4 U RNAse H at 37°C for 30 min. For PCR amplification, each 20 μl reaction contained 2 μl RT aliquot, 1, 8 μM each dNTP, 1 μl (α-^{35}S) dATP (1000 Ci/mmol), 1 μM oligo(dT$_{12}$VN) primers, 0, 2 μM arbitrary upstream primer, 1.5 mM $MgCl_2$ and 2.5 units AmpliTaq DNA polymerase (Perkin Elmer). Following an initial denaturation step of 5 min at 95°C, amplification consisted of 40 cycles of 45 s at 94°C, 1 min at 40°C, 1 min at 72°C followed by a final extension step of 5 min at 72°C. Amplification was performed in a Geneamp PCR system 2400 (Perkin Elmer). PCR products were analyzed on a 6% DNA sequencing gel and exposed for autoradiography.

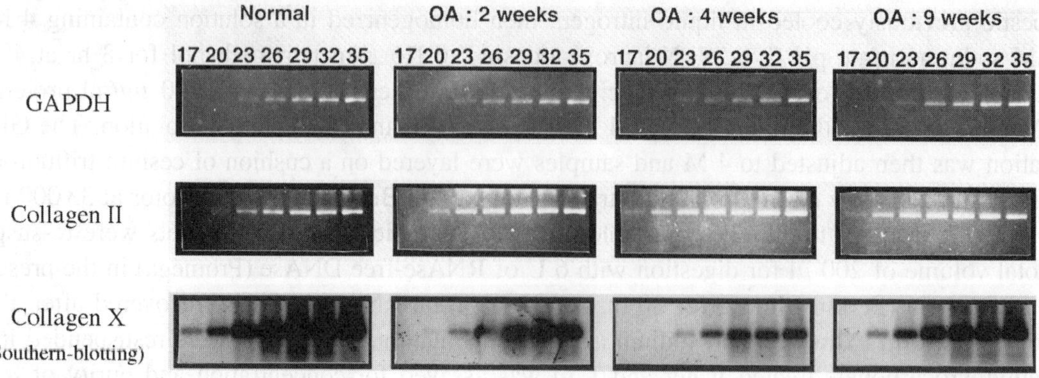

Fig. 1. Examples of semiquantitative mRNA analysis from articular cartilage in normal rabbit and during development of experimental OA. The gene products analyzed are indicated on the left. After several numbers of cycles (indicated on the top), amplicons are shown after ethidium bromide staining for GAPDH and type II collagen or after Southern blotting followed by autoradiography for type X collagen.

Table 1

Comparison of gene expression in normal rabbit articular cartilage and in articular cartilage from rabbits sacrificed 2, 4 and 9 weeks following ACL transection

Gene products	Normal (cycles)	OA: 2 weeks	OA: 4 weeks	OA: 9 weeks
GAPDH	23	→	→	→
MMP-1	40	↗↗	↗	↗↗
MMP-3	32	↗↗	↗	↗
MMP-13	40	↗↗↗↗↗	↗↗↗↗	↗↗↗↗
TIMP-1	26	↗	→	→
Aggrecanase-1*	29	→	↘	→
Aggrecanase-2*	32	→	→	→
Type II collagen	20	→	↗	↗
Type X collagen*	17	↘↘	↘↘	↘
Osteonectin	26	→	↘	↘
Aggrecan	29	→	→	↘
βig-h3	26	↗	→	→
BiP	26	→	→	→

The cycle number at which the PCR product was first detected on the agarose gel or by autoradiography* is indicated for the markers analyzed in the normal cartilage. → Represents the same number of cycles as in the normal cartilage. One ↗ represents an increase in the level of gene expression corresponding to a three PCR cycle difference with the normal cartilage. One ↘ represents a decrease in the level of gene expression corresponding to a three PCR cycle difference with the normal cartilage.

2.5. Determination of enzymatic activity

Cartilage fragments were homogenized at 4°C with a Polytron homogenizer. The homogenate was centrifuged at 5000*g* for 5 min and the supernatant was used as a crude enzymatic fraction. Collagenolytic and caseinolytic activities were determined after APMA activation by digestion of ^{3}H-labeled type II collagen and casein fluorescein isothiocyanate, respectively, as described previously [18]. PLA2 activity was measured using unlabeled L-α-phosphatidyl-ethanolamine and labeled 1-palmitoyl-2-($1^{14}C$)-linoleyl-L-3-phosphatidyl-ethanolamine (NEN) as a substrate [45]. The activity of glycosyltransferases (xylosyl-, galactosyl-, glucuronyl- and N-acetylgalactosaminyl-transferase was measured as described previously [12].

3. Results and discussion

In all operated knee joints, reproducible evolution of lesions was observed at different time points following ACL surgery [5]: articular cartilage was histologically normal during the first week. Chondrocytes proliferated between 2 and 3 weeks, as attested by numerous clusters of cells throughout the matrix. Between 4 and 6 weeks the cartilage swelled in appearance and the surface was irregular and fibrillated. After 8 weeks, erosion of the tissue was evident with progressive decrease in cartilage thickness and bone exposure in the late stage.

Semiquantitative RT–PCR was used to follow expression of several relevant molecules (type II and X collagens, aggrecan, osteonectin, βig-h3, BiP, TIMP1, MMP-1, -3, -13, aggrecanase-1, -2, GAPDH) during development of OA in rabbit articular cartilage. As rabbit cDNA sequences for type X collagen, aggrecanase-1 and -2, osteonectin and BiP were not available, we cloned partial cDNA sequences coding for these proteins. These sequences are shown in Figs 2–5. Comparison of these rabbit sequences with previously published sequences of other species are the following: the rabbit type X collagen sequence (563 bp) exhibited 86% and 90% nucleotide identity with the bovine and human sequence, respectively (Fig. 2); the rabbit aggrecanase-1 sequence (317 bp) showed 90% nucleotide identity with the human sequence and the rabbit aggrecanase-2 sequence (323 bp) showed 86%, 87% and 90% nucleotide identity with murine, human and bovine sequences, respectively (Fig. 3); the rabbit osteonectin sequence (840 bp) exhibited 89% and 91% nucleotide identity with the bovine and human sequence, respectively (Fig. 4) and the rabbit BiP sequence (279 bp) showed 90% and 92% nucleotide identity with the murine and the human sequence, respectively (Fig. 5).

The methods employed here cannot detect small differences in gene expression as previously discussed [5] and we present here (Table 1) a recapitulative view of our results, considering the level of

```
           F  I  P  Y  A  I  K  S  K  G  I  P  V  R  G  E  Q  G  I  P  G  P  P  G  P  A  G  P  R  G  H  P  G
Rabbit : CTTCATTCCCTATGCCATAAAGAGTAAAGGTATACCAGTGAGAGGAGAACAAGGTATTCCTGGACCGCCAGGCCCTGCTGGACCCCGAGGGCACCCAGGT
Human  : ............CA...................G...A.......G......C.....T..A................T..............
Bovine : ......C..G............G.......G..T..C..........G.....C..C.....T..A........C.......T.............C

          P  S  G  P  P  G  K  P  G  Y  G  T  P  G  L  Q  G  E  P  G  L  P  G  P  P  G  L  S  A  T  G  K  P
Rabbit : CCTTCTGGACCACCAGGAAAACCAGGCTATGGAACCCCGGGACTGCAAGGAGAGCCAGGGTTGCCAGGACCTCCAGGACTATCAGCCACTGGGAAGCCAG
Human  : ............................C....GT..T.....C.........................A..G....C......TGTA.....A....
Bovine : ..G..A..............G..C...ACC....GT..T..G.CC......C....................A..G....CG..................

         G  L  P  G  L  P  G  K  Q  G  E  R  G  P  S  G  P  K  G  D  I  G  P  A  G  L  P  G  P  R  G  P  P  G
Rabbit : GTTTGCCAGGACTCCCAGGAAAACAAGGGGAGAGAGGACCAAGTGGACCGAAAGGAGACATTGGGCCAGCTGGTCTACCAGGACCCAGGGGCCCACCAGG
Human  : ..G...................C...A...........TA......A........TG.....A........C............C............
Bovine : ..................................T..A.......A........T.....A.........T..........AC............

           P  P  G  I  P  G  P  A  G  I  S  V  P  G  K  P  G  Q  Q  G  L  T  G  A  P  G  P  R  G  F  P  G  E
Rabbit : GCCACCTGGAATCCCAGGACCAGCTGGCATTTCTGTGCCAGGAAAACCTGGACAACAGGGACTTACAGGAGCCCCAGGCCCCCGCGGCTTTCCTGGAGAA
Human  : A..............T.....G.....A.................................CC..............A...A.G..............
Bovine : A.....C.........C..C........A........T................G.C...A....C........AA.....T...A.G............

          K  G  A  P  G  A  P  G  V  S  G  Q  K  G  E  T  G  Y  G  A  P  G  R  P  G  E  R  G  L  P  G  P  Q
Rabbit : AAGGGTGCACCAGGAGCCCCTGGTGTGAGTGGACAGAAAGGGGAAACAGGATATGGTGCTCCTGGCCGCCCAGGTGAGAGGGGCCTTCCAGGCCCTCAGG
Human  : ................T.......A...A.................TG.................T..T.............T...............
Bovine : ......A.GT....G.T......AC...A.................TG..GC..T..A.....T.......................C.....T......

         G  P  M  G  P  P  G  P  P  G  V  G  K  R  G  E  N  G  F  P  G
Rabbit : GTCCCATGGGACCACCTGGCCCTCCTGGAGTGGGGAAAAGAGGTGAGAATGGGTTTCCAGGGC
Human  : ......CA......T...................A..........A......G.......A.
Bovine : ......C............................................A........A.....A.
```

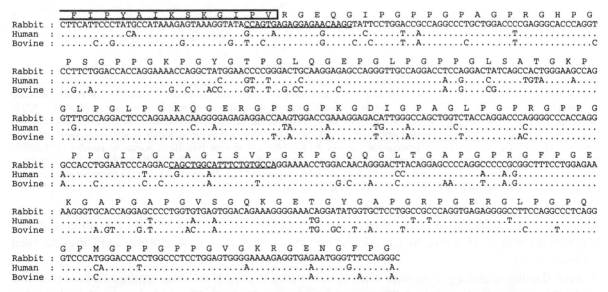

Fig. 2. Partial nucleotide sequence of rabbit cDNA for proα1(X) collagen mRNA. The rabbit sequence (GenBank accession number AF247705) is compared with the corresponding human (EMBL accession number X60382) and bovine (EMBL accession number X53556) sequences. A dot (.) indicates that the nucleotide is identical to the rabbit sequence. Nucleotide sequences of the primers used for the construction of the probe for Southern-blotting are underlined. The top line shows the translated amino acids of the rabbit sequence. The N-terminal non-collagenous domain (NC2) is indicated by a box and is followed by the triple-helical domain.

A

```
            D  C  T  R  P  V  P  R  N  G  G  K  Y  C
Rabbit : GGATTGCACGAGGCCCGTCCCCCGGAACGGTGGCAAGTATTGT
Human  : A..C.........T..........T..........C...

          E  G  R  R  T  R  F  R  S  C  N  T  E  N
Rabbit : GAGGGCCGCCGGACTCGCTTCCGCTCCTGCAACACGGAGAACT
Human  : ..........T..C...................T...G...

         [C] P  H  G  S  A  L  T  F  R  E  E  Q  C  A
Rabbit : GCCCACACGGCTCAGCATTGACCTTCCGTGAAGAGCAGTGTGC
Human  : .....ACT........CC..........C..G..........

          A  Y  N  H  R  T  D  L  F  K  S  F  P  G
Rabbit : TGCCTACAACCACCGAACCGACCTCTTCAAGAGCTTTCCAGGG
Human  : ...............C.....................C.....

          P  M  D  W  V  P  R  Y  T  G  V  A  P  R
Rabbit : CCCATGGACTGGGTTCCGCGCTACACAGGTGTGGCCCCTCGAG
Human  : ...................T...........C........C.AG.

         D  Q  C  K  L  T  C  Q  A  R  A  L  G  Y  Y
Rabbit : ACCAATGCAAACTCACCTGCCAGGCCCGGGCACTGGGCTACTA
Human  : ....G.....................................

          Y  V  L  E  P  R  V  A  D  G  T  P  C  S
Rabbit : CTACGTATTGGAGCCCCGGGTGGCAGATGGGACTCCCTGCTCC
Human  : ...T..GC.......A......T.........C.....T...

          P  D  T  S  S
Rabbit : CCAGACACCTCCTCT
Human  : ..G....G......G
```

B

```
          C  N  L  T  F  G  P  E  Y  T  V  C  P  G
Rabbit : TGCAACTTGACGTTCGGGCCTGAGTACACCGTGTGTCCCGGC
Bovine : ......C...C........A.....T...............
Human  : ......C...A..............T...............
Murine : ..........A..T............T.G.....C..T...

          M  D  V  C  A  R  L  W  C  A  V  V  R  Q
Rabbit : ATGGACGTGTGTGCCCGCCTGTGGTGTGCGGTGGTGCGCCAG
Bovine : ....................C.......C...........
Human  : .....T..C....T.............T....A.....
Murine : .....T..C....G..G...........T..........A

          G  Q  M  V  C  L  T  K  K  L  P  A  V  E
Rabbit : GGCCAGATGGTGTGCCTCACCAAGAAGCTGCCCGCCGTGGAG
Bovine : ............T..G...............G........A
Human  : ..........C..T..G...............T..G....A
Murine : ....A........T..G...............G..T.....

          G  T  P  C  G  K  G  R  I  C  L  Q  G  K
Rabbit : GGCACACCCTGTGGAAAGGCAGGATCTGCCTGCAGGGCAAG
Bovine : ..A....A..........G.....................
Human  : ..G..G..T.....A..G..G..A.................A
Murine : ....T...........G..A..AG.......T..A.....A

          C  A  D  K  T  K  K  K  Y  Y  S  T  S  S
Rabbit : TGTGCGGACAAGACCAAGAAGAAGTATTACTCAACATCAAGT
Bovine : ....T......A.........A..A....T..G........C
Human  : ....T......A.........A..A....T.....G.....C
Murine : ....T......A..T.....A..A........G........C

          H [G  N  W  G  S  W  G  P  W  G  Q  C  S
Rabbit : CATGGAAACTGGGGGTCCTGGGGTCCCTGGGGCCAGTGTTCT
Bovine : ..C..C....................GT...............
Human  : .....C.........A..T....AT.................
Murine : ........T..................C.......T.......

          R  S  C  G  G  G  V  Q  F  A  Y  R  H  C
Rabbit : CGCTCTTGTGGGGGAGGAGTACAGTTTGCCTATCGCCACTGC
Bovine : .....G........C.....G............C........
Human  : .....A.....A.........G................T.....T
Murine : .....A..C...........G............C.....T..T

          N  N  P  A  P  R  N  S][G  P
Rabbit : AATAACCCTGCACCCAGAAACAGTGGCCG
Bovine : ........A..G....AG....AC.....
Human  : ............T..........AC..A..
Murine : ...............TC.............
```

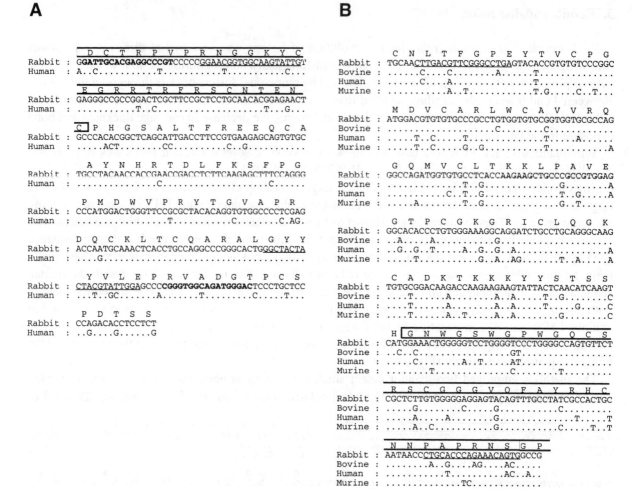

Fig. 3. Partial nucleotide sequences of rabbit cDNA for aggrecanase-1 (A) and aggrecanase-2 (B). (A) The rabbit sequence (GenBank accession number AF247707) is compared with the corresponding human (GenBank accession number AF148213) sequence. Nucleotide sequences of the primers used for the construction of the probe for Southern-blotting are underlined. (B) The rabbit sequence (GenBank accession number AF247708) is compared with the corresponding bovine (GenBank accession number AF192771), human (GenBank accession number AF142099) and murine (GenBank accession number AF140673) sequences. Nucleotide sequences of the primers used for the construction of the probe for Southern-blotting are underlined. (A, B) The top line shows the translated amino acids of the rabbit sequence with the box indicating the thrombospondin type I motif. A dot (.) indicates that the nucleotide is identical to the rabbit sequence.

gene expression as increased or decreased if it corresponds at least to a three PCR cycle difference with the normal cartilage. However, we consider here as truly significant only differences in PCR cycle equal or higher to six.

Type II collagen and aggrecan were actively transcribed in normal articular cartilage and not modified at the OA time intervals examined. The young age of our animals (9–10 months) can explain these data as well as their discrepancy with those of Matyas et al. [29] who showed elevated type II collagen and aggrecan mRNA levels in dog OA knee after ACL transection.

Type X collagen was demonstrated at the surface of normal human, porcine and rat articular cartilage [35], and in human osteoarthritic cartilage [46] and thus is not restricted to the growth plate as

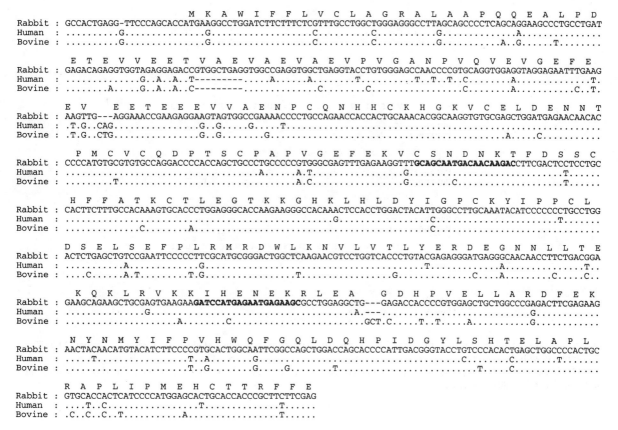

```
                                M  K  A  W  I  F  F  L  V  C  L  A  G  R  A  L  A  A  P  Q  Q  E  A  L  P  D
Rabbit : GCCACTGAGG-TTCCCAGCACCATGAAGGCCTGGATCTTCTTTCTCGTTTGCCTGGCTGGGAGGGCCTTAGCAGCCCCTCAGCAGGAAGCCCTGCCTGAT
Human  : ..........G...............G...................C..........C...........G..............A...............
Bovine : ..........G...............G...................C..........C...........G...........A..G......T........

          E  T  E  V  V  E  E  T  V  A  E  V  A  E  V  A  E  V  P  V  G  A  N  P  V  Q  V  E  V  G  E  F  E
Rabbit : GAGACAGAGGTGGTAGAGGAGACCGTGGCTGAGGTGGCCGAGGTGGCTGAGGTACCTGTGGGAGCCAACCCCGTGCAGGTGGAGGTAGGAGAATTTGAAG
Human  : ..............G..A..A..T---------.....A......A........T..........T..T..T..C........A..............T.
Bovine : ........A.....G..A..A..C---------..............C........C.................C........A...........C..T.

         E  V     E  E  T  E  E  E  V  V  A  E  N  P  C  Q  N  H  H  C  K  H  G  K  V  C  E  L  D  E  N  N  T
Rabbit : AAGTTG---AGGAAACCGAAGAGGAAGTAGTGGCCGAAAACCCCTGCCAGAACCACCACTGCAAACACGGCAAGGTGTGCGAGCTGGATGAGAACAACAC
Human  : .T.G..CAG................G..G.....G.....T...........................................................
Bovine : .T.G..CTG................G..G........G............................................A.....C...........

           P  M  C  V  C  Q  D  P  T  S  C  P  A  P  V  G  E  F  E  K  V  C  S  N  D  N  K  T  F  D  S  S  C
Rabbit : CCCCATGTGCGTGTGCCAGGACCCCACCAGCTGCCCTGCCCCCGTGGGCGAGTTTGAGAAGGTTTGCAGCAATGACAACAAGACCTTCGACTCCTCCTGC
Human  : ....................................A......A.T.................G.............................T......
Bovine : .........T.................................A.C.................G........C....................T......

          H  F  F  A  T  K  C  T  L  E  G  T  K  K  G  H  K  L  H  L  D  Y  I  G  P  C  K  Y  I  P  P  C  L
Rabbit : CACTTCTTTGCCACAAAGTGCACCCTGGAGGGCACCAAGAAGGGCCACAAACTCCACCTGGACTACATTGGGCCTTGCAAATACATCCCCCCCTGCCTGG
Human  : ..................................................G.................C.......................T.......
Bovine : ..............C........A............................................C...............................

         D  S  E  L  S  E  F  P  L  R  M  R  D  W  L  K  N  V  L  V  T  L  Y  E  R  D  E  G  N  N  L  L  T  E
Rabbit : ACTCTGAGCTGTCCGAATTCCCCCTTCGCATGCGGGACTGGCTCAAGAACGTCCTGGTCACCCTGTACGAGAGGGATGAGGGCAACAACCTTCTGACGGA
Human  : ...........A.............G.........................................T.............A...............T..
Bovine : ....C......A.T.........T.G.................T.................G..............C....A.........C.....C..

           K  Q  K  L  R  V  K  K  I  H  E  N  E  K  R  L  E  A     G  D  H  P  V  E  L  L  A  R  D  F  E  K
Rabbit : GAAGCAGAAGCTGCGAGTGAAGAAGATCCATGAGAATGAGAAGCGCCTGGAGGCTG---GAGACCACCCCGTGGAGCTGCTGGCCCGAGACTTCGAGAAG
Human  : ...............G......................................A.---............................G............
Bovine : .....................A........C.........................GCT.C.....T..T.....A...........G............

          N  Y  N  M  Y  I  F  P  V  H  W  Q  F  G  Q  L  D  Q  H  P  I  D  G  Y  L  S  H  T  E  L  A  P  L
Rabbit : AACTACAACATGTACATCTTCCCCGTGCACTGGCAATTCGGCCAGCTGGACCAGCACCCCATTGACGGGTACCTGTCCCACACTGAGCTGGCCCCACTGC
Human  : .....T.................T..A........G......................................C........C........T.......
Bovine : .......................T..G........G.....G........T..........................T.....C................

         R  A  P  L  I  P  M  E  H  C  T  T  R  F  F  E
Rabbit : GTGCACCACTCATCCCCATGGAGCACTGCACCACCCGCTTCTTCGAG
Human  : ....T..C.................T..............T......
Bovine : .C..C..C..T...........A.................T......
```

Fig. 4. Partial nucleotide sequence of rabbit cDNA for osteonectin. The rabbit sequence (GenBank accession number AF247647) is compared with the corresponding human (GenBank accession number J03040) and bovine (Genbank accession number J03233) sequences. A dot (.) indicates that the nucleotide is identical to the rabbit sequence. The top line shows the translated amino acids of the rabbit sequence.

originally thought [36]. Interestingly, the present work also provides evidence of type X collagen transcription in normal and OA rabbit articular cartilage, with a decrease 2 weeks after ACL transection. This decrease could be related to the mitogenic activity of the chondrocytes observed in our histological sections, since chondrocyte proliferation is not compatible with full expression of type X collagen [26].

No significant variation in gene expression was observed for osteonectin, βig-h3 (TGF-β induced gene-3, a collagen-associated protein) and BiP (immunoglobulin heavy chain binding protein, a collagen chaperone) [15] related to HSP70 [39].

The most important results obtained in this study are related to the gene expression of matrix metalloproteinases (MMPs), aggrecanases and one inhibitor (TIMP1) in control and OA cartilage. Several studies have demonstrated that MMPs and aggrecanases play roles in the turnover of aggrecan and type II collagen in normal cartilage [4,10,23] and osteoarthritis is induced by an imbalance between synthesis and degradation of the extracellular matrix of articular cartilage [27]. In our study (Table 1), we observed a very large increase in MMP-13 (collagenase-3) gene expression early after ACL transection followed by a constant high level thereafter. Increase in MMP-1 (collagenase-1) and MMP-3 gene expression was also noted associated with an absence of variation for TIMP-1. These data were confirmed by the analysis of the total collagenase and caseinase activities present in the corresponding cartilage extracts (Table 2). A significant increase in collagenase activity was noted 3 weeks after surgery, still persistent

```
            A  A  A  I  A  Y  G  L  D  K  R  E  G  E
Rabbit : TGCAGCTGCTATTGCTTATGGTCTGGATAAGCGGGAGGGAGAG
Human :  G....................C........A........G...
Murine : A..............A.....C........A.A..........

          K  N  I  L  V  F  D  L  G  G  G  T  F  D
Rabbit : AAGAACATCCTGGTGTTTGACCTGGGTGGTGGAACCTTTGATG
Human :  .............................C........C....
Murine : ...........T.................C..C.....C....

         V  S  L  L  T  I  D  N  G  V  F  E  V  V  A
Rabbit : TGTCTCTTCTCACCATTGACAATGGTGTCTTCGAAGTCGTGGC
Human :  .....................................T.....
Murine : ...............................T.....G.....

           T  N  G  D  T  H  L  G  G  E  D  F  D  Q
Rabbit : CACTAATGGAGACACTCACCTGGGCGGGGAAGACTTCGACCAG
Human :  ............T.....T.....T..A........T......
Murine : ............T...........T...........T..T...

          R  V  M  E  H  F  I  K  L  Y  K  K  K  T
Rabbit : CGTGTCATGGAGCACTTCATCAAGCTGTACAAAAAGAAGACTG
Human :  ...........A...........A.................G.
Murine : ..G........A............T.............A....

         G  K  D  V  R  K  D  N  R  A  V  Q  K  L  R
Rabbit : GCAAAGATGTCAGGAAAGACAACAGAGCTGTGCAGAAGCTCCG
Human :  ................G.....T..............A.....
Murine : .T........T...........................A.....

           R  E  V  E  K  A  K
Rabbit : GCGTGAAGTGGAAAAGGCCAAG
Human :  ...C..G..A............
Murine : ......G..A........T...
```

Fig. 5. Partial nucleotide sequence of rabbit cDNA for BiP. The rabbit sequence (GenBank accession number AF247706) is compared with the corresponding human (GenBank accession number AF188611) and murine (EMBL accession number AJ002387) sequences. A dot (.) indicates that the nucleotide is identical to the rabbit sequence. The top line shows the translated amino acids of the rabbit sequence.

Table 2

Determination of different enzyme activities in rabbit cartilage extracts at different times following ACL transection

	Collagenase (+ APMA) (dpm/h/mg prot.)	Caseinase (+ APMA) (UF/h/mg prot.)	PLA_2 (dpm/h/mg prot.)
Normal	35 000 ± 4359	2450 ± 397	8750 ± 902
OA: 1 week	31 000 ± 3605	2700 ± 458	8100 ± 656
OA: 3 weeks	52 000 ± 8888**	3200* ± 360	9200 ± 1114
OA: 9 weeks	45 000 ± 6245*	2800 ± 530	8300 ± 1212
OA: 12 weeks	36 000 ± 5568	2400 ± 360	9500 ± 1353

	Xylosyl-transferase (dpm/h/mg prot.)	Galactosyl-transferase (dpm/h/mg prot.)	Glucuronyl-transferase (dpm/h/mg prot.)	N-acetyl-galactosaminyl-transferase (dpm/h/mg prot.)
Normal	1250 ± 305	880 ± 214	5160 ± 986	1640 ± 275
OA: 1 week	1420 ± 204	940 ± 171	4220 ± 620	1850 ± 360
OA: 3 weeks	1150 ± 265	605 ± 161	3900 ± 625	1760 ± 275
OA: 9 weeks	850 ± 278	850 ± 263	4940 ± 675	1720 ± 160
OA: 12 weeks	1020 ± 157	680 ± 246	4310 ± 508	1680 ± 322

Data represent mean ± SD of triplicate assays; Student's t-test: $^{**}P < 0.04$, $^{*}P < 0.07$.

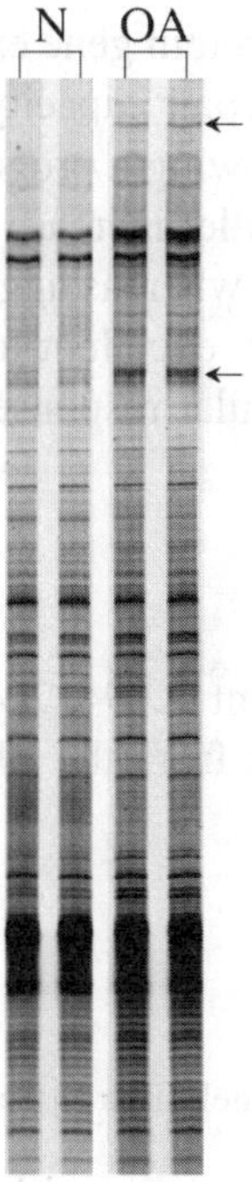

Fig. 6. Example of differential display RT–PCR analysis of normal rabbit articular cartilage (N) and articular cartilage from rabbit sacrificed 4 weeks following ACL surgery (OA). The upstream arbitrary primer was GGAAGCAGCT and the downstream primer was an oligo($dT_{12}VN$) (V represents A, G or C and N represents A, T, G or C). The PCR reaction was repeated twice for the normal or OA sample and the duplicates are shown under brackets. Arrows indicate cDNA candidates that appear to be differentially expressed in OA cartilage.

after 9 weeks. A parallel increase in MMP-13 mRNA and protein levels was noted in synovium and meniscus in the same model [19]. All these results are in good accordance with previous data obtained in human OA [25,30,44,47] and experimental OA [13,22]. The increase in MMP-1 and -3 synthesis observed in the early phase of OA induced after rupture of the ACL in dog [13] is also consistent with our data. Furthermore, as MMP-3 is capable of activating pro-MMP-1 [9] and pro-MMP-13 [28] by cleavage of their propeptide [11], the parallel enhanced gene expression of MMP -1, -3 and -13 observed in our rabbit model of OA is consistent with the general increase in MMP activities observed at the same time post surgery.

In contrast, levels of gene expression for aggrecanase-1 or -2 remained stable during development of OA in our rabbit model. It reveals a differential expression of MMPs and aggrecanases. As human cartilage explants exposed to retinoic acid or to interleukin-1 showed a marked increase in the release of aggrecanase-generated aggrecan catabolites with no concomitant increase in aggrecanase-1 mRNA levels [14], it is possible that aggrecanase activity may be predominantly regulated by post-transcriptional mechanisms [3].

Finally, we have undertaken a search for differentially expressed genes by differential display RT–PCR. We found that a treatment by proteinase K in 1 M GIT prior to CsTFA ultracentrifugation was crucial to increase the yield and purity of RNA extracted from cartilage matrix. This protocol led indeed in reproducible patterns of differential display RT–PCR and should be useful for identifying genes differentially expressed by chondrocytes *in situ* (Fig. 6).

In summary, our data indicate that matrix protein gene expression remains stable or varies little during development of OA, which suggests a poor repair capacity of the rabbit articular cartilage. This is corroborated by our histological observations showing a progressive destruction of extracellular matrix [5]. We have also found an early (2–3 weeks) parallel increase in gene expression of MMP-1, -3 and-13 and in global collagenase and caseinase activities, whereas aggrecanase gene expression remains stable with time. This animal model of OA appears as an excellent tool to study the influence of an experimental modification in joint biomechanics on the cellular response in articular cartilage and meniscus.

Acknowledgements

This work was funded by an Emergence grant (L094120402) with a fellowship to G.B., a "Programme Thématique Prioritaire" (085113) both grants from the Rhône-Alpes region and by a CNRS-INSERM grant "Ingénierie Tissulaire" (2000-03).

References

[1] M.E. Adams, D.H. Huang, L.Y. Yao and L.J. Sandell, Extraction and isolation of mRNA from adult articular cartilage, *Analyt. Biochem.* **202** (1992), 89–95.
[2] R.D. Altman and D. Dean, Osteoarthritis research: animal models, *Semin. Arthritis Rheum.* **19** (1990), 21–25.
[3] E.C. Arner, M.A. Pratta, J.M. Trzaskos, C.P. Decicco and M.D. Tortorella, Generation and characterization of aggrecanase: A soluble, cartilage-derived aggrecan-degrading activity, *J. Biol. Chem.* **274** (1999), 6594–6601.
[4] R.C. Billinghurst, L. Dahlberg, M. Ionescu, A. Reiner, R. Bourne, C. Rorabeck, P. Mitchell, J. Hambor, O. Diekmann, H. Tschesche, J. Chen, H. Van Wart and A.R. Poole, Enhanced cleavage of type II collagen by collagenases in osteoarthritic articular cartilage, *J. Clin. Invest.* **99** (1997), 1534–1545.
[5] G. Bluteau, T. Conrozier, P. Mathieu, E. Vignon, D. Herbage and F. Mallein-Gerin, Matrix metalloproteinase-1-2-13 and aggrecanase-1 and -2 are differentially expressed in experimental osteoarthritis, *Biochim. Biophys. Acta* **1526** (2001), 147–158.
[6] G. Bluteau, L. Labourdette, M.C. Ronziere, T. Conrozier, P. Mathieu, D. Herbage and F. Mallein-Gerin, Type X collagen in rabbit and human meniscus, *Osteoarthritis Cartilage* **7** (1999), 496–501.
[7] P. Bornstein, Diversity of function is inherent in matricellular proteins: an appraisal of thrombospondin 1, *J. Cell Biol.* **130** (1995), 503–506.
[8] K.D. Brandt, E.M. Braunstein, D.M. Visco, B. O'Connor, D. Heck and M. Albrecht, Anterior (cranial) cruciate ligament transection in the dog: a bona fide model of osteoarthritis, not merely of cartilage injury and repair, *J. Rheumatol.* **18** (1991), 436–446.
[9] C.E. Brinckerhoff, K. Suzuki, T.I. Mitchell, F. Oram, C.I. Coon, R.D. Palmiter and H. Nagase, Rabbit procollagenase synthesized and secreted by a high-yield mammalian expression vector requires stromelysin (matrix metalloproteinase-3) for maximal activation, *J. Biol. Chem.* **265** (1990), 22 262–22 269.
[10] S. Chubinskaya, K.E. Kuettner and A.A. Cole, Expression of matrix metalloproteinases in normal and damaged articular cartilage from human knee and ankle joints, *Lab. Invest.* **79** (1999), 1669–1999.
[11] S. Curran and G.I. Murray, Matrix metalloproteinases in tumour invasion and metastasis, *J. Pathol.* **189** (1999), 300–308.
[12] M.J. David, E. Vignon, M.J. Peschard, P. Louisot and M. Richard, Effect of non-steroidal anti-inflammatory drugs (NSAIDS) on glycosyltransferase activity from human osteoarthritic cartilage, *Brit. J. Rheumatol.* **31** (1992), 13–17.
[13] J.C. Fernandes, J. Martel-Pelletier, V. Lascau-Coman, F. Moldovan, D. Jovanovic, J.-P. Raynauld and J.-P. Pelletier, Collagenase-1 and collagenase-3 synthesis in normal and early experimental osteoarthritic canine cartilage: an immunohistochemical study, *J. Rheumatol.* **25** (1998), 1585–1594.
[14] C.R. Flannery, C.B. Little, C.E. Hughes and B. Caterson, Expression of ADAMTS homologues in articular cartilage, *Biochem. Biophys. Res. Commun.* **260** (1999), 318–322.
[15] A.-M. Freyria, M.-C. Ronzière, M.-M. Boutillon and D. Herbage, Effect of retinoic acid on protein synthesis by foetal bovine chondrocytes in high-density culture: down-regulation of the glucose-regulated protein, GRP-78, and type II collagen, *Biochem. J.* **305** (1995), 391–396.
[16] F. Guilak, R.L.Y. Sah and L.A. Setton, Physical regulation of cartilage metabolism in: *Basic Orthopaedic Biomechanics*, V.C. Mow and W.C. Hayes, eds, Lippincott-Raven, Philadelphia, 1997, pp. 179–207.

[17] F. Guilak and V.C. Mow, The mechanical environment of the chondrocyte: a biphasic finite element model of cell–matrix interactions in articular cartilage, *J. Biomechanics* **33** (2000), 1663–1673.

[18] M.-P. Hellio, M.J. Peschard, C. Cohen, M. Richard and E. Vignon, Calcitonin inhibits phospholipase A2 and collagenase activity of human osteoarthritc chondrocytes, *Osteoarthritis Cartilage* **5** (1997), 121–128.

[19] M.-P. Hellio Le Graverand, J. Eggerer, P. Sciore, C. Reno, E. Vignon, I. Otterness and D.A. Hart, Matrix metalloproteinase-13 expression in rabbit knee joint connective tissues: influence of maturation and response to injury, *Matrix Biol.* **19** (2000), 431–441.

[20] M.-P. Hellio Le Graverand, E. Vignon, I. Otterness and D.A. Hart, Early changes in lapine menisci during osteoarthritis development: Part I: Cellular and matrix alterations, *Osteoarthritis Cartilage* **9** (2001), 56–64.

[21] M.-P. Hellio Le Graverand, E. Vignon, I. Otterness and D.A. Hart, Early changes in lapine menisci during osteoarthritis development: Part II: Molecular alterations, *Osteoarthritis Cartilage* **9** (2001), 65–72.

[22] J.L. Huebner, I.G. Otterness, E.M. Freund, B. Caterson and V.B. Kraus, Collagenase 1 and collagenase 3 expression in a guinea pig model of osteoarthritis, *Arthritis Rheum.* **41** (1998), 877–890.

[23] M.W. Lark, E.K. Bayne, J. Flanagan, C.F. Harper, L.A. Hoerrner, N.I. Hutchinson, I.I. Singer, S.A. Donatelli, J.R. Weidner, H.R. Williams, R.A. Mumford and L.S. Lohmander, Aggrecan degradation in human cartilage, *J. Clin. Invest.* **100** (1997), 93–106.

[24] P. Liang and A.B. Pardee, Differential display of eukaryotic messenger RNA by means of the polymerase chain reaction, *Science* **257** (1992), 967–971.

[25] O. Lindy, Y.T. Konttinen, T. Sorsa, Y. Ding, S. Santavirta, A. Ceponis and C. Lopez-Otin, Matrix metalloproteinase 13 (collagenase 3) in human rheumatoid synovium, *Arthritis Rheum.* **40** (1997), 1391–1399.

[26] F. Mallein-Gerin and B.R. Olsen, Expression of simian virus 40 large T (tumor) oncogene in mouse chondrocytes induces cell proliferation without loss of the differentiated phenotype, *Proc. Natl. Acad. Sci. USA* **90** (1993), 3289–3293.

[27] J. Martel-Pelletier, R. Mc Collum, N. Fujimoto, K. Obata, J.M. Cloutier and J.P. Pelletier, Excess of metalloproteases over tissue inhibitor of metalloprotease may contribute to cartilage degradation in osteoarthritis and rheumatoid arthritis, *Lab. Invest.* **70** (1994), 807–815.

[28] L.M. Matrisian, Metalloproteinases and their inhibitors in matrix remodeling, *Trends Gen.* **6** (1990), 121–125.

[29] J.R. Matyas, P.F. Ehlers, D. Huang and M.E. Adams, The early molecular history of experimental osteoarthritis. I. Progressive discoordinate expression of aggrecan and type II procollagen messenger RNA in the articular cartilage of adult animals, *Arthritis Rheum.* **42** (1999), 993–1002.

[30] P.G. Mitchell, H.A. Magna, L.M. Reeves, L.L. Lopresti-Morrow, S.A. Yocum, P.J. Rosner, K.F. Geoghegan and J.E. Hambor, Cloning, expression, and type II collagenolytic activity of matrix metalloproteinase-13 from human osteoarthritic cartilage, *J. Clin. Invest.* **97** (1996), 761–768.

[31] R.W. Moskowitz, D.S. Howell, V.M. Goldberg, O. Muniz and J. Pita, Cartilage proteoglycan alterations in an experimentally induced model of rabbit osteoarthritis, *Arthritis Rheum.* **22** (1979), 155–163.

[32] V.C. Mow, C.C. Wang and C.T. Hung, The extracellular matrix, interstitial fluid and ions as a mechanical signal transducer in articular cartilage, *Osteoarthritis Cartilage* **7** (1999), 41–58.

[33] P.C. Pastoureau, A.C. Chomel and J. Bonnet, Evidence of early subchondral bone changes in the meniscectomized guinea pig. A densitometric study using dual-energy X-ray absorptiometry subregional analysis, *Osteoarthritis Cartilage* **7** (1999), 466–473.

[34] J.P. Pelletier, J. Martel-Pelletier, R.D. Altman, L. Ghandur-Mnaymneh, D.S. Howell and J.F. Woessner, Collagenolytic activity and collagen matrix breakdown of the articular cartilage in the Pond-Nuki dog model of osteoarthritis, *Arthritis Rheum.* **26** (1983), 866–874.

[35] G.M. Rucklidge, G. Milne and S.P. Robins, Collagen type X: a component of the surface of normal human, pig, and rat articular cartilage, *Biochem. Biophys. Res. Commun.* **224** (1996), 297–302.

[36] T.M. Schmid and T.F. Linsenmayer, Immunohistochemical localization of short chain cartilage collagen (type X) in avian tissues, *J. Cell Biol.* **100** (1985), 598–605.

[37] R. Sironen, M. Elo, K. Kaarniranta, H.J. Helminen and M.J. Lammi, Transcriptional activation in chondrocytes submitted to hydrostatic pressure, *Biorheology* **37** (2000), 85–93.

[38] R.A. Stockwell, M.E.J. Billingham and H. Muir, Ultrastructural changes in articular cartilage after experimental section of the anterior cruciate ligament of the dog knee, *J. Anat.* **136** (1983), 425–439.

[39] K. Takahashi, T. Kubo, Y. Arai, J. Imanishi, M. Kawata and Y. Hirasawa, Localization of heat shock protein in osteoarthritic cartilage, *Scan. J. Rheumatol.* **26** (1997), 368–375.

[40] J.P.G. Urban, Present perspectives on cartilage and chondrocyte mechanobiology, *Biorheology* **37** (2000), 185–190.

[41] J.B.J. van Meurs, P.L.E.M. van Lent, L.A.B. Joosten, P.M. van der Kraan and W.B. van den Berg, Quantification of mRNA levels in joint capsule and articular cartilage of the murine knee joint by RT–PCR: kinetics of stromelysin and IL-1 mRNA levels during arthritis, *Rheumatol. Int.* **16** (1997), 197–205.

[42] E. Vignon, J. Bejui, P. Mathieu, D.J. Hartmann, G. Ville, J.C. Evreux and J. Descotes, Histological cartilage changes in a rabbit model of osteoarthritis, *J. Rheumatol.* **14** (1987), 104–106.

[43] E. Vignon, O. Gateau, A. Martin, D. Hartmann, J. Bejui, M.C. Biol, M.T. Vanier, P. Louisot and M. Richard, Screening of degradative enzymes from articular cartilage in experimental osteoarthritis, *Clinic. Rheumatol.* **6** (1987), 208–204.
[44] E. Vignon, D.J. Hartmann, G. Vignon, B. Moyen, M. Arlot and G. Ville, Cartilage destruction in experimentally induced osteoarthritis, *J. Rheumatol.* **11** (1984), 202–207.
[45] E. Vignon, P. Mathieu, P. Louisot, J. Vilamitjana, M.F. Harmand and M. Richard, Phospholipase A2 activity in human osteoarthritic cartilage, *J. Rheumatol.* **16**(Suppl. 18) (1989), 35–38.
[46] K. von der Mark, T. Kirsch, A. Nerlich, A. Kuss, G. Weseloh, K. Gluckert and H. Stoss, Type X collagen synthesis in human osteoarthritic cartilage, *Arthritis Rheum.* **35** (1992), 806–811.
[47] D. Wernicke, C. Seyfert, B. Hinzmann and E. Gromnica-Ihle, Cloning of collagenase 3 from the synovial membrane and its expression in rheumatoid arthritis and osteoarthritis, *J. Rheumatol.* **23** (1996), 590–595.
[48] R.J. Wilkins, J.A. Browning and J.P.G. Urban, Chondrocyte regulation by mechanical load, *Biorheology* **37** (2000), 67–74.
[49] Z. Yang, T. Kyriakides and P. Bornstein, Matricellular proteins as modulators of cell–matrix interactions: adhesive defect in thrombospondin 2-null fibroblasts is a consequence of increased levels of MMP-2, *Mol. Biol. Cell* **11** (2000), 3353–3364.

Biorheology 39 (2002) 259–268
IOS Press

Bioreactor studies of native and tissue engineered cartilage

G. Vunjak-Novakovic [a,*] B. Obradovic [b], I. Martin [c] and L.E. Freed [a]

[a] *Division of Health Sciences and Technology, Massachusetts Institute of Technology, Cambridge, MA 02139, USA*

[b] *Department of Chemical Engineering, Belgrade University, Belgrade, Yugoslavia*

[c] *Department of Surgery, Research Division, University of Basel, Switzerland*

Abstract. Functional tissue engineering of cartilage involves the use of bioreactors designed to provide a controlled *in vitro* environment that embodies some of the biochemical and physical signals known to regulate chondrogenesis. Hydrodynamic conditions can affect *in vitro* tissue formation in at least two ways: by direct effects of hydrodynamic forces on cell morphology and function, and by indirect flow-induced changes in mass transfer of nutrients and metabolites. In the present work, we discuss the effects of three different *in vitro* environments: static flasks (tissues fixed in place, static medium), mixed flasks (tissues fixed in place, unidirectional turbulent flow) and rotating bioreactors (tissues dynamically suspended in laminar flow) on engineered cartilage constructs and native cartilage explants. As compared to static and mixed flasks, dynamic laminar flow in rotating bioreactors resulted in the most rapid tissue growth and the highest final fractions of glycosaminoglycans and total collagen in both tissues. Mechanical properties (equilibrium modulus, dynamic stiffness, hydraulic permeability) of engineered constructs and explanted cartilage correlated with the wet weight fractions of glycosaminoglycans and collagen. Current research needs in the area of cartilage tissue engineering include the utilization of additional physiologically relevant regulatory signals, and the development of predictive mathematical models that enable optimization of the conditions and duration of tissue culture.

1. Introduction

Tissue engineering of articular cartilage has been motivated by the need to develop novel treatment options for the repair of articular cartilage lesions that are a significant clinical problem. Current therapies, which include methods for facilitating cartilage healing, transplantation of cartilage or isolated cells, and implantation of prosthetic devices, have limitations and have not been shown to predictably restore a durable articular surface [4,10]. One approach to cartilage tissue engineering is to grow functional cartilaginous constructs by the integrated use of three components: (i) chondrogenic cells that can be expanded or transfected to overexpress genes of interest, (ii) polymer scaffolds that can provide a defined 3-dimensional structure for tissue development and biodegrade at a controlled rate, and (iii) bioreactors that can provide a controlled *in vitro* environment designed to promote chondrogenesis [23–25]. The use of tissue-like constructs, as compared to the use of cells or biomaterial scaffolds alone, can potentially improve the localization of cell delivery and promote graft fixation and survival, thereby enhancing tissue repair.

Functional cartilaginous constructs could also serve as a controlled physiological model system for quantitative studies of tissue development and function that are designed to distinguish the effects of

*Address for correspondence: G. Vunjak-Novakovic, MIT E25-330, 45 Carleton St., Cambridge, MA 02139, USA. Tel.: +1 617 452 2593; Fax: +1 617 258 8827; E-mail: gordana@mit.edu.

specific biochemical and physical signals from other systemic effects present *in vivo*. Ideally, a bioreactor should be designed as a controlled system that can establish a spatially uniform initial cell distribution on polymer scaffolds, control the factors of interest in the immediate cell environment (e.g., temperature, pH, concentration of oxygen), and expose the developing tissue to biochemical and physical signals known to regulate chondrogenesis. Bioreactors can be utilized to enhance mass transport between a gas phase and the culture medium (to maintain oxygen and pH levels) and between the medium and the cultured tissue (to promote the cellular exchange of nutrients, oxygen and metabolites). However, the flow and mixing utilized to facilitate mass transport can also directly affect cell function and tissue development [20,25,37,39]. We have previously shown that dynamic (rather than steady) and laminar (rather than turbulent) flow patterns were associated with rapid *in vitro* assembly of functional cartilaginous matrix containing glycosaminoglycans (GAG) and type II collagen by articular chondrocytes cultured on biodegradable scaffolds [21,39].

In order to assess the effects of bioreactor hydrodynamics on *in vitro* chondrogenesis, we conducted comparative studies of engineered cartilage constructs and native cartilage explants in three different systems: static flasks (tissues fixed in place, static medium), mixed flasks (tissues fixed in place, unidirectional turbulent flow) and rotating vessels (tissues dynamically suspended in laminar flow) over a period of 6 weeks. The mechanical properties of engineered and freshly explanted cartilage were correlated to the respective tissue compositions.

2. Materials and methods

2.1. *Tissue harvest, cell isolation and culture*

Full-thickness articular cartilage was harvested aseptically from the femoropatellar grooves of 2–3 week old bovine calves within 8 hours of slaughter. *For explant cultures*, 5 mm diameter × 2 mm thick discs were cut from middle sections of full-thickness cartilage plugs and rinsed in phosphate buffered saline (PBS) supplemented with 50 U/ml penicillin and 50 μg/ml streptomycin. *For engineered constructs*, chondrocytes were isolated using type II collagenase (Worthington, Freehold, NJ) and resuspended in culture medium (Dulbecco's Modified Eagle Medium (DMEM) containing 4.5 g/l glucose and supplemented with 10% Fetal Bovine Serum, 10 mM N-2-HydroxyEthylPiperazine N′-2-EthaneSulfonicAcid (HEPES), 50 U/ml penicillin, 50 μg/ml streptomycin, 0.1 mM nonessential amino acids, 0.4 mM proline, 50 μg/ml ascorbic acid, 0.5 μg/ml fungizone) [5]. Cells were seeded onto biodegradable polyglycolic acid scaffolds (5 mm diameter × 2 mm thick discs, void volume 96% [16]) for 3 days using well-mixed spinner flasks and an initial seeding density of 5×10^6 cells per scaffold [38].

Cell–polymer constructs and freshly harvested cartilage explants were subdivided into three groups each and transferred into static flasks, mixed flasks or rotating bioreactors (Fig. 1). Culture vessels containing 11–12 constructs or explants in 110–120 ml of culture medium were placed in 37°C/10% CO_2 incubators. In static and mixed flasks, tissue samples were fixed in place and exposed to either static or turbulently mixed culture medium; gas was exchanged by surface aeration. In rotating vessels, tissue samples were freely suspended in culture medium in the annular space between an outer and an inner cylinder, the latter of which served as a gas exchange membrane. Each bioreactor was attached to a base that provided vessel rotation and gas supply. Vessel rotation speed was adjusted throughout the period of cultivation to maintain each growing tissue settling in a relatively steady position within the vessel. In all

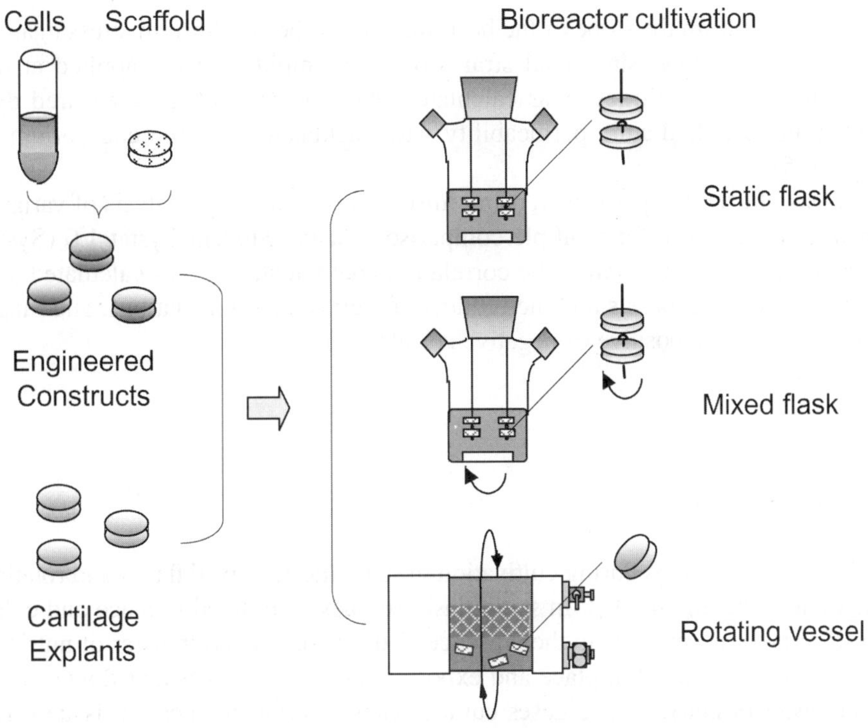

Fig. 1. Model system. Cells isolated from bovine articular cartilage were seeded for 3 days on biodegradable scaffolds (polyglycolic acid mesh, 5 mm diameter $\times$ 2 mm thick discs) in well mixed flasks [38]. The resulting cell–polymer constructs and 5 mm diameter $\times$ 2 mm thick explants of native bovine calf cartilage were cultured for 6 weeks in static flasks (tissues fixed in place, static medium), mixed flasks (tissues fixed in place, unidirectional turbulent flow) and rotating bioreactor vessels (tissues dynamically suspended in dynamic laminar flow) [39].

groups, culture medium was replaced at a rate of 50% every 2–3 days, i.e., 3 cm^3 per tissue sample per day [39].

2.2. *Structural and functional assessments*

Histological samples were fixed in 10% neutral buffered formalin, embedded in paraffin, sectioned (8 μm thick), and stained with safranin-O/fast green for GAG. Samples for biochemical analyses were frozen, lyophilized, and digested in a buffered solution of proteinase K [21]. The amount of GAG was determined spectrophotometrically [13]. Total collagen was determined from the measured hydroxyproline content [40], using a ratio of hydroxyproline to collagen of 0.1 [27]. Medium pH, partial pressure of oxygen (p_{O_2}) and partial pressure of CO_2 (p_{CO_2}) were determined immediately after sampling using a Blood-Gas analyzer (model 1610, Instrumentation Laboratory, Lexington, MA) with an accuracy of 0.1% for pH and 2% for each p_{O_2} and p_{CO_2}. Medium concentrations of glucose, lactate and ammonia (NH_3) were determined spectrophotometrically in duplicate using commercially available enzymatic assays (Sigma, St. Louis, MO). Medium concentrations of GAG were measured spectrophotometrically using bovine chondroitin sulfate in medium as a standard [13].

Mechanical properties were measured as previously described [30,39]. In brief, 3 mm diameter $\times$ 2 mm thick discs were sectioned from the central regions of engineered and native cartilage tissues and compressed at sequential increments of 10% strain up to a maximum of 40% strain [14]. The equilibrium modulus was determined from the slope of the best linear fit of the equilibrium stress against the applied strain. At a static offset of 30%, sinusoidal strains of 0.5% amplitude were applied at frequencies of 0.025–1 Hz and the dynamic stiffness was calculated from the ratio of the measured dynamic stress and the applied strain. The hydraulic permeability was calculated using the equilibrium modulus and dynamic stiffness [15].

Statistical analysis of data was performed using fully factorial one-way analysis of variance (ANOVA) in conjunction with Tukey's test for multiple comparisons, using Student Systat 1.0 (Systat, Evanston, IL). For structure–function relationships, the correlation coefficient, ρ_{xy} was calculated as the ratio between the covariance of two data sets and the product of their standard deviations, and ranged from zero (no correlation) to ± 1 (strong positive or negative correlation).

3. Results

3.1. Bioreactor hydrodynamics

The flow and mixing conditions during cultivation in static flasks, mixed flasks and rotating vessels are summarized in Table 1 (based on [39]). In static flasks, tissues were fixed in place and cultured without hydrodynamic shear at tissue surfaces in the presence of diffusional mass transfer of nutrients and gases. In mixed flasks, tissues were fixed in place and exposed to the steady turbulent flow of medium, which enhanced mass transfer of nutrients and gases but also caused turbulent shear at tissue surfaces [37]. In rotating vessels, tissues were dynamically suspended in a laminar, rotational flow field, and mass transfer was enhanced by laminar convection due to construct settling [17]. The composition of culture medium changed with cultivation time due to cell metabolism and batchwise medium replacement [33]. Average

Table 1

Hydrodynamic conditions in tissue culture vessels

Cultivation vessel	Static flask (ST)	Mixed flask (MIX)	Rotating vessel (RV)
Operating conditions			
Medium volume (cm^3)	120	120	110
Tissue constructs or explants (5 mm diameter × 2 mm thick discs)	Fixed in place; $n \leqslant 12$ per vessel	Fixed in place; $n \leqslant 12$ per vessel	Freely settling; $n \leqslant 12$ per vessel
Medium exchange	Batch-wise (3 cm^3 per tissue per day)	Batch-wise (3 cm^3 per tissue per day)	Batch-wise (3 cm^3 per tissue per day)
Gas exchange	Continuous, via surface aeration	Continuous, via surface aeration	Continuous, via an internal membrane
Mixing mechanism	None	Magnetic stirring	Settling in rotational flow
Fluid flow	None	Turbulent(1)	Laminar(2)
Mass transfer in bulk medium	Molecular diffusion	Turbulent convection (due to medium stirring)	Laminar convection (due to tissue settling)
References	[18,20–25,30,33,37,39]	[18,20–26,30,33,37–39]	[17–25,29,30,37,39]

(1)The smallest turbulent eddies had a diameter of 250 μm and velocity of 0.4 cm/s; estimated according to Cherry and Papoutsakis [11] by Vunjak-Novakovic et al. [37].

(2)Tissues were settling in a laminar tumble-slide regimen in a rotational field; estimated according to Clift et al. [12] by Freed and Vunjak-Novakovic [17] and Neitzel et al. [32].

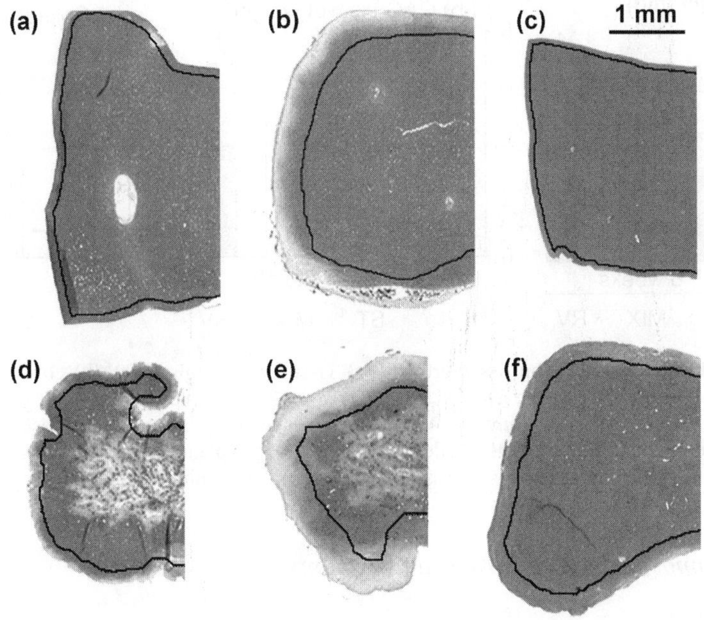

Fig. 2. Bioreactor hydrodynamics: effects on tissue morphology. Histological sections of native cartilage explants ((a), (b), (c)) and engineered cartilage constructs ((d), (e), (f)) cultured for 6 weeks in static flasks ((a), (d)), mixed flasks ((b), (e)), and rotating vessels ((c), (f)). Safranin-O stained cross-sections were bisected, and one representative half is shown for each group [29].

steady state levels of p_{O_2}, ammonia, pH and GAG were generally comparable for the corresponding cultures of constructs and explants, and lower in static than in either mixed culture. Yields of lactate on glucose ($\sim$1.6 mol/mol in static cultures, $\sim$1.3 mol/mol in mixed flasks and rotating vessels) suggested that cell metabolism was more anaerobic in static cultures than in either mixed flasks or rotating vessels.

3.2. Histomorphology of constructs and explants

Histomorphology of cultured constructs and explants depended on *in vitro* culture environment [29]. In static construct cultures, GAG accumulated mostly at the periphery (Fig. 2(a), (d)), presumably due to diffusionally constrained mass transfer. In mixed flasks, mixing enhanced mass transport throughout culture medium and at the tissue surfaces, but the associated turbulent shear was associated with the formation of an outer fibrous capsule at construct and explant surfaces (Fig. 2(b), (e)). Only in rotating vessels were GAG concentrations in constructs and explants high and spatially uniform (Fig. 2(c), (f)). All tissues contained an external region with high concentration of elongated cells and low GAG content (Fig. 2, solid line) which was relatively thin (70–265 μm) for constructs and explants cultured in static flasks and rotating vessels and thicker ($\sim$450 μm) for constructs and explants cultured in mixed flasks [29]. The corresponding gradients in GAG level at the tissue surfaces were consistent with the measured GAG release into the culture medium [33]. Over 6 weeks of culture, the fractional release of newly synthesized GAG by constructs and explants was 10–30% in static flasks and rotating vessels, as compared to 40–60% in mixed flasks.

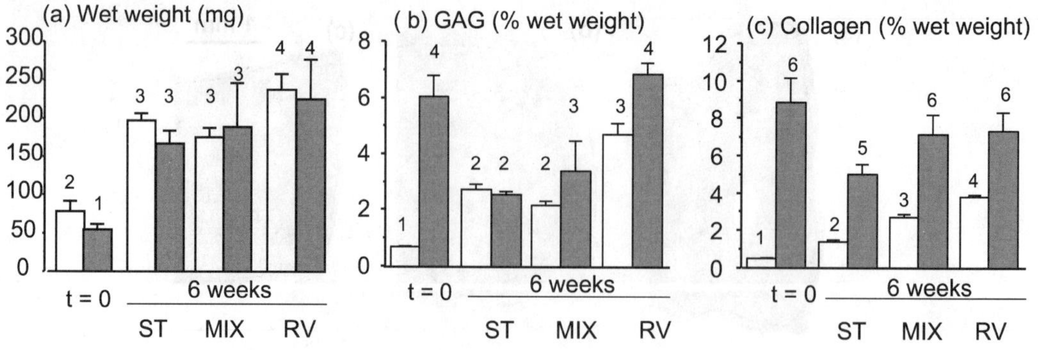

Fig. 3. Bioreactor hydrodynamics: effects on tissue growth. (a) Tissue wet weight (mg), (b) GAG (% wet weight), (c) Total collagen (% wet weight). Data represents the average ± SD for $n = 3$–6 samples per group at the time of transfer into the culture vessel ($t = 0$) and after 6 weeks of culture. ST, MIX and RV, respectively, refer to static flasks, mixed flasks and rotating vessels. Open bars refer to engineered constructs, shaded bars to cartilage explants. Numbers above bars indicate significantly different levels of a given variable among groups, as assessed by ANOVA in conjunction with Tukey's HSD test ($p < 0.05$).

3.3. Biochemical compositions of constructs and explants

After 6 weeks of *in vitro* culture, the wet weights of cartilage constructs and explants were 4–5-fold larger than at the beginning of cultivation, due to the progressive accumulation of GAG and collagen (Fig. 3(a), open and shaded bars, respectively). In both groups, wet weight fractions of GAG (Fig. 3(b)) and collagen (Fig. 3c) were markedly higher rotating vessels than in static culture. In 6-week constructs from rotating vessels, the fractions of both GAG and total collagen were five times as high as in the initial constructs and twice as high as in constructs from static flasks (Fig. 3(b) and (c), open bars), but remained subnormal, at 78% and 43% of the respective values measured for freshly explanted cartilage. In 6-week explants from rotating vessels, the fractions of GAG and total collagen were maintained at their initial levels, in contrast to static flasks where both fractions decreased to ∼50% of initial (Fig. 3(b) and (c)), shaded bars). Total amounts of biochemically determined tissue components (cells + GAG + collagen % of the wet weight) were 4.63 ± 0.13, 5.88 ± 0.33 and 9.04 ± 0.53 for 6-week constructs cultured in static flasks, mixed flasks and rotating vessels, respectively, and 8.05 ± 0.68, 11.26 ± 2.07 and 14.89 ± 0.71, for explants cultured in static flasks, mixed flasks and rotating vessels, respectively, as compared to 15.93 ± 2.45 for freshly explanted cartilage.

3.4. Structure–function correlations for engineered cartilage

Confined-compression equilibrium modulus, dynamic stiffness (measured at 30% strain, 1 Hz frequency) and hydraulic permeability (calculated at the 30% strain using combined data for static and dynamic compression) correlated with the wet weight fractions of GAG (Fig. 4(a), (c), (e)) and total collagen (Fig. 4(b), (d), (f)). In particular, equilibrium modulus was strongly positively correlated with wet weight fractions of GAG and total collagen (Fig. 4(a), (b)) and dynamic stiffness was strongly positively correlated with wet weight fractions of GAG and total collagen (Fig. 4(c), (d)). Consistently, hydraulic permeability was negatively correlated with wet weight fractions of GAG and total collagen (Fig. 4(e), (f)).

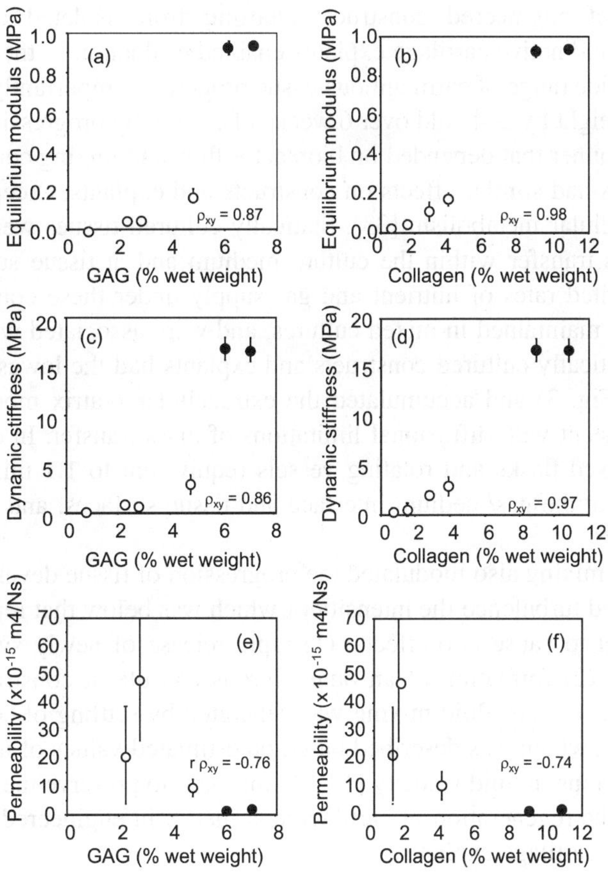

Fig. 4. Structure–function correlations. Equilibrium modulus vs. (a) GAG (% wet weight) and (b) total collagen (% wet weight); dynamic stiffness (30% strain, 1 Hz) vs. (c) GAG (% wet weight) and (d) total collagen (% wet weight); hydraulic permeability (30% strain) vs. (e) GAG (% wet weight) and (f) total collagen (% wet weight). Data represents the Average ± SD for $n = 3$–6 samples per point. Data for engineered cartilage are indicated with open symbols; data for native cartilage are indicated with closed symbols. Correlation coefficients are denoted by ρ_{xy}.

4. Discussion

We report some of the *in vitro* factors relevant for tissue engineering of functional cartilaginous constructs that can be used for controlled studies of chondrogenesis and potentially for *in vivo* joint repair. Ideally, engineered constructs should resemble native articular cartilage with respect to architectural features, biochemical composition and functional properties. Because the primary function of cartilage *in vivo* is mechanical, the mechanical properties of engineered cartilage are key for the success of any tissue engineering approach. It has been shown that the mechanical properties of cartilage, both native and engineered, are largely determined by the composition of extracellular matrix [24,31], and that the composition and mechanical properties of engineered cartilage can be modulated by the conditions and duration of bioreactor cultivation [5,30,39]. In the present work, engineered and native cartilage were studied to assess the effects of hydrodynamic conditions on *in vitro* chondrogenesis, identify correlations between the tissue composition and mechanical properties, and relate progression of chondrogenesis to the conditions of bioreactor cultivation.

Comparative studies of engineered constructs (starting from isolated chondrocytes seeded on biodegradable scaffolds) and native cartilage explants enabled evaluation of the effects of hydrodynamic factors over a relatively wide range of cartilaginous tissue properties. Importantly, both constructs and explants increased in wet weight by 2–4-fold over 6 weeks of culture by progressive accumulation of GAG and total collagen, in a manner that depended on bioreactor flow and mixing conditions (Table 1, Fig. 3). Hydrodynamic conditions had similar effects on constructs and explants, except that the constructs exhibited higher rates of cellular metabolism [33]. Statically cultured tissues were not exposed to hydrodynamic shear, and mass transfer within the culture medium and at tissue surfaces was governed by molecular diffusion. Limited rates of nutrient and gas supply under these conditions reduced p_{O_2} and pH as compared to levels maintained in mixed cultures, and were associated with predominantly anaerobic cell metabolism. Statically cultured constructs and explants had the lowest wet weights, GAG and total collagen contents (Fig. 3) and accumulated the extracellular matrix mostly at the tissue periphery (Fig. 2(a), (d)), consistent with diffusional limitations of mass transfer. In contrast, efficient mixing of culture medium in mixed flasks and rotating vessels (equivalent to 1.1 mixed tanks in series [20]) promoted mass transport at the gas/medium interface and tissue surfaces, and resulted in more aerobic metabolism.

The nature of flow and mixing also modulated the progression of tissue development. In mixed flasks, magnetic stirring generated turbulence the intensity of which was below that reported to cause cell damage [11,37], but sufficient to cause two effects: (1) rapid release of newly synthesized GAG into tissue culture medium, and (2) formation of an outer fibrous capsule at construct and explant surfaces (Fig. 2(b), (d)). In rotating vessels, fluid mixing was generated by settling of constructs and explants in the "tumble-slide" regime, which was described based on estimated values of inertia and drag ([32], Table 1). Associated fluctuations in fluid velocity [12,18] appeared to preserve the differentiated phenotype of chondrocytes, permit the regeneration of cartilaginous matrix in engineered constructs and maintain the matrix composition in cartilage explants.

The absence of physiologically relevant mechanical loading reportedly caused cartilage matrix degradation, *in vitro* [6] and *in vivo* [3]. The stimulation of *in vitro* development of engineered tissues by hydrodynamic forces in rotating vessels, as compared to static culture [39] is analogous to the enhancement of GAG and collagen synthesis by dynamic compression of agarose-immobilized chondrocytes, as compared to static loading [6], and by intermittent hydrostatic pressure applied to cell–polymer constructs, as compared to non-pressurized controls [9]. Steady hydrodynamic shear enhanced GAG synthesis in chondrocyte monolayers [36], while intermittent motion of medium in roller bottles stimulated chondrocytes to form cartilaginous nodules [28].

The utilization of dynamic laminar flow during *in vitro* cultivation resulted in constructs consisting of a continuous cartilaginous matrix with GAG fractions ranging from $\sim$5% wet weight after 6 weeks of culture [21,39] to 8.8% wet weight after 7 months of culture [19] and high and stable fractions of type II collagen (75–90% of the total collagen of the construct [21]). Under the same conditions, cartilage explants increased their wet weights by approximately 4 times while maintaining physiological wet weight fractions of GAG and collagen (Fig. 3). In both groups, static cultures supported tissue growth at the outer periphery and not in the inner tissue phase, turbulent mixing caused the formation of an outer fibrous capsule, and only dynamic laminar flow supported the development of uniformly cartilaginous constructs and explants (Fig. 2). It may be that dynamic changes in hydrodynamic forces acting on freely settling tissues in rotating flow resemble some aspects of dynamic loading of cartilage *in vivo*. However, hydrodynamic forces are different in nature and several orders of magnitude lower than compressive forces in articular joints [2,18,22,36,37]. Studies of cartilaginous cells and tissues under physiological levels of dynamic loadings [7,8,34] would allow a more direct comparison of tissue responses to forces acting *in vitro* and *in vivo*.

Mechanical properties of engineered and freshly explanted cartilage correlated with wet weight fractions of GAG and total collagen (Fig. 4). As compared to chondrocytes cultured for 5 weeks in agarose gels [5], constructs cultured for 6 weeks in rotating vessels had 3 times higher GAG fractions and 2 times higher equilibrium moduli and dynamic stiffnesses. As compared to native cartilage [30,39], the 6-week constructs had subnormal compositions and mechanical properties. Confined-compression equilibrium moduli and dynamic stiffness correlated strongly with the wet weight fractions of GAG and total collagen (Fig. 4(a)–(d)), presumably due to the functional assembly of cartilaginous tissue matrix. The hydraulic permeability decreased as expected with an increase in either GAG or collagen content (Fig. 4(e), (f)). Correlations between biochemical compositions and mechanical properties detected in the present study were qualitatively similar to those previously reported for native cartilage immediately after explant [1], engineered constructs [26] and cultured explants [35]. Subnormal construct compositions and mechanical properties after 6 weeks of culture stress the need to utilize bioreactors that can provide additional physiologically relevant regulatory signals, and develop mathematical models that can describe structure–function relationships in terms of the underlying phenomena on the cellular and molecular levels.

In summary, this study demonstrated that the cell metabolism, morphology and composition of engineered and native cartilage all depend on hydrodynamic conditions present during *in vitro* cultivation. Rotating bioreactors provided efficient mass transfer essential for cell proliferation and synthesis of matrix components and dynamic laminar flow patterns that promoted cell differentiation, retention of newly synthesized macromolecules, and maintenance of cartilaginous tissue.

Acknowledgements

We would like to thank Sue Kangiser for her help with the manuscript preparation. This work was supported by NASA grant NCC8-174.

References

[1] C.G. Armstrong and V.C. Mow, Variations in the intrinsic mechanical properties of human articular cartilage with age, degeneration, and water content, *J. Bone Joint Surg.* **44A** (1982), 88–94.
[2] F. Berthiaume and J. Frangos, Effects of flow on anchorage-dependent mammalian cells-secreted products, in: *Physical Forces and The Mammalian Cell*, J. Frangos, ed., Academic Press, San Diego, 1993, pp. 139.
[3] J.A. Buckwalter and H.J. Mankin, Articular cartilage, part II: degeneration and osteoarthrosis, repair, regeneration, and transplantation, *J. Bone Joint Surg.* **79A** (1997), 612–632.
[4] J.A. Buckwalter and H.J. Mankin, Articular cartilage repair and transplantation, *Arthritis Rheum.* **41** (1998), 1331–1342.
[5] M.D. Buschmann et al., Chondrocytes in agarose culture synthesize a mechanically functional extracellular matrix, *J. Orthop. Res.* **10** (1992), 745–752.
[6] M.D. Buschmann et al., Mechanical compression modulates matrix biosynthesis in chondrocyte/agarose culture, *J. Cell Sci.* **108** (1995), 1497–1508.
[7] S.E. Carver and C.A. Heath, Influence of intermittent pressure, fluid flow, and mixing on the regenerative properties of articular chondrocytes, *Biotechnol. Bioeng.* **65** (1999), 274–281.
[8] S.E. Carver and C.A. Heath, Increasing extracellular matrix production in regenerating cartilage with intermittent physiological pressure, *Biotechnol. Bioeng.* **62** (1999), 166–174.
[9] S.E. Carver and C.A. Heath, Semi-continuous perfusion system for delivering intermittent physiological pressure to regenerating cartilage, *Tiss. Eng.* **5** (1999), 1–11.
[10] F.S. Chen et al., Repair of articular cartilage defects: Part II. Treatment options, *Am. J. Orthop.* **28** (1999), 88–96.

[11] R.S. Cherry and T. Papoutsakis, Physical mechanisms of cell damage in microcarrier cell culture bioreactors, *Biotechnol. Bioeng.* **32** (1988), 1001–1014.
[12] R. Clift et al., *Bubbles, Drops, and Particles*, Academic Press, New York, 1978.
[13] R.W. Farndale et al., Improved quantitation and discrimination of sulphated glycosaminoglycans by the use of dimethylmethylene blue, *Biochim. Biophys. Acta* **883** (1986), 173–177.
[14] E.H. Frank and A.J. Grodzinsky, Cartilage electromechanics I. Electrokinetic transduction and the effects of electrolyte pH and ionic strength, *J. Biomech.* **20** (1987), 615–627.
[15] E.H. Frank and A.J. Grodzinsky, Cartilage electromechanics II. A continuum model of cartilage electrokinetics and correlation with experiments, *J. Biomech.* **20** (1987), 629–639.
[16] L.E. Freed et al., Biodegradable polymer scaffolds for tissue engineering, *Bio/Technology* **12** (1994), 689–693.
[17] L.E. Freed and G. Vunjak-Novakovic, Cultivation of cell–polymer constructs in simulated microgravity, *Biotechnol. Bioeng.* **46** (1995), 306–313.
[18] L.E. Freed and G. Vunjak-Novakovic, Tissue engineering of cartilage, in: *Biomedical Engineering Handbook*, J.D. Bronzino, ed., CRC Press, Boca Raton, 1995, pp. 1788–1807.
[19] L.E. Freed et al., Tissue engineering of cartilage in space, *Proc. Natl. Acad. Sci. USA* **94** (1997), 13 885–13 890.
[20] L.E. Freed and G. Vunjak-Novakovic, Tissue culture bioreactors: chondrogenesis as a model system, in: *Principles of Tissue Engineering*, R. Langer and W. Chick, eds, R.G. Landes, Austin, 1997, pp. 150–158.
[21] L.E. Freed et al., Chondrogenesis in a cell–polymer-bioreactor system, *Exp. Cell Res.* **240** (1998), 58–65.
[22] L.E. Freed et al., Frontiers in tissue engineering: in vitro modulation of chondrogenesis, *Clin. Orthop.* **367S** (1999), S46–S58.
[23] L.E. Freed and G. Vunjak-Novakovic, Tissue engineering bioreactors, in: *Principles of Tissue Engineering*, 2nd edn., R.P. Lanza et al., eds, Academic Press, San Diego, 2000, pp. 143–156.
[24] L.E. Freed and G. Vunjak-Novakovic, Tissue engineering of cartilage, in: *The Biomedical Engineering Handbook*, Vol. II, 2nd edn., J.D. Bronzino, ed., CRC Press, Boca Raton, 2000, pp. 124-121–124-126.
[25] L.E. Freed and G. Vunjak-Novakovic, Culture environments: Cell–polymer-bioreactor systems, in: *Methods of Tissue Engineering*, A. Atala and R.P. Lanza, eds, Academic Press, San Diego, 2002, pp. 97–111.
[26] K.J. Gooch et al., IGF-I and mechanical environment interact to modulate engineered cartilage development, *Biochem. Bioph. Res. Co.* **286** (2001), 909–915.
[27] A.P. Hollander et al., Increased damage to type II collagen in osteoarthritic articular cartilage detected by a new immunoassay, *J. Clin. Invest.* **93** (1994), 1722–1732.
[28] K.E. Kuettner et al., Synthesis of cartilage matrix by mammalian chondrocytes in vitro. II. Maintenance of collagen and proteoglycan phenotype, *J. Cell Biol.* **93** (1982), 751–757.
[29] I. Martin et al., A method for quantitative analysis of glycosaminoglycan distribution in cultured natural and engineered cartilage, *Ann. Biomed. Eng.* **27** (1999), 656–662.
[30] I. Martin et al., Modulation of the mechanical properties of tissue engineered cartilage, *Biorheology* **37** (2000), 141–147.
[31] V.C. Mow et al., Cartilage and diarthrodial joints as paradigms for hierarchical materials and structures, *Biomaterials* **13** (1992), 67–97.
[32] G.P. Neitzel et al., Cell function and tissue growth in bioreactors: fluid mechanical and chemical environments, *J. Japan. Soc. Microgr. Appl.* **15** (1998), 602–607.
[33] B. Obradovic et al., Bioreactor studies of natural and tissue engineered cartilage, *Ortopedia Traumatologia Rehabilitacja* **3** (2001), 181–189.
[34] J.J. Parkkinen et al., Effects of cyclic hydrostatic pressure on proteoglycan synthesis in cultured chondrocytes and articular cartilage explants, *Arch. Biochem. Biophys.* **300** (1993), 458–465.
[35] R.L. Sah et al., Differential effects of serum, insulin-like growth factor-I, and fibroblast growth factor-2 on the maintenance of cartilage physical properties during long-term culture, *J. Orthop. Res.* **14** (1996), 44–52.
[36] R.L. Smith et al., Effect of fluid-induced shear on articular chondrocyte morphology and metabolism in vitro, *J. Orthop. Res.* **13** (1995), 824–831.
[37] G. Vunjak-Novakovic et al., Effects of mixing on the composition and morphology of tissue-engineered cartilage, *AIChE J.* **42** (1996), 850–860.
[38] G. Vunjak-Novakovic et al., Dynamic cell seeding of polymer scaffolds for cartilage tissue engineering, *Biotechnol. Prog.* **14** (1998), 193–202.
[39] G. Vunjak-Novakovic et al., Bioreactor cultivation conditions modulate the composition and mechanical properties of tissue engineered cartilage, *J. Orthop. Res.* **17** (1999), 130–138.
[40] J.F. Woessner, The determination of hydroxyproline in tissue and protein samples containing small proportions of this imino acid, *Arch. Biochem. Biophys.* **93** (1961), 440–447.

Biorheology 39 (2002) 269–276
IOS Press

The biochemical content of articular cartilage: An original MRI approach

Damien Loeuille [a,b,1], Pierre Olivier [a,1], Astrid Watrin [a], Laurent Grossin [a], Patrick Gonord [c], Geneviève Guillot [c], Stéphanie Etienne [a], Alain Blum [a], Patrick Netter [a,*] and Pierre Gillet [a]
[a] *UMR 7561 CNRS - Université Nancy I, France*
[b] *Clinique Rhumatologique CHU Nancy, France*
[c] *U2R2M, ESA CNRS 8081, Orsay, Paris, France*

Abstract. The MR aspect of articular cartilage, that reflects the interactions between protons and macromolecular constituents, is affected by the intrinsic tissue structure (water content, the content of matrix constituents, collagen network organization), imager characteristics, and acquisition parameters. On the T1-weighted sequences, the bovine articular cartilage appears as an homogeneous tissue in high signal intensity, whatever the age of animals considered, whereas on the T2-weighted sequences, the articular bovine cartilage presents variations of its imaging pattern (laminar appearance) well correlated to the variations of its histological and biochemical structure. The T2 relaxation time measurement (T2 mapping), which reflects quantitatively the signal intensity variations observed on T2 weighted sequences, is a way to evaluate more precisely the modifications of cartilage structure during the aging and maturation processes (rat's study). This technique so far confined to experimental micro-imagers is now developed on clinical imagers. Consequently, it may permit to depict the early stages of osteoarthritic disease (OA) or to evaluate the chondroprotective effect of drugs.

Keywords: Cartilage, MRI, aging, extracellular matrix, biochemistry, histological correlations

1. Introduction

Osteoarthritic disease (OA) is a common disease, affecting the elderly population. Its therapy was so far mostly symptomatic, aimed at reducing the inflammatory process and the joint pains. Diagnosis of OA is based on clinics and confirmed by the X-ray radiographs. Cartilage alterations are indirectly appreciated by X-rays radiographs that visualize joint space narrowing and modifications of underlying bone (osteophytes, cysts, subchondral bone) [7]. Compared to X-ray, arthroscopy and MRI, which offer a direct approach of the cartilage tissue, permit to characterize chondral lesions in terms of localization, surface and severity [9,10,19]. However, recent works suggest that OA therapy would be more efficient if applied earlier in the evolution of the disease, i.e., before morphological lesions appear. Such an approach permits to characterize precisely the early changes of cartilage tissue, and more specifically the disorganization of the collagen network and the loss of PGs content in the upper part of the cartilage. The signal that supports MR images, is related to water protons mobility which depends on the biochemical content and organization of the tissue studied. The articular cartilage is composed of a single cell type, the chondrocytes, dispersed throughout a large volume of extracellular matrix composed of water, type II collagen, proteoglycan molecules, and hyaluronic acid. This tissue is characterized by age-related

[1] D.L. and P.O. made equal contributions to this work.
*Address for correspondence: Laboratoire de Pharmacologie, UMR 7561, CNRS-Université Henri Poincaré, Nancy I, "Physiopathologie et Pharmacologie Articulaires", Faculté de Médecine, BP 184, Avenue de la Forêt de Haye, F 54505 Vandoeuvre Les Nancy, France. Tel.: +33 3 83 59 26 22; Fax: +33 3 83 59 26 21; E-mail: Patrick.Netter@medecine.uhp.nancy.fr.

modifications such as variations in cell morphology and density, variations in thickness, and variations in biochemical composition and physical properties of the extracellular matrix [6,29,30]. Thus, several authors have evaluated the potentialities of MR to evaluate physiological and/or pathological changes of the cartilage structure [12,16–18,20–22,25,27,28,32,34].

2. Basics and limitations of hydrogen imaging

MRI is a non-invasive imaging technique based on the NMR process of absorption and emission of energy by proton. The protons that initially precess along an external magnetic field B_0 are submitted to a brief radio frequency (RF) pulse. This RF pulse will give to these protons an higher energy state that will make them precess around an axis different from B_0. After the RF pulse, the protons progressively recover their steady state, the restituted energy being collected by a reception coil to constitute the MR signal. This MR signal depends on imaging parameters (TR-TE) but also on the intrinsic characteristics of the tissue. Three different types of sequences are defined according to these MR parameters: T1-weighted sequences (short TE and TR values), T2-weighted sequences (long TE and TR values), and proton density sequences (short TE and long TR) [18].

Several mechanisms have been identified as potentially able to lead to the artefactual aspects of cartilage not related to any anatomic substratum. The magic angle phenomenon observed in anisotropic tissues may lead to the variations of the MR signal inside the cartilage secondary to variations of the collagen network orientation according to B_0 [4,33]. The truncature artefact may be observed in articular cartilage due to the abrupt transition of the MR signal between this tissue and the adjacent structures (synovial fluid and subchondral bone) [11,13]. This artefact, which is affected by the parameters supporting the spatial resolution (FOV, matrix size), is more frequently observed in the T1-weighted sequences. The chemical shift artefact is mainly observed in the cartilage/subchondral bone interface and is secondary to the differences in proton mobility between water and fat. This artefact is substantially reduced by the MR sequences in which fat signal was suppressed. Finally, the last artefact that must be considered is the partial volume effect that could lead to an average of the MR signal between the cartilage and the adjacent structures. Thus, all these potential artefacts should be taken into account before considering any correlation between MR imaging and the cartilage structure.

In human, the correlations between MRI and structural characteristics are impossible except the studies on cadeveric specimen usually restricted to severe OA lesions. Conversely, there are no limitations for *ex vivo* animal studies, with especially, the possibility to consider animals issued from different age-related groups. Moreover, such *ex vivo* studies permit to perform optimal MRI procedures with an optimal spatial resolution, without limitation regarding the duration of acquisitions, or the possibility of movement.

3. MR aspect of bovine cartilage: Correlations with biochemistry and histology

The bovine patellar cartilage has a thickness similar to the one of human patellar cartilage (approximately 3 to 4 mm) [24]. The patella specimen issued from three-month, three-year and thirteen-year-old animals were studied. The acquisitions were performed with a 1.5 T clinical scan and a 3 inch surface coil with a resultant spatial resolution of $375 \times 375\ \mu m^2$. Giving this resolution, the total cartilage thickness is displayed by 8 to 10 pixels, which permits an accurate study despite the partial volume effect.

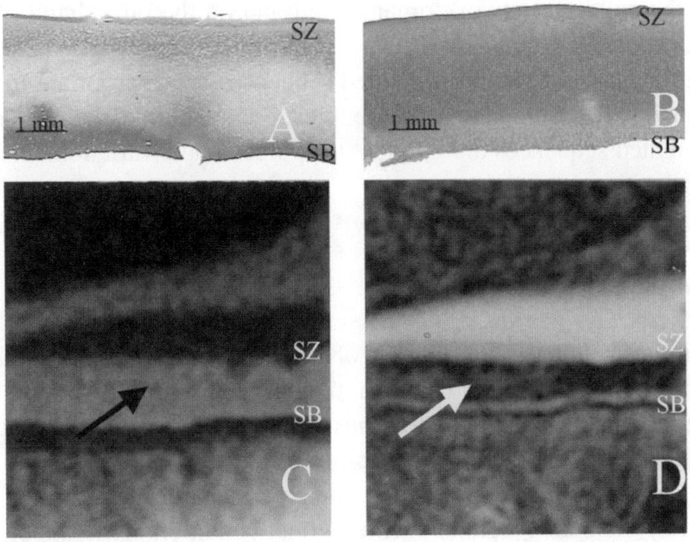

Fig. 1. The histological aspects and the MR pattern of the calf patellar cartilage. On histological sections, the cartilage is characterized by the variations of proteoglycan (PGs) (A) and collagen (B) contents through its depth (SZ: superficial zone; SB: subchondral bone). On T1-weighted sequences (C) the cartilage (black arrow) appears homogeneous, in high signal intensity, whereas cartilage presents a variation of its internal structure (laminar appearance, white arrow) on the T2-weighted and proton sequences (D).

The patella were always placed in the same position regarding the magnetic field B_0, such a precaution was intended to limite the impact of the magic angle phenomenon. In order to match MR imaging with histological and biochemical studies accurately, cylindrical calibrated lesions (2 mm diameter) were performed perpendicularly to the cartilage surface.

In the T1-weighted sequences, the articular cartilage, whose thickness decreases with age, is depicted as an homogeneous structure in high signal intensity, clearly delineated from adjacent structures in low signal intensity (subchondral bone and synovial fluid) (Fig. 1). These T1-weighted sequences offer the possibility to perform fast 3-D acquisitions characterized by a good spatial resolution and a high signal-to-noise ratio. The MR signal profiles performed on cartilage depth are globally flat whereas histological and biochemical data demonstrated variations with the cartilage depth whatever the age of the animals. The cartilage may present a false laminar appearance related to truncature artefact. In such a situation, the cartilage is striped with thin bands in low signal intensity oriented parallel to the articular surface, bands the number of which varies with the parameters that affect the spatial resolution.

Conversely on T2-weighted sequences, the cartilage presents a variation of its internal structure. This heterogeneity is especially marked in young specimens with a plurilaminar aspect. There is a parallel evolution between variations of MR signal and, on one hand, the collagen network organization, and on the other hand, the variations with depth of the matrix constituents. In the three-year-old animal, the articular cartilage becomes more homogeneous and is depicted in an intermediate signal intensity with a thick deep region in low signal intensity. The histological and biochemical data demonstrate a mature organization with less marked variations of matrix constituents.

On the proton density sequences, the cartilage pattern is globally similar to the one observed on the T2-weighted sequences but its heterogeneous pattern is less pronounced and the deep zone more clearly visualized than with the T2-weighted sequences. There are similarities between the variations of MR signal: on one hand, the collagen network organization and on the other hand, the variations with the depth of the matrix constituents.

Thus, in bovine cartilage, while the T1-weighted sequences are appropriate to characterize morphological abnormalities and to determine the cartilage thickness or volume, the T2-weighted sequences and the proton density-weighted sequences are efficient to appreciate the variations in matrix content and the collagen network organization.

4. T2 mapping in rat patellar cartilage

T2 mapping permits a quantitative evaluation of the tissue structure in which each voxel is characterized by a T2 relaxation time value. This quantitative method offers a more robust approach than T2-weighted imaging in which MR signal variations are highly dependent on study conditions (type of scanner and coil, acquisition parameters, signal-to-noise ratio). T2 mapping was developed on experimental high field imagers with specific acquisition and processing characteristics [35]. Such imagers are well adapted to the study of small specimens with optimal conditions in terms of spatial resolution and contrast. We applied this technique to the rat patella, this animal being a model used on one hand in rheumatic experimental osteo-articular diseases and on the other hand to evaluate the action of drugs. The MRI acquisitions of *ex vivo* specimens, were performed on a 8.5 T microimager in the axial plan with a 4×4 mm field of view and a 128×128 matrix. For each patella studied, eight spin echo sequences with constant TR and variable TE (5 to 30 ms) were performed on its middle part (1 mm slice thickness). These eight different sequences were aimed at calculating the T2 relaxation time values for each voxel (Fig. 2).

A total of 24 rat patellae issued from 4 age groups (4 weeks to more than 6 months) were investigated for T2 mapping, histological data and biochemical content. We demonstrated a progressive and significant decrease of T2 global values of cartilage with aging from 12.10 ± 0.13 ms in four-week-old rats to 7.95 ± 0.65 ms in old rats. During maturation and aging processes, we noted a decrease of PGs content concomitant to an increase of collagen content. The collagen content ($r = -0.76$; $p < 0.01$) may rely

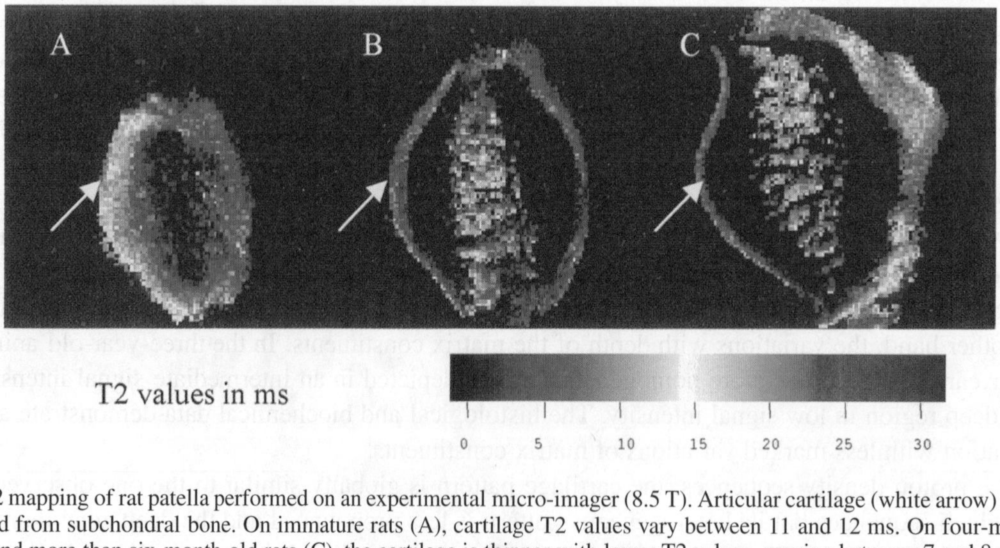

Fig. 2. T2 mapping of rat patella performed on an experimental micro-imager (8.5 T). Articular cartilage (white arrow) is clearly delineated from subchondral bone. On immature rats (A), cartilage T2 values vary between 11 and 12 ms. On four-month-old rats (B) and more than six-month-old rats (C), the cartilage is thinner with lower T2 values, varying between 7 and 8 ms.

more on the MR signal than the PG content ($r = 0.63$; $p < 0.01$). Moreover, whatever the age, a zonal variation was present, with a high T2 value in the superficial layer and a lower T2 value in the deepest layer. The variations in the collagen network organization we observed on polarized light may account for these zonal variations.

5. T2 mapping in clinical practice

Such an ability of T2 mapping to depict the variations of the cartilage internal structure suggests it could be able to detect the early stages of OA diseases before the occurrence of macroscopical abnormalities. Until a recent past, T2 mapping was confined to experimental studies on small size specimens, performed on dedicated high field imagers, with specific characteristics in terms of acquisition sequences and processing [14,15]. The T2 mapping technique is now available in clinical practice to be performed in association with classical osteo-articular MR sequences. Recently, Dardzinski et al. reported a monotonic increase in T2 values in normal human patellar cartilage, from the deep zone to the surface on a 3.0 T imager. They also noted an increase of T2 values in the upper part of fibrillar cartilage [8]. In human patellar cartilage, Mosher et al. [23] reported some variations of zonal T2 values with aging, affecting the transitional zone. Using a 1.5 T imager, we observed a relatively flat profile of T2 values from the deepest zone to the transitional zone with values ranging from 20 to 30 ms and then a slight increase of these values in the superficial zone (40 ms) (Fig. 3). In OA patellar cartilage, the T2 profile was also flat as far as the transitional zone (60 ms), but was marked by an abrupt increase from this transitional zone to the superficial zone (100 ms) where fibrillar lesions were arthroscopically observed. If these results were confirmed, this T2 mapping technique would constitute an interesting approach to differentiate the changes related to the aging process from the early stages of OA disease. The main limitation of T2 mapping in the clinics is the duration of the specific acquisition required (e.g., about 30 min for a T2 mapping of the human patellar cartilage) coming in addition to the duration of usual sequences. Another limitation is the fact that the T2 mapping developed in clinics is less accurate than the T2 mapping performed on experimental scans.

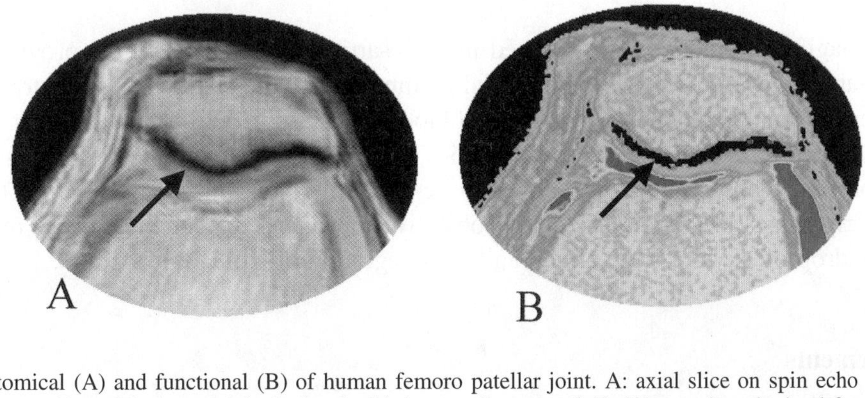

Fig. 3. MR anatomical (A) and functional (B) of human femoro patellar joint. A: axial slice on spin echo T2-weighted sequences: articular cartilage (black arrow) is depicted with intermediate signal. B: T2 mapping obtained from four spin echo acquisitions with a constant TR value = 3500 ms, and variable TE values (15, 30, 45, 60 ms). T2 relaxation time value of each voxel is coded with a color scale. T2 relaxation time values of normal articular cartilage (black arrow) vary from 20 to 40 ms.

6. Biochemical MR approach: Other techniques

Other MR applications are developed in order to characterize the biochemical content of cartilage. Variations in diffusion of ionic contrast agents by the T1 mapping technique reflect the variations of proteoglycans content in cartilage ($R = 0.90$). The loss of proteoglycans can lead to a loss of the fixed negative charge density which can be mapped either positively or negatively according to contrast agents used [2,3]. These contrast agents are injected either in the joint or intravenously. When injected in the joint, the contrast agent will diffuse in the articular cartilage from its superficial to its deepest zones. This technique is invasive and its use is restricted to some countries.

When injected intravenously, there are two ways for the contrast agent to reach the cartilage [31]. The first way is a diffusion from the synovial fluid after a passive transfer through the synovial membrane. The other way is a diffusion from the highly vascularized subchondral bone. The reproducibility of this approach is limited since the intracartilagenous concentration of the agent will vary with local conditions of inflammation that affect the diffusion of the agent through the subchondral bone and the synovial membrane. Another limitation of these techniques is related to the difficulties for matching images performed before injection with those performed several hours after injection. Finally, the use of such techniques in clinics is limited by the fact they are time consuming.

MR magnetization transfer techniques [1] permit to evaluate the interactions between free protons and macromolecules. For cartilage, this magnetization transfer is mainly affected by the collagen network integrity and to a least degree by the PGs molecules. Thus, this method showed great promise to detect early cartilage abnormalities but it was so far confined to experimental studies or to evaluate cartilage volume in the knee [26].

Finally, Borthakur et al. [5] demonstrated that besides hydrogen imaging, sodium imaging was another promising technique that permits to accurately detect PGs loss through concomitant changes in sodium signal ($R^2 = 0.85$; $p < 0.01$). PGs play an important role in determining the tissue integrity and their negatively charged molecules will interact with positive charged sodium ions present in the synovial fluid. The main limitation of this technique is represented by the long duration of acquisitions required to compensate the low signal-to-noise ratio.

7. Conclusion

MR is an imaging technique routinely used in osteo-articular diseases, able to provide high resolution multiplanar images of the bone, the synovial membrane and the abarticular structures (tendons and menisci). These recent developments regarding a biochemical approach in MRI suggest that this technique, so far limited to morphological abnormalities, could be extended to the cartilage structure itself. This new approach, whose validation remains to be performed in clinics, could permit to better characterize the early stages of OA disease and to objectively evaluate the action of drugs on the cartilage structure (chondroprotective or toxic effect) and the integration of biomaterials grafts.

Acknowledgements

This work was granted by Projet Hospitalier de Recherche Clinique 1998, the “Pole Européen de Santé”, and GDR CNRS 2237. Thanks are due to Mrs Cabaret for English speaking.

References

[1] R.S. Balaban and T.L. Cekler, Magnetization contrast in magnetic resonance imaging, *Magn. Reson. Q* **8** (1992), 116–137.

[2] A. Bashir, M.L. Gray and D. Burstein, Gd-DTPA2- as a measure of cartilage degradation, *Magn. Reson. Med.* **36** (1996), 665–673.

[3] A. Bashir, M.L. Gray, J. Hartke and D. Burstein, Nondestructive imaging of human cartilage glycomaminoglycan concentration by MRI, *Magn. Reson. Med.* **41** (1999), 857–865.

[4] A. Benninghoff, Form und bau der gelenkknorpfel in ihren beziehungen zur funktion. I. Die modellierenden und formerhaltenden faktorendes knorpelreliefs, *Z. Gesamte Anat.* **76** (1925), 43–63.

[5] A. Borthakur, E.M. Shapiro, J. Beers, S. Kudchodkar, J.B. Kneeland and R. Reddy, Sensitivity of MRI to proteoglycan depletion in cartilage: comparison of sodium and proton MRI, *Osteoarthritis Cartilage* **8** (2000), 288–293.

[6] J.S.A. Buckwalter, L.L. Rosenberg and E.B. Hunzicker, Articular cartilage composition, structure, response to injury, and methods of facilitating repair, in: *Articular Cartilage and Knee Joint Function: Basis Science and Arthroscopy*, J.W. Ewing, ed., Raven, New York, 1990, pp. 19–56.

[7] W.P. Chan, P. Lang, M.P. Stevens, K. Sack, S. Majumdar, D.W. Stoller et al., Osteoarthritis of the knee, comparison of radiography, CT and MR imaging to assess extent and severity, *Am. J. Roentgenol.* **157** (1991), 799–806.

[8] B.J. Dardzinski, T.J. Mosher, S. Li, M.A. Van Slyke and M.B. Smith, Spatial variation of T2 in human articular cartilage, *Radiology* **205** (1997), 546–550.

[9] M. Dougados, X. Ayral, V. Listrat, A. Gueguen, J. Bahuaud, P. Beaufils, J.A. Beguin, J.P. Bonvarlet, T. Boyer, H. Coudane et al., The SFA systemfor assessing articular cartilage lesions at arthroscopy of the knee, *Arthroscopy* **10** (1994), 69–77.

[10] J.L. Drape, E. Pessis, G.R. Auteley, A. Chevrot, M. Dougados and X. Ayral, Quantitative MR imaging evaluation of chondropathy in osteoarthritic knee, *Radiology* **6** (1998), 160–166.

[11] S.J. Erickson, J.G. Waldschmidt, L.F. Czervionke and R.W. Prost, Hyaline cartilage: truncation artefact as a cause of trilaminar appearance with fat-suppressed three dimensional spoiled gradient-recalled sequences, *Radiology* **201** (1996), 260–264.

[12] E. Fragonas, V. Mlynarik, V. Jellus, F. Micali, A. Piras, R. Toffanin et al., Correlation between biochemical composition and magnetic resonance appearance of articular cartilage, *Osteoarthritis Cartilage* **6** (1998), 24–32.

[13] L.R. Frank, J. Brossmann, R.B. Buxton and D. Resnick, MR imaging truncation artefacts can create a false laminar appearance in cartilage, *AJR (Am. J. Roentgenol.)* **168** (1997), 547–554.

[14] L.R. Frank, E.C. Wong, W.M. Luh, J.M. Ahn and D. Resnick, Articular cartilage in the knee: mapping of the physiologic parameters at MR imaging with a local gradient coil-preliminary results, *Radiology* **210** (1999), 241–246.

[15] D.W. Goodwin, Y.Z. Wadghiri and J.F. Dunn, Microimaging of articular cartilage: T2 proton density, and the magic angle effect, *Acad. Radiol.* **5** (1998), 790–798.

[16] D.J. Kim, J.S. Suh, E.K. Jeong, K.H. Shin and W.I. Yang, Correlation of laminated MR appearance of articular cartilage with histology, ascertained by artificial landmarks on the cartilage, *J. Magn. Reson. Imaging* **10** (1999), 57–64.

[17] K.B. Lehner, H.P. Rechl, J.K. Gmeinwieser, A.F. Heuck, H.P. Lukas and H.P. Kohl, Structure, function, and degeneration of bovine hyaline cartilage: assessment with MR imaging in vitro, *Radiology* **170** (1989), 495–499.

[18] D. Loeuille, P. Olivier, D. Mainard, P. Gillet, P. Netter and A. Blum, Magnetic resonance imaging of normal and osteoarthritic cartilage, *Arthritis Rheum.* **41** (1998), 963–975.

[19] D. Loeuille, D. Perard, P. Pere, D. Mainard, P. Gillet, P. Netter, A. Blum and A. Gaucher, MRI quantitative score of femoro-patellar and femoro-tibial artcicular cartilage, *Arthritis and Rheumatism* **41** (1998), S144 (abstract).

[20] V. Mlynarik, A. Degrassi, R. Toffanin, F. Vittur, M. Cova and R.S. Pozzi-Mucelli, Investigation of laminar appearance of articular cartilage by means of magnetic resonance microscopy, *Magn. Reson. Imaging.* **14** (1996), 435–442.

[21] V. Mlynarik, S. Trattnig, M. Huber, A. Zembsch and H. Imhof, The role of relaxation times in monitoring proteoglycan depletion in articular cartilage, *J. Magn. Reson. Imaging* **10** (1999), 497–502.

[22] J.M. Modl, L.A. Sether, V.M. Haughton and J.B. Kneeland, Articular cartilage: correlation of histologic zones with signal intensity at MR imaging, *Radiology* **181** (1999), 853–855.

[23] T.J. Mosher, B.J. Dardzinski and M.B. Smith, Human articular cartilage: influence of aging and early symptomatic degeneration on the spatial variation of T2 – preliminary findings at 3 T, *Radiology* **214** (2000), 259–266.

[24] P. Olivier, D. Loeuille, A. Watrin, F. Walter, S. Etienne, Netter, P. Gillet and A. Blum, Structural evaluation of articular cartilage: potential contribution of MR techniques used in clinical practice, *Arthritis and Rheumatism* **44** (2001), 2285.

[25] P.K. Paul, E. O'Byrne, V. Blancuzzi, D. Wilson, D. Gunson, F.L. Douglas, J.Z. Wang and R.S. Mezrich, Magnetic resonance imaging reflects cartilage proteoglycan degradation in the rabbit knee, *Skeletal Radiol.* **20** (1991), 31–36.

[26] C.G. Peterfy, C.F. van Dijke, Y. Lu, A. Nguyen, T.J. Connick, J.B. Kneeland, P.F. Tirman et al., Quantification of the volume of articular cartilage in the metacarpophalangeal joints of the hand: accuracy and precision of three-dimensional MR imaging, *AJR (Am. J. Roentgenol.)* **165**(2) (1995), 371–375.

[27] J. Rubenstein, M. Recht, D.G. Disler, J. Kim and R.M. Henkelman, Laminar structures on MR images of articular cartilage, *Radiology* **20** (1997), 15–16.

[28] J.D. Rubenstein, J.K. Kim, I. Morova-Protzner, P.L. Stanchev and R.M. Henkelman, Effects of collagen orientation on MR imaging characteristics of bovine articular cartilage, *Radiology* **188** (1993), 219–226.

[29] R.A. Stockwell, Cell density, cell size, and cartilage thickness in adult mammalian articular cartilage, *J. Anat.* **108** (1971), 584–588.

[30] E.J. Thonar, J.A. Buckwalter and K.E. Kuettner, Maturation-related differences in the structure and composition of proteoglycans synthesized by chondrocytes from bovine articular cartilage, *J. Biol. Chem.* **261** (1986), 2467–2474.

[31] S. Trattnig, V. Mlynarik, M. Breitenseher, M. Huber, A. Zembsch, T. Rand et al., MRI visualization of proteoglycan depletion in articular cartilage via intravenous administration of Gd-DTPA, *Magn. Reson. Imaging* **17** (1999), 577–583.

[32] M. Uhl, C. Ihling, K.H. Allmann, J. Laubenberger, U. Tauer, C.P. Adler et al., Human articular cartilage: in vitro correlation of MRI and histologic findings, *Eur. Radiol.* **8** (1998), 1123–1129.

[33] F.K. Wacker, X. Bolze, D. Felsenberg and K.J. Wolf, Orientation-dependent changes in MR signal intensity of articular cartilage: a manifestation of the "magic angle" effect, *Skeletal Radiol.* **27** (1998), 306–310.

[34] J.G. Waldschmidt, R.J. Rilling, A.A. Kajdacsy-Balla, M.D. Boynton and S.J. Erickson, In vitro and in vivo MR imaging of hyaline cartilage: zonal anatomy, imaging pitfalls, and pathologic conditions, *Radiographics* **17** (1997), 1387–1402.

[35] A. Watrin, J.P. Ruaud, P. Olivier, N. Guingamp, P. Gonord, P. Netter et al., T2 map of rat patellar cartilage, *Radiology* **219** (2001), 395–402.

Biorheology 39 (2002) 277–285
IOS Press

In vitro study of intracellular IL-1β production and β1 integrins expression in stimulated chondrocytes – Effect of rhein

C. Gigant-Huselstein [a], D. Dumas [a], E. Payan [b], S. Muller [a], D. Bensoussan [a], P. Netter [b] and J.F. Stoltz [a,*]

[a] *Mécanique et Ingénierie Cellulaire et Tissulaire, LEMTA UMR INPL CNRS 7563 et IFR 111 Bioingénierie, Faculté de Médecine, 54505 Vandoeuvre-lès-Nancy, France*

[b] *Physiopathologie et Pharmacologie Articulaires – UMR CNRS 7561 et IFR 111 Bioingénierie, Faculté de médecine, 54504 Vandoeuvre-lès-Nancy, France*

Abstract. The purpose of the present study was to investigate the intracellular IL-1β production and β1 integrins (α4/β1 and α5/β1) expression on chondrocytes.

Chondroytes monolayer (human chondrosarcoma cell line HEM-C55) were incubated for 12, 24 and 48 hours in the presence of Tumor Necrosis Factor-α (TNF-α, Sigma, France) or recombinant human IL-1α (*rh*-IL1α, Becton Dickinson, France). After direct immunolabelling, cells were either analyzed on FACScan flow cytometer (Becton Dickinson, France), or observed under an epi-fluorescence inverted microscope equipped with the CellScan EPR™ optical scanning acquisition system (IPLab-Scanalytics, USA). We found that the IL-1β mean fluorescence intensity in flow cytometry and in 3D microscopy was increased in the presence of TNF-α or rh-IL-1α, and α4/β1 or α5/β1 expression was higher on stimulated cells than on control cells.

On the other hand, we have evaluated the *in vitro* effects of rhein (10^{-5} M, Negma, France), an active metabolite of diacerein, on the intracellular IL-1β and β1 integrins expressed by stimulated or no-stimulated chondrocytes. The results indicated that rhein leads to a reduction of IL-1β synthesis whereas a weak decrease of β1 integrins receptors expression is observed.

From this study, it seems that rhein partially reduce cytokine-induced intracellular IL-1β production, and it has a weak action on α4/β1 or α5/β1 receptors.

1. Introduction

Osteoarthritis (OA) is a degenerative disease of diarthrodial joints in which degradative and repair processes in articular cartilage, subchondral bone and synovium occur concurrently. The appearance of fibrillations, matrix depletion, cell clusters and changes in matrix composition reflect the aberrant behavior of resident chondrocytes. In OA, there is evidence supporting the involvement of proinflammatory cytokines such as interleukin 1 (IL-1) [1], and tumor necrosis factor-α (TNF-α) [2]. IL-1 was shown to be capable not only of inducing expression of matrix metalloproteases (MMPs) but also of suppressing synthesis of the major cartilage matrix molecules, collagen and aggrecan [3,4]. This cytokine exists in two forms IL-1α and IL-1β which interact with the same cellular receptors and exhibit analogous function effects. The TNF-α appeared to exert similar effects as IL-1 [5]. Biomechanical factors are, also,

*Address for correspondence: Jean Françis Stoltz, Mécanique et Ingénierie Cellulaire et Tissulaire, LEMTA UMR CNRS 7563, Faculté de Médecine - B.P. 184, 54505 Vandoeuvre-lès-Nancy Cedex, France. Tel.: +33 03.83.59.26.41; Fax.: +33 03.83.59.26.43; E-mail: stoltz@hemato.u-nancy.fr.

strongly implicated in OA. *In vivo* studies have shown that the mechanical environment can influence the structure and function of the cartilage [6,7].

The signal transduction mechanisms from extracellular matrix to the interior of the cells are poorly understood. Recent reports suggest that integrins act as mechanoreceptors in a variety of different cell types, including chondrocytes and bone cells [8–10]. Integrins receptors are heterodimers composed of two transmembrane polypeptide subunits (α and β) each having large extracellular domains that join together to form binding sites for specific extracellular matrix ligands and F-actin cytoskeleton [11]. It has been shown that normal human articular chondrocytes expressed integrin subunits $\alpha 1, \alpha 5, \alpha V, \beta 1, \beta 4$ and $\beta 5$ whereas in OA cartilage, a neo-expression of $\alpha 2, \alpha 4$ and $\beta 2$ was observed [12]. Several studies support the idea that integrins, $\alpha 5\beta 1$ integrin in particular, can transduce a mechanical signal into cellular response in both normal and OA [9,13]. In fact, Clancy et al. have recently reported that nitric oxide (NO) and NO-producing cytokines, including IL-1α, can exert profound effects on fibronectin $\alpha 5\beta 1$ signalling in bovine chondrocytes [14]. Moreover, Arner et al. shown that signal transduction through chondrocytes integrin receptors up-regulated MMPs expression and that this was likely mediated through induction of interleukin-1 (IL-1) [15].

Diacerhein is one of the proposed therapeutic agents to regulate the inflammatory process. It has shown anti-OA effects in experimental animal models [16,17] and is used in the treatment of OA patients, with beneficial improvements of clinical symptoms [18,19]. Diacerein has been shown, *in vitro*, to inhibit collagenase production by cultured rabbit articular chondrocytes and partially reverse IL-1-induced inhibition of proteoglycan synthesis [20,21]. The drug, and its active metabolite rhein, have been reported to inhibit superoxide release from human neutrophils and to reduce the phagocytic ability of mouse peritoneal macrophages [22,23]. These data, when taken together, suggest that rhein could interfere with the cytokines which play a role in the imbalance of cartilage homeostasis during OA but also with the β1 integrin receptors.

The present preliminary study was designed to quantify the receptors β1 integrin $\alpha 4/\beta 1$ (VLA-4 or CD49d/CD29) and $\alpha 5/\beta 1$ (VLA-5 or CD49e/CD29) and the intracellular IL-1β in stimulated chondrocytes by TNF-α and *rh*IL-1α. Then, we evaluated the *in vitro* effects of rhein on intracellular IL-1β and on β1 integrin VLA-4 and VLA-5.

2. Materials and methods

2.1. Culture conditions

A cell line of human chondrosarcoma (H-EM-CSS, ECACC, United Kingdom) were seeded at high density in tissue culture flasks (Fisher, France) and cultured in Dulbecco's Modified Eagle's Medium/Ham's F-12 (Life Technologies, France) containing 10% decomplemented fetal calf serum (FCS, Life Technologies), 2 mM L-glutamine (Life Technologies), 100 units/ml penicillin and 100 units/ml streptomycin at 37°C in 5% CO_2 and 95% air. At confluence, the cells were detached and passaged once, then seeded at high density either in 24 well culture plates (Becton Dickinson, France), or on glass cover plates (Polylabo, France) previously coated with gelatin (1%, Sigma, France) and were allowed to grow until the confluence was reached.

For experimental studies, chondrocytes were incubated for 24, 48 and 72 h at 37°C in a humidified atmosphere of 5% CO_2 and 95% air under the following conditions: (i) no additive, (ii) 10^{-5} M rhein (Negma Laboratories, France), (iii) 2 ng/ml TNF-α (Sigma, France), (iv) 2 ng/ml recombinant human

IL-1α (*rh*IL-1α, Pharmingen, France), (v) TNF-α and rhein 2 ng/ml each, (vi) *rh*Il-1α and rhein 2 ng/ml each. Since rhein was dissolved in dimethylsulfoxide (DMSO), parallel controls experiments were set up to ensure that DMSO did not affect cell viability or results (data not shown).

2.2. Intracellular IL-1β production

Chondrocytes were fixed (paraformaldehyde, 1%, Sigma) and permeabilised (Triton X100, 0.5%), then incubated with fluorescein isothiocyanate (FITC)-conjugated monoclonal antibody (mAb) IL-1β (Becton Dickinson, France) for 30 minutes at room temperature in the dark. At the end of incubation period, cells were washed with PBS and analyzed by FACScan flow cytometer (Becton Dickinson, San Jose, USA).

2.3. β1 integrin expression

Adhesion receptor expression was determined by a direct immunofluorescence assay. MAbs used in this study included FITC-conjugated CD49d (α4/β1) and CD49e (α5/β1) (Immunotech, France). The quantification of CD49d and CD49e expression was performed using Quantum Simply Cellular calibrator (Sigma, France). Calibration beads as described previously were used to transform the conventional mean fluorescence intensity obtained in flow cytometry as arbitrary units into an absolute number of antibody molecules bound per cell [24].

2.4. Analysis by flow cytometry

List mode data was acquired on a flow cytometer using PCLysis software. The flow cytometer was equipped with a 488 nm argon laser and the serial filter configuration. Dead cells were excluded by forward and side scatter gating. At least, 10 000 cells were counted. List mode files were then analyzed using WinMDI software.

2.5. 3D conventional optical scanning microscopy/deconvolution studies

Optical sections were obtained with an Olympus IX-70 epi-fluorescence inverted microscope equipped with the CellScan EPR^{TM} optical scanning acquisition system (IPLab-Scanalytics, USA) and a 60$\times$ PSF/1.25 NA oil immersion objective (Olympus, France) according to Dumas et al. [25].

2.6. Statistical analysis

Results were obtained in at least two different experiments, in triplicate. Results are expressed as mean $\pm$ standard deviation (SD). An unpaired Student's t-test was performed with SPSS for statistical analyses. Significant differences were confirmed only with a probability of less than or equal to 5%.

3. Results

3.1. Expression of intracellular IL-1β (Figs 1 and 2)

When cells were treated only with rhein for 48 h, any significant variation of intracellular IL-1β expression was observed. Whereas, a stimulation with *rh*IL-1α or TNF-α resulted in a significant increase of mean fluorescence intensity ($p < 0.01$).

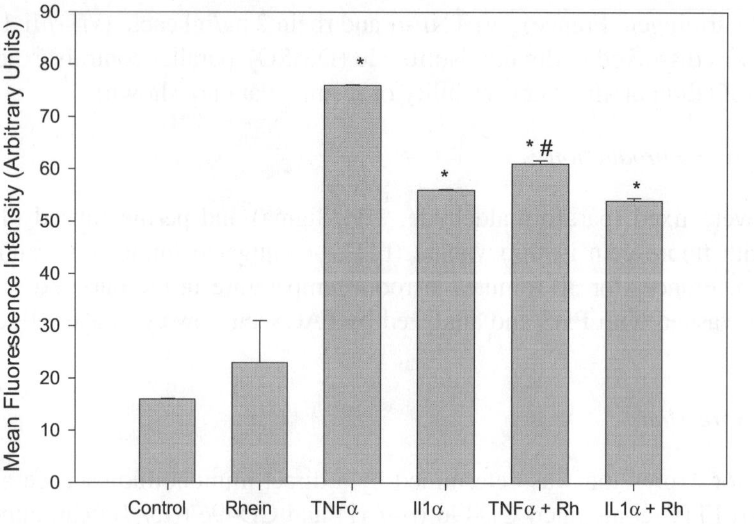

Fig. 1. Effect of rhein on intracellular IL-1β expression using flow cytometry. Confluent chondrocytes were incubated for 48 hours with control medium or medium containing cytokines (IL-1α or TNF-α, at 2 ng/ml) or with rhein (10^{-5} M) and cytokines. Values are means $\pm$ SD. Statistically significant differences ($p < 0.05$) are earmarked: *: control versus IL-1α or TNF-α; #: cytokines versus rhein + cytokines.

On the other hand, incubation of cells with cytokines (*rh*IL-1α or TNF-α) and rhein resulted in a significant decrease of mean fluorescence intensity ($p < 0.05$).

Cells observation with 3D fluorescence microscopy showed an increase of intracellular IL-1β in stimulated cells (number and density of spot). We also observed that in the presence of rhein (simultaneous addition), the number and size of spots decreased (Fig. 2).

3.2. Quantitative expression of α4/β1 and α5/β1 receptors on chondrocytes (Figs 3 and 4)

The α5/β1 receptors were highly expressed on control chondrocytes than α4/β1 receptors (between 3300 and 8000 α4/β1 receptors and between 6500 and 11 500 α5/β1 receptors). Moreover, any quantitative variation of β1 integrins was observed in the presence of rhein ($p > 0.05$).

The stimulation of chondrocytes by cytokines (TNF-α or rhIL-1α at both 2 ng/ml) induces quantitative variation of α4/β1 and α5/β1. In fact, until incubation of 24 h, the α4/β1 receptor density was significantly greater on TNF-α stimulated cells than on control cells ($p < 0.05$, 12 h: $+53 \pm 27\%$ and 24 h: $+6 \pm 3\%$). The α4/β1 receptor number significantly increased ($+11 \pm 4\%$) after an incubation in the presence of *rh*IL-1α for 24 hours (Fig. 3).

The α5/β1 receptor number was significantly increased when cells were stimulated either with TNF-α or *rh*IL-1α, from 24 hours of incubation ($p < 0.05$, TNF-α: 24 h: $+37 \pm 15\%$ and 48 h: $+30 \pm 8\%$, *rh*IL-1α: 24 h: $+71 \pm 40\%$ and 48 h: $+106 \pm 25\%$) (Fig. 4).

On the other hand, in the presence of TNF-α and rhein, a significantly decrease of α4/β1 expression was observed compared to stimulated cells without rhein ($p < 0.05$, for 12 h: $-31 \pm 2\%$ and for 24 h: $-16 \pm 5\%$). Conversely, any significantly variation of this receptor was observed when cells were incubated with *rh*IL-1α and rhein.

Finally, the α5/β1 receptor number on chondrocytes, after treatment with *rh*IL-1α and rhein, significantly decreased for 48 hours of incubation ($p < 0.05$, $-25 \pm 13\%$).

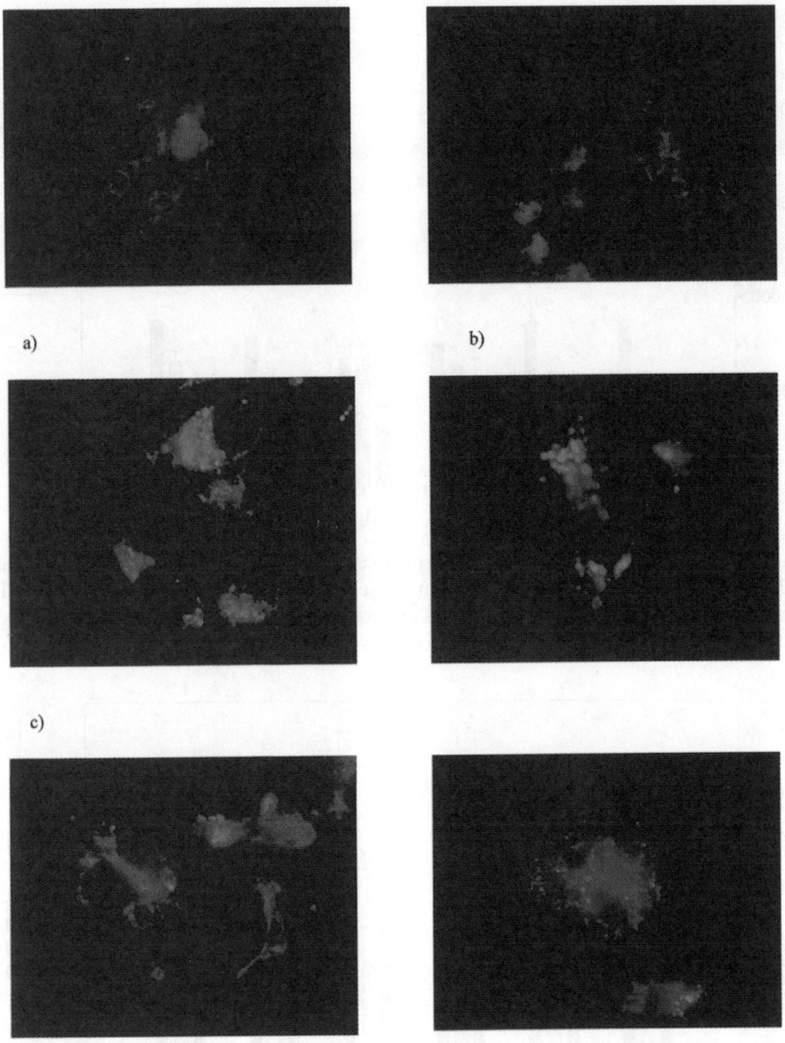

a) b)

c)

d)

Fig. 2. Effect of rhein on intracellular IL-1β expression (observation with 3D microscopy). Confluent chondrocytes were incubated for 48 hours with control medium or medium containing cytokines or with rhein (10^{-5} M) and cytokines. (a): Intracellular IL-1β expression in control cells; (b): intracellular IL-1β expression in cells treated with rhein; (c): intracellular IL-1β expression in stimulated chondrocytes; (d): intracellular IL-1β expression after stimulation and treatment with rhein.

4. Discussion

Cartilage destruction is a crucial feature of OA and is generally considered irreversible. Accordingly, drugs that block the destruction of cartilage will be of therapeutic value. Diacerein has shown to be effective and well tolerated in the long term treatment of OA [18] and it is currently under evaluation as a disease modifying OA drug in randomized placebo-controlled clinical trials of patients with OA of the hip and knee. The efficacy of diacerein has been well documented in animal models but its precise mechanism of action on chondroprotection is unclear.

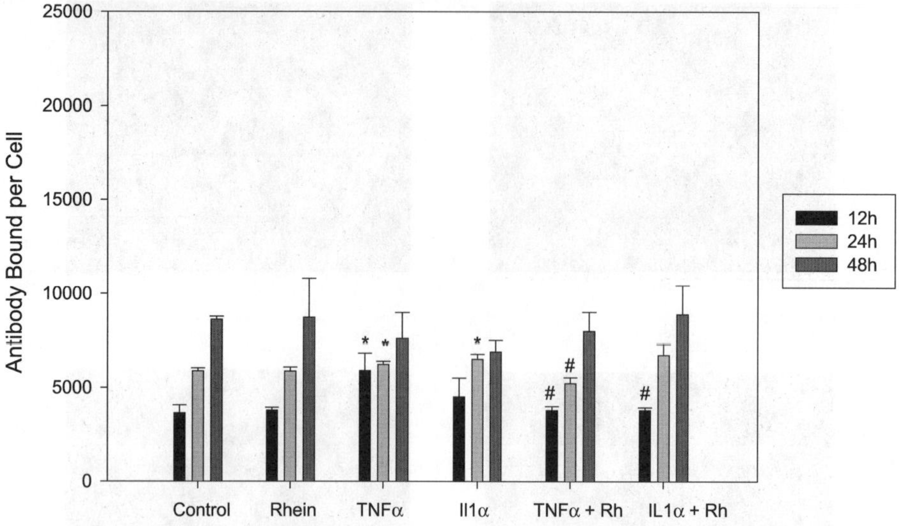

Fig. 3. Effect of rhein on quantitative expression of integrin $\alpha4\beta1$ receptors using flow cytometry. Confluent chondrocytes were incubated for 12, 24 and 48 hours with control medium or medium containing cytokines (IL-1α or TNF-α, at 2 ng/ml) or with rhein (10^{-5} M) and cytokines. Results are expressed as antibody bound per cell ± SD. Statistically significant differences ($p < 0.05$) are earmarked: *: control versus IL-1α or TNF-α, #: cytokines versus rhein + cytokines.

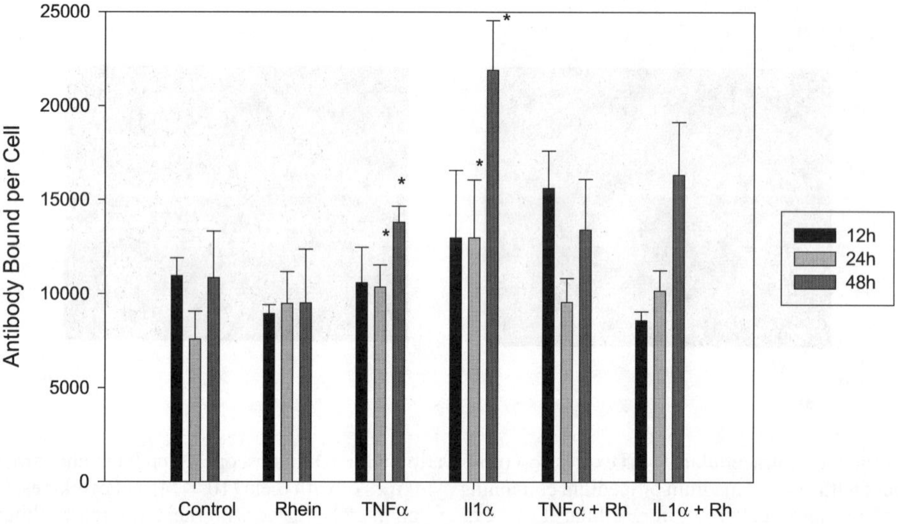

Fig. 4. Effect of rhein on quantitative expression of integrin $\alpha5\beta1$ receptors by flow cytometry. Confluent chondrocytes were incubated for 12, 24 and 48 hours with control medium or medium containing cytokines (IL-1α or TNF-α, at 2 ng/ml) or with rhein (10^{-5} M) and cytokines. Results are expressed as antibody bound per cell ± SD. Statistically significant differences ($p < 0.05$) are earmarked: *: control versus IL-1α or TNF-α, #: cytokines versus rhein + cytokines.

In order to investigate the articular chondrocytes activation by cytokines *rh*IL-1α and TNF-α, we estimated by flow cytometry the intracellular IL-1β production and the expression of $\beta1$ integrins receptors ($\alpha4/\beta1$ and $\alpha5/\beta1$). Then, we studied the effect of rhein on intracellular IL-1β production and on $\beta1$ integrin expression.

Flow cytometry and 3D microscopy showed that cells stimulation by TNF-α and *rh*IL-1α induce a significant increase of intracellular IL-1β production. Whereas, the addition of rhein leads to a reduction

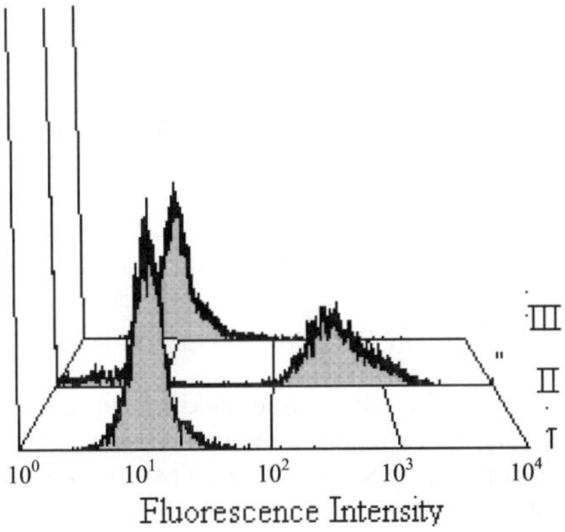

Fig. 5. Effect of rhein on intracellular IL-1β produced by TNF-α stimulated neutrophils by flow cytometry. Cells were incubated during 1 hour either with TNF-α either with TNF-α and rhein. I: Intracellular IL-1β expression in control cells; II: intracellular IL-1β expression after cells stimulation by TNF-α; III: intracellular IL-1β expression after cells treatment with TNF-α and rhein.

of intracellular IL-1β production. These results are in agreement with recent data describing a reduction of IL-1 synthesis by a selective inhibition of the IL-1 (cytokine and specific receptor) in relation with TNF-α [2,26].

The immunolabelling calibrator use allowed us to produce results as antigen density. This method is applied to quantify adhesion molecules expression on chondrocytes in order to establish a relationship between their quantitative expression and cell functionality. The quantitative study of α4/β1 and α5/β1 receptors showed that normal chondrocytes expressed more strongly α5/β1 receptor than α4/β1 receptor. Moreover, stimulated cells by proinflammatory cytokines induced a significant increase of α5/β1 from 24 hours of incubation, suggesting a role of this receptor in transduction signal generated by *rh*Il-1α or TNF-α. Millward-Sadler et al. reported that the difference in response of OA chondrocytes could appear to be the result of altered activation of a different intracellular signal cascade via α5/β1 [13]. Conformational changes in the extracellular domain of integrins which result in changes in affinity can be generated by activating signals from growth factors, cytokines and neuropeptides, a process termed "inside–out" signalling [27]. Modulation of "outside–in" integrin signalling may also occur as a result of growth factor regulation of intracellular molecules such as Rho [28]. A weak increase of α4β1 receptor expression was also observed from 12 to 24 hours. When chondrocytes are treated with cytokines and rhein, a weak decrease of β1 integrins receptors expression was observed, suggesting an action of rhein on these integrins receptors expression. Actually, diacerein is reported to be able of stimulating TGF-β production [29]. This growth factors may play an important role as it can enhance matrix production via α5β1 integrin [30,31]. TGF-β may also counter the IL-1-induced effects on expression of MMPs and matrix molecules and down-regulate the expression of IL-1 receptor in articular chondrocytes [32].

From this study, it seems that rhein has a weak action on α4β1 or α5β1 receptors. Moreover, it was shown that rhein partially reduce TNF-α-induced intracellular IL-1β synthesis. Other kind of cells implicated in the inflammatory process of OA joint (such as the synoviocytes but also the neutrophils infiltrating synovial fluid) also produce cytokines [22,33,34]. Works in progress in our laboratory show also the effect of rhein on neutrophils IL-1 production (preliminary results are shown in Fig. 5). These pharmacological activities of rhein, observed at doses devoided of cell toxicity and closed to human therapeutic doses, contribute to protect the human OA cartilage.

References

[1] C.A. Towle, H.H. Hung, L.J. Bonassar, B.V. Treadwell and D.C. Mangham, Detection of interleukin 1 in the cartilage of patients with osteoarthritis: a possible autocrine/paracrine role in pathogenesis, *Osteoarthritis Cartilage* **5** (1997), 293–300.

[2] J. Martel-Pelletier, F. Mineau, F.C. Jolicoeur, J.M. Cloutier and J.P. Pelletier, In vitro effects of diacerhein and rhein on interleukin-1 and tumor necrosis factor-α systems in human osteoarthritic synovium and chondrocytes, *J. Rheum.* **25** (1998), 753–762.

[3] H.P. Benton and J.A. Tyler, Inhibition of cartilage proteoglycan synthesis by interleukin-1, *Biochem. Biophys. Res. Commun.* **154** (1988), 421–428.

[4] M.B. Goldring, L. Birkead, T. Kimura and S.M. Krane, Interleukin-1 suppresses expression of cartilage specific types II and IX collagens and increases types I and III collagens in human chondrocytes, *J. Clin. Invest.* **82** (1988), 2026–2037.

[5] J.R. Kammermann, S.A. Kincaid, P.F. Rumph, D.K. Baird and D.M. Visco, Tumor necrosis factor-alpha (TNF-alpha) in canine osteoarthritis: Immunolocalization of TNF-alpha, stromelysin and TNF receptors in canine osteoarthritic cartilage, *Osteoarthritis Cartilage* **4** (1996), 23–34.

[6] Y.J. Kim, R.L. Sah, A.J. Grodzinsky, A.H. Plaas and J.D. Sandy, Mechanical regulation of cartilage biosynthetic behavior: physical stimuli, *Arch. Biochem. Biophys.* **311** (1994), 1–12.

[7] J.H. Kimura, O.D. Schipplein, K.E. Keuttner and T.P. Andriacchi, Effects of hydrostatic loading on extracellular matrix formation, *Trans. Orthop. Res. Soc.* **10** (1985), 365.

[8] D.M. Salter, J.E. Robb and M.O. Wright, Electrophysiological responses of human bone cells to mechanical stimulation: evidence for specific integrin function in mechanotransduction, *J. Bone Min. Res.* **12** (1997), 1133–1141.

[9] M.O. Wright, J.L. Godolphin, E. Dunne, C. Bavington, P. Jopanbutra, G. Nuki et al., Hyperpolarization of cultured human chondrocytes following cyclical pressurisation involves $\alpha5\beta1$ integrin and integrin-associated intracellular pathways, *J. Orthop. Res.* **15** (1997), 742–747.

[10] R.L. Juliano and S. Haskill, Signal transduction from the extracellular matrix, *J. Cell. Biol.* **120** (1993), 577–585.

[11] V.L. Woods, Jr, P.L. Screck, D.S. Gesink, H.O. Pacheco, D. Amiel, W.H. Akeson and M. Lotz, Integrin expression by human articular chondrocytes, *Arthritis Rheum.* **37** (1994), 537–544.

[12] K. Ostergaard, D. Salter, J. Petersen, K. Bendtzen, J. Hvolris and C.B. Andersen, Expression of a and b subunits of the integrin superfamily in articular cartilage from macroscopically normal and osteoarthritic human femoral heads, *Ann. Rheum. Dis.* **57** (1998), 303–308.

[13] S.J. Millward-Sadler, M.O. Wright, H.S. Lee, H. Caldwell, G. Nuki and D.M. Salter, Altered electrophysiological responses to mechanical stimulation and abnormal signalling through $\alpha5\beta1$ integrin in chondrocytes from osteoarthritic cartilage, *Osteoarthritis Cartilage* **8** (2000), 272–278.

[14] R.M. Clancy, J. Rediske, X. Tang, N. Nijher, S. Frenkel, M. Philips et al., Outside-in signaling in the chondrocyte. Nitric oxyde disrupts fibrnectin-induced assembly of a subplasmalemmal actin/rho A/Focal adhesion kinase signaling complex, *J. Clin. Invest.* **100** (1997), 1789–1796.

[15] E.C. Arner and M.D. Tortorella, Signal transduction through chondrocyte integrin receptors induces matrix metalloproteinase synthesis and synergizes with interleukin-1, *Arthritis Rheum.* **38** (1995), 1304–1314.

[16] A.R. Moore, K.J. Greenslade, C.A. Alam and D.A. Willoughby, Effects of diacerein on granuloma induced cartilage breakdown in the mouse, *Osteoarthritis Cartilage* **1** (1998), 19–23.

[17] K.D. Brandt, G. Smith, S.Y. Kang, S. Myers, B. O'Connor and M. Albrecht, Effect of diacerein an accelerated canine model of osteoarthrosis, *Osteoarthritis Cartilage* **5** (1997), 438–449.

[18] M. Nguyen, M. Dougados, L. Berdah and B. Amor, Diacerein in the treatment of osteoarthritis of the hip, *Arthritis Rheum.* **37** (1994), 529–536.

[19] M. Dougados, M. Nguyen, L. Berdah, M. Lequesne, B. Mazieres and E. Vignon, Methodes d'évaluation de l'arthrose: à propos de l'étude ECHODIAH, *Rev. Prat.* **46** (1996), S53–S56.

[20] J.P. Pujol, Collagenolytic enzymes and interleukin-1: their role in inflammation and cartilage degradation; the antagonistic effects of diacerein on IL-1 action on cartilage matrix components, *Osteoarthritis Cartilage* **1** (1993), 82.

[21] M. Boittin, F. Redini, G. Loyau and J.P. Pujol, Effect of diacetylrhein on matrix synthesis and collagenase release in cultures rabbit articular chondrocytes, *Rev. Rheum.* **60** (1993), 68S–76S.

[22] M. Mian, S. Brunelleschi, S. Tarli, A. Rubino, D. Benetti and R. Fantozzi, Rhein: an anthraquinone that modultes superoxide anion production from human neutrophil, *J. Pharm. Pharmacol.* **39** (1987), 845–847.

[23] M. Mian, D. Benetti, S. Rosini and R. Fantozzi, Effects of diacerein on the quantity and phagocytic activity of thioglycollate-elicited mouse peritoneal macrophages, *Pharmacology* **39** (1989), 362–366.

[24] P. Poncelet, F. George, S. Papa and F. Lanza, Quantitation of hemopoietic cell antigens in flow cytometry, *Eur. J. Histochem.* **40** (1996), 15–32.

[25] D. Dumas, C. Gigant, N. Presle, C. Cipolletta, G. Miralles, E. Payan, J.Y. Jouzeau, D. Mainard, B. Terlain, P. Netter and J.F. Stoltz, The role of 3D-microscopy in the study of chondrocyte–matrix interaction (alginate bead or sponge, rat femoral

head cap, human osteoarthritic cartilage) and pharmacological application, *Biorheology* **37** (2000), 165–176.
[26] D. Dumas, K. Bordji, E. Payan, P. Gillet, P. Netter and J.F. Stoltz, Imagerie ultrastructurale du chondrocyt, *Rev. Prat.* **49** (1999), S9–SS14.
[27] K.M. Yamada, Integrin signalling, *Matrix Biol.* **16** (1997), 137–141.
[28] S. Dedhar and G.E. Hannigan, Integrin cytoplasmic interactions and bidirectional transmembrane signalling, *Curr. Opin. Cell. Biol.* **8** (1996), 657–669.
[29] J.P. Pujol, N. Felisaz, K. Boumediene, C. Ghayor, J.F. Herrouin, P. Bogdanowicz and P. Galera, Effects of diacerein on biosynthesis activities of chondrocytes in culture, *Biorheology* **37** (2000), 177–184.
[30] J.P. Pujol, P. Galera, S. Pronost, Boumediene, D. Vivien, M. Macro, W. Min, F. Redini, H. Penfornis, M. Daireaux et al., Transforming growth factor-β (TGF-β) and articular chondrocytes, *Ann. Endocrinol.* **55** (1994), 109–120.
[31] R.F. Loeser, Chondrocyte integrin expression and function, *Biorheology* **37** (2000), 109–116.
[32] F. Rédini, A. Mauviel, S. Pronost, G. Loyau and J.P. Pujol, Transforming growth factor-b exerts opposite effects from inteleukin-1b (IL-1b) on cultured rabbit articular chondrocytes through reduction of IL-1 receptor expression, *Arthritis Rheum.* **36** (1993), 44–50.
[33] K. Fassbender, S. Kaptur, P. Becker, J. Groschl and Hennerici, Adhesion molecules in tissue injury: Kinetics of expression and shedding and association with cytokine release in humans, *Clin. Immunol. Immunopathol.* **89** (1998), 54–60.
[34] D.D. Wood, E.J. Ihrie and D. Hamerman, Release of interleukin-1 from human synovial tissue in vitro, *Arthritis Rheum.* **28** (1985), 853–862.

Biorheology 39 (2002) 287–288
IOS Press

Concluding remarks and perspectives on cartilage and chondrocyte mechanobiology

J.F. Stoltz
Faculty of Medicine, UMR 7563 LEMTA and IFR 111 Bioingénierie, 54500 Vandoeuvre lès Nancy, France

Research on the influence of physiological forces on cartilage and chondrocyte metabolism has increased during the last 10 years. For instance, it is known today that the chondrocyte is sensitive not only to the intensity of mechanical applied forces, but also to the frequency of its application. Thus, cyclic application of loads stimulates matrix synthesis in cartilage explants, whereas as for other tissues, static loads depress synthesis (vessels, bone). These results lead to the notion of tissue remodeling that Wolff had already empirically described for bone in the late 19th century.

After the first symposium of Nice-Sainte Maxime in 1999, organized by the French Society of Biorheology with the support of Negma-Lerads Laboratories, the purpose of this second meeting was to draw up a balance, certainly not exhaustive, of the metabolic remodeling of cartilage induced by forces, understanding the possibilities of OA and the development of new substitution tissues.

In the first session, different communications have developed the theoretical, mechanical and physical approaches on cartilage and chondrocyte. In fact, as for all porous media, cartilage exhibits a coupling between mechanics and electrokinetics with the appearance of a streaming potential. This is how the poro-elasticity theories can be used. During these two days, Dr. Leping Li proposed a representation of cartilage using a 3-phase model composed of: fibrillar matrix (collagen), non-fibrillar matrix (solid excluding collagen) and water. The electrical phenomena related to the deformation by compression were also the subject of interesting approaches in the papers of Grodzinsky, Lai and Wohlrab. Thus, the results presented by M. Lai show that inside the tissue, in addition to the streaming potential caused by fluid movement, there is also a "diffusion potential" caused by cation and anion concentration gradients that are induced by the gradient of fixed charge density inside the tissue. These electrical phenomena coupled with the ionic channels should certainly be taken into account in the interpretation of signal transduction mechanisms.

In the second session, the group of communications was more focused on the mechanisms of mechanotransduction. In fact, although the physiological effects of compression are now well described, the transcription pathways remain hypothetical. Thus, the deformation of chondrocytes is believed to be important in mechanotransduction with the possible involvement of stretch activated ion channels within the membrane (Bader et al.). The role of the tensegrity architecture of the cytoskeleton should also be mentioned. The work of Lammi et al. on cDNA array experiments revealed some interesting candidate genes whose expression levels may be changed during cyclic loading (IL-1β, stromelysin, tissue inhibitor of metalloproteinases 3). The expression of these genes may be important for the functions of cartilage.

The determining role of the mechanical forces in the development of cartilage was particularly demonstrated by Van de Lest and colleagues in growing foals who observed, in the training group of animals, the ability of the cartilage to increase proteoglycan synthesis.

However, how may these experimental approaches provide information on the pathology of cartilage, and in particular on the appearance of osteoarthritis (OA)? Several studies on TNF (Westacott) and IL-1β (d'Andrea), have demonstrated the primary role of the cytokeines in the phenomena of degeneration, and the differential responses of the normal and OA chondrocytes (Salter, Fernandes and Bluteau) to mechanical stimulation. The role of Cox-1 and Cox-2 in articular joint disease was also underlined (Colville-Nash).

Because it is difficult to obtain joint tissues from humans with OA, the use of animal models provides the only useful way to examine the processes involved in OA and to study the pharmacological agents. Kenneth Brandt has developed these approaches through different examples.

Finally, recent progress in tissue engineering can provide functional cartilaginous constructs for controlled studies of chondrogenesis *in vitro* and *in vivo* and for joint repair. Cartilage tissue engineering was discussed with respect to the various parameters that can be used to modulate the growth and functions of the constructs (Levenston). Of course, these substitutes must respond to functional characteristic of the cartilages and their integrations in the surrounding tissue (Van Osch).

To conclude, and with regard to prospective work, I would like to repeat the general outline of the investigations on mechanotransduction in cartilage that was suggested two years ago by J. Urban during the 1st symposium in Ste-Maxime, that summarizes so well the different questions to be solved (*Biorheology* **37** (2000), 186). Many pathways remain to be explored with the knowledge that, as for other tissues, the mechanisms of mechanotransduction can be different according to the properties studied, and it is not known how intracellular changes influence transcriptional activation. However, it is clear that many genes are up or down regulated by mechanical forces. It is also clear that chondrocyte responses to mechanical signals are very complicated. In fact, the response depends not only on the duration and amplitude of the forces, but also on their variations in time. On the other hand, it is fundamental to have in mind that one of the main objectives of these investigations is also to understand the mechanisms that lead to arthrosis, and how to treat a pathological cartilage. Studies on the role of mechanical forces in the aetiology of osteoarthritis are necessary, as well as on the effects of the pharmacological agents.

In order to provide some new answers to these questions, I am looking forward to meeting you again for the third symposium in two or three years.

I would like to thank Negma-Lerads Laboratories for its continuous cooperation and support. Our collaboration is an outstanding example of how pharmaceutical companies and scientific societies should work together in a constructive way. Finally, I extend my thanks to Dr Martine Burger, who gave so generously of her time in the preparation of this symposium.

Paris, 28 April 2001